Sintering Technology

Sintering Technology

edited by

Randall M. German
Gary L. Messing
Robert G. Cornwall

The Pennsylvania State University
University Park, Pennsylvania

Marcel Dekker, Inc.   New York • Basel • Hong Kong

Library of Congress Cataloging-in-Publication Data

Sintering technology / edited by Randall M. German, Gary L. Messing, Robert G. Cornwall.
    p. cm.
    Includes index.
    ISBN 0-8247-9775-2 (hardcover : alk. paper)
    1. Sintering—Congresses. I. German, Randall M. II. Messing, Gary L.
III. Cornwall, Robert G.
TN695.S594 1996
671.3'73—dc20   96-26522
                                                              CIP

The publisher offers discounts on this book when ordered in bulk quantities. For more information, write to Special Sales/Professional Marketing at the address below.

This book is printed on acid-free paper.

MARCEL DEKKER, INC.
270 Madison Avenue, New York, New York 10016

Current printing (last digit):
10 9 8 7 6 5 4 3 2 1

PRINTED IN THE UNITED STATES OF AMERICA

Preface

Over the past 50 years there has been a series of important research conferences aimed at documenting the status of sintering theory and practice. Previous meetings were organized by the Tokyo Institute of Technology, Notre Dame University, the University of British Columbia, and the International Institute for the Science of Sintering in Yugoslavia. In September 1995, a similar meeting was held at the Scanticon Conference Center on the campus of The Pennsylvania State University. The meeting and these proceedings were organized by the Particulate Materials Center and P/M Lab of Penn State. We were fortunate to have the world's experts participate in critical discussions and analyses of our current understanding of sintering concepts.

The conference consisted of nearly 170 invited, submitted, and poster presentations, as well as table-top displays and discussion periods. Participants came from 27 countries and included most of the major universities involved in sintering research. Industrial participants provided valuable information on new applications, materials, and technical problems encountered in sintering practice. An important aspect was the intermixing of academic, industrial research, production, engineering, and technological views of sintering. Accordingly, the meeting explored many developments in sintering as applied to engineered materials. This included new models for particle bonding and improved simulations of densification and microstructure development during sintering. Interlaced were discussions on new materials, processes, and applications that provided an update of the technology. Thus, *Sintering 1995* constituted an important forum for the sintering community. This book contains the major contributions and is an important benchmark for assessing progress and for identifying needed research. It consists of 63 selected papers from the presentations made at the meeting. All the papers were reviewed by the editors and at least two other experts in the field. On the basis of the review process, we have assembled this compilation to document the status of sintering concepts as a stepping-stone for future investigators.

Randall M. German
Gary L. Messing
Robert G. Cornwall

Contents

Liquid Phase Sintering

Alumina-Sintering and Grain Growth

Grain Growth

Novel and Reactive Sintering Processes

Constrained Densification

Contributors

John Ågren   *Royal Institute of Technology, Stockholm, Sweden*

T. Aizawa   *University of Tokyo, Tokyo, Japan*

D. J. Aldrich   *Colorado School of Mines, Golden, Colorado*

F. Jorge Alves   *Lehigh University, Bethlehem, Pennsylvania*

C. Argento   *E. I. Dupont de Nemours & Co., Inc., Wilmington, Delaware*

R. S. Averback   *University of Illinois at Urbana-Champaign, Urbana, Illinois*

P. K. Bagdi   *Indian Institute of Technology, Kanpur, India*

Youngmin Baik   *McGill University, Montreal, Quebec, Canada*

H. Balmori-Ramírez *National Polytechnic Institute, Mexico City, Mexico*

J. A. Belward *The University of Queensland, Queensland, Australia*

I. Bennett *Eindhoven University of Technology, Eindhoven, The Netherlands*

Rajendra K. Bordia *University of Washington, Seattle, Washington*

Animesh Bose *Parmatech Corporation, Petaluma, California*

Didier Bouvard *Institut National Polytechnique de Grenoble, Saint Martin d'Hères, France*

R. C. Bradt *The University of Alabama, Tuscaloosa, Alabama*

Jan Brandt *University of Linköping, Linköping, Sweden*

Richard J. Brook *University of Oxford, Oxford, United Kingdom*

Jeffrey W. Bullard *University of Illinois at Urbana-Champaign, Urbana, Illinois*

S. L. Burkett *The University of Alabama, Tuscaloosa, Alabama*

K. Burrage *The University of Queensland, Queensland, Australia*

Hugo S. Caram *Lehigh University, Bethlehem, Pennsylvania*

C. A. Carmichael *Oak Ridge National Laboratory, Oak Ridge, Tennessee*

V. M. Carreño *National Polytechnic Institute, Mexico City, Mexico*

C. Carry *Université Paris Sud, Orsay, France*

W. Craig Carter *National Institute of Standards and Technology, Gaithersburg, Maryland*

J. M. Chabala *The University of Chicago, Chicago, Illinois*

Helen M. Chan *Lehigh University, Bethlehem, Pennsylvania*

Horng-Yi Chang *National Tsing-Hua University, Hsinchu, Taiwan, R.O.C.*

H. C. Chen *National Taiwan University, Taipei, Taiwan, R.O.C.*

Long-Qing Chen *The Pennsylvania State University, University Park, Pennsylvania*

G. Cisneros-Gonzalez *National Polytechnic Institute, Mexico City, Mexico*

Herbert Danninger *Technische Universität Wien, Vienna, Austria*

Burtron H. Davis *Center for Applied Energy Research, Lexington, Kentucky*

L. C. De Jonghe *University of California, Berkeley, California*

M. Demartin *Swiss Federal Institute of Technology, Lausanne, Switzerland*

G. de With *Eindhoven University of Technology, Eindhoven, The Netherlands*

Graham P. Dransfield *Tioxide Specialties, Cleveland, United Kingdom*

G. Drazic *University of Ljubljana, Ljubljana, Slovenia*

Robin A. L. Drew *McGill University, Montreal, Quebec, Canada*

J. Duszczyk *Delft University of Technology, The Netherlands*

J. Dutta *Ecole Polytechnique Fédérale de Lausanne, Lausanne, Switzerland*

J. L. Estrada-Haén *National Polytechnic Institute, Mexico City, Mexico*

Danan Fan *The Pennsylvania State University, University Park, Pennsylvania*

Jianxin Fang *Lehigh University, Bethlehem, Pennsylvania*

Brian D. Flinn *University of Washington, Seattle, Washington*

Randall M. German *The Pennsylvania State University, University Park, Pennsylvania*

H. Collin Gill *University of Sunderland, Sunderland, United Kingdom*

A. M. Glaeser *Lawrence Berkeley National Laboratory, Berkeley, California*

W. S. Hackenberger *The Pennsylvania State University, University Park, Pennsylvania*

Sven Haglund *Royal Institute of Technology, Stockholm, Sweden*

Charles E. Hamrin, Jr. *University of Kentucky, Lexington, Kentucky*

Martin P. Harmer *Lehigh University, Bethlehem, Pennsylvania*

M. Herrmann *Fraunhofer-Institute of Ceramic Technologies and Sintered Materials, Dresden, Germany*

H. Hofmann *Ecole Polytechnique Fédérale de Lausanne, Lausanne, Switzerland*

Seong Hyeon Hong *The Pennsylvania State University, University Park, Pennsylvania*

Debra Horn *The Pennsylvania State University, University Park, Pennsylvania*

K. S. Hwang *National Taiwan University, Taipei, Taiwan, R.O.C.*

Masaru Inoue *Toyo Kohan Co., Ltd., Kudamatsu, Japan*

A. Jagota *E. I. Dupont de Nemours & Co., Inc., Wilmington, Delaware*

D. Jaramillo-Vigueras *National Polytechnic Institute, Mexico City, Mexico*

John L. Johnson *The Pennsylvania State University, University Park, Pennsylvania*

Suk-Joong L. Kang *Korea Advanced Institute of Science and Technology, Taejon, Korea*

Ingrid Kerscht *The Pennsylvania State University, University Park, Pennsylvania*

J. Kihara   *University of Tokyo, Tokyo, Japan*

Kwan-Hyeong Kim   *Samsung Electromechanics Co., Suwon, Korea*

M. K. Kim   *Kyungpook National University, Taegu, Korea*

M. Kitayama   *University of California, Berkeley, California*

D. Kolar   *University of Ljubljana, Ljubljana, Slovenia*

Masao Komai   *Toyo Kohan Co., Ltd., Kudamatsu, Japan*

M. Kosec   *University of Ljubljana, Ljubljana, Slovenia*

O. J. Kwon   *Kyungpook National University, Taegu, Korea*

F. F. Lange   *University of California, Santa Barbara, California*

Simon Lawson   *British Nuclear Fuels, Cumbria, United Kingdom*

Sung-Min Lee   *Korea Advanced Institute of Science and Technology, Taejon, Korea*

Mitchell D. Lehigh   *The University of Pittsburgh, Pittsburgh, Pennsylvania*

E. R. Leite   *DQ-UFSCar, São Carlos, Brazil*

Gert Leitner   *Fraunhofer-Institut für Keramische Technologien und Sinterwerkstoffe, Dresden, Germany*

R. Levi-Setti   *The University of Chicago, Chicago, Illinois*

I-Nan Lin *National Tsing-Hua University, Hsinchu, Taiwan, R.O.C.*

Per Lindskog *Sandvik AB, Stockholm, Sweden*

Yixiong Liu *The Pennsylvania State University, University Park, Pennsylvania*

Elson Longo *Universidade Estadual Paulista, Araraquara, Brazil*

Y. C. Lu *National Taiwan University, Taipei, Taiwan, R.O.C.*

B. Malič *University of Ljubljana, Ljubljana, Slovenia*

Keiich Masuyama *Toyohashi University of Technology, Toyohashi, Japan*

Hideaki Matsubara *Japan Fine Ceramics Center, Nagoya, Japan*

Katsura Matsubara *NTK Technical Ceramics, Komaki, Japan*

Satoru Matsuo *Toyo Kohan Co., Ltd., Kudamatsu, Japan*

S. Mazur *E. I. Dupont de Nemours & Co., Inc., Wilmington, Delaware*

Robert M. McMeeking *University of California, Santa Barbara, California*

Gary L. Messing *The Pennsylvania State University, University Park, Pennsylvania*

Z. A. Munir *University of California, Davis, California*

Kazunori Nakano *Toyo Kohan Co., Ltd., Kudamatsu, Japan*

Ian Nettleship *The University of Pittsburgh, Pittsburgh, Pennsylvania*

Zoran S. Nikolic *University of Nish, Nish, Yugoslavia*

M. A. L. Nobre *DQ-UFSCar, São Carlos, Brazil*

S. I. Nunes *The University of Alabama, Tuscaloosa, Alabama*

Shinya Ozaki *Toyo Kohan Co., Ltd., Kudamatsu, Japan*

Leinig Perazolli *Universidade Estadual Paulista, Araraquara, Brazil*

I. Podolsky *The University of Queensland, Queensland, Australia*

J. D. Powers *University of California, Berkeley, California*

M. N. Rahaman *University of Missouri, Rolla, Missouri*

D. W. Readey *Colorado School of Mines, Golden, Colorado*

S. Rimlinger *Quartz and Silice, Nemours, France*

Momcilo M. Ristic *Serbian Academy of Sciences and Arts, Belgrade, Yugoslavia*

M. A. Ritland *Golden Technologies Company, Inc., Golden, Colorado*

Jürgen Rödel *Technische Hochschule, Darmstadt, Germany*

E. Rupp *Fraunhofer-Institute of Ceramic Technologies and Sintered Materials, Dresden, Germany*

Samuel M. Salamone *University of Washington, Seattle, Washington*

G. B. Schaffer *The University of Queensland, Queensland, Australia*

J. H. Schneibel *Oak Ridge National Laboratory, Oak Ridge, Tennessee*

Chr. Schubert *Fraunhofer-Institute of Ceramic Technologies and Sintered Materials, Dresden, Germany*

Matthew Seabaugh *The Pennsylvania State University, University Park, Pennsylvania*

T. Senda *The University of Alabama, Tuscaloosa, Alabama*

T. B. Sercombe *The University of Queensland, Queensland, Australia*

Shinya Shiga *Niihama National College of Technology, Niihama, Japan*

Soon-Gi Shin *Japan Fine Ceramics Center, Nagoya, Japan*

T. R. Shrout *The Pennsylvania State University, University Park, Pennsylvania*

J. Malcolm Smith *University of Sunderland, Sunderland, United Kingdom*

K. K. Soni *The University of Chicago, Chicago, Illinois*

R. F. Speyer *Georgia Institute of Technology, Atlanta, Georgia*

Richard M. Spriggs *Alfred University, Alfred, New York*

Laura C. Stearns   *Lehigh University, Bethlehem, Pennsylvania*

Suiying Su   *Center for Applied Energy Research, Lexington, Kentucky*

M. S. Suh   *Hyundai Motor Company, Yangjungdong, Ulsan, Korea*

H. Suzuki   *The University of Alabama, Tuscaloosa, Alabama*

Ken-ichi Takagi   *Toyo Kohan Co., Ltd., Tokyo, Japan*

K. Sing Tan   *University of Sunderland, Sunderland, United Kingdom*

A. M. Thompson   *GE Corporate Research and Development, Schenectady, New York*

Gerhard Tomandl   *TU Bergakademie Freiberg, Freiberg, Germany*

F. Tsumori   *University of Tokyo, Tokyo, Japan*

Björn Uhrenius   *Sandvik Hard Materials, Stockholm, Sweden*

Minoru Umemoto   *Toyohashi University of Technology, Toyohashi, Japan*

G. S. Upadhyaya   *Indian Institute of Technology, Kanpur, India*

P. G. Th. van der Varst   *Eindhoven University of Technology, Eindhoven, The Netherlands*

José A. Varela   *Universidade Estadual Paulista, Araraquara, Brazil*

Péter Varkoly *Institute of Ceramic Materials, Freiberg, Germany*

Masakazu Watanabe *NTK Technical Ceramics, Komaki, Japan*

D. B. Williams *Lehigh University, Bethlehem, Pennsylvania*

Suxing Wu *Lehigh University, Bethlehem, Pennsylvania*

Yuji Yamasaki *Toyo Kohan Co., Ltd., Kudamatsu, Japan*

Kazuo Yamazaki *University of California at Davis, Davis, California*

Huilong Zhu *University of Illinois at Urbana-Champaign, Urbana, Illinois*

De-sintering, A Phenomenon Concurrent with Densification Within Powder Compacts: A Review

F. F. Lange

Materials Department, University of California, Santa Barbara, CA 93106

Abstract

A morphological instability characterized by the separation of grain pairs and the disappearance of grain boundaries is a common feature in polycrystalline bodies that are constrained from shrinking during mass transport. This instability, know as de-sintering, occurs during grain coarsening as sequentially observed for thin films with columnar microstructures constrained by a substrate, fibers with a 'bamboo' microstructure constrained by a matrix, and powder matrices within composites. De-sintering is a common phenomena in partially dense, polycrystalline bodies; it occurs concurrently with densification phenomena and is emphasized in composites where the partially dense matrix is constrained from shrinking by a reinforcement network. De-sintering occurs in any lower density region were shrinkage is constrained by the average shrinkage of the body.

1. Introduction

The reader must pull out their Introduction to Ceramics (Kingery et al., 1976) and turn to pages 482 and 483 to study the consecutive micrographs, supplied by C. Greskovich and K. Lay, showing the microstructural development of alumina oxide during densification. A quick glance should tell even the causal student that grain growth can be concurrent with densification. The four micrographs on these two pages clearly show that the particle size of the initial powder compact, and the size of the grains for the partially dense compacts after heat treatments at 1700 °C for 1, 2.5 and 6 minutes. The gain size increases by about an order of magnitude between each pair of consecutive micrographs. Concurrent to this enormous amount of grain growth is a comparable growth of pores. If the large amount of grain growth does not bother the serious student, who expects to

1

understand densification, the concurrent growth of the pores should trigger a lack of comprehension. That is, although the student can not read about a comprehensive link between grain growth and densification, at least the text explains grain growth in terms of either grain boundary motion or coarsening, i.e., the mass transfer between grain surfaces. On the other hand, since our Ceramic texts only show why and how pores in a powder compact can disappear during densification, it should be incomprehensible to the serious student to see them grow as clearly illustrated in the sequential micrographs supplied by Greskovich and Lay.

The objective of this review is to explain the why and how of pore growth during densification. The why and how have simple answers, but the remaining paper will detail their explanation. Why is answered by recognizing that the growth of a void can reduce the surface to volume ratio of the solid, and thus, best reduce the free energy of the system under some circumstances. How is answered by stating that just as voids can become smaller by the formation necks between touching particles, a phenomenon know as sintering, voids can also become larger (e.g., link together) when the neck and grain boundary between two grains disappear, a phenomenon described as de-sintering.

As it will be seen, the explanations to these simple answers require some constraint to shrinkage during the densification process. That is, within the body, a portion of the partially dense material needs to be partially constrained from shrinking by surrounding material. Since the voids within these regions are constrained from decreasing their free energy by shrinking,, i.e., decreasing their surface to volume ratio, they, instead, decrease their free energy by coalescing to become larger. Coalescence requires grain pairs, which had previously formed necks, to de-sinter. It will also be shown that de-sintering is generally, but not exclusively, associated with grain growth. Namely, de-sintering can also occur when grain pairs are pulled apart as one region, linked to another via the grain pairs, move in opposite directions. Although the linking together of voids via de-sintering does not decrease the free energy as much as if they were to disappear, any reduction of free energy appears kinetically expedient.

The review initiates with conditions where constraint to shrinkage is more obvious, viz., where grain growth occurs in dense, polycrystalline fibers and films which are fully constrained from shrinking. It then moves to composites, where the powder matrix is either partially or fully constrained from shrinking by a reinforcement phase. The reader is then asked to return to the four micrographs mensioned above to discover that the reason for pore growth in powder compacts is now more obvious.

2. Morphological Instability of Polycrystalline Fibers

The de-sintering phenomena associated with the instability of polycrystalline fiber is the easiest to see and explain, and has great similarity to the instability phenomena associated with grain pairs in partially dense networks. In its own right, the thermal stability of fibers is an important concern in composite materials technology. At high temperatures, the fibers can undergo grain growth. As shown elsewhere (Miller and Lange, 1989), mass transport can dramatically change the morphology of fibers, such that they become a collection of isolated grains that replaces the original fiber. This phenomenon is caused by the reduction of free energy (minimizing the interfacial energy per unit volume) that is concurrent with grain growth. Plateau (1873) was the first to show that when the length of a cylinder exceeds a critical value, its surface energy per unit volume can be decreased by breaking into spheres. Rayleigh (1945) expanded this observation into his theory of liquid jet instability to show that a cylindrical liquid jet of radius r could continuously decrease its surface energy by breaking into droplets, when subjected to symmetrical disturbances with wavelengths greater than $2\pi r$. Nichols and Mullins (1965) further extended this concept to solid rods within a solid body. These experimentally verified theories rely on the fact that large amplitude diametrical perturbations can grow by mass transport along the cylinder axis. Miller and Lange (1989) showed through theory and experiments that large amplitude perturbations in a polycrystalline cylinder, developed by grain boundary grooving, will break the fiber into individual grains provided that the mass centers of the grains are fixed, and the grain size to fiber diameter ratio exceeds a critical value.

Figure 1 illustrates a polycrystalline fiber that has under gone grain growth until the grains, assumed to be nearly identical, have equilibrium shapes, i.e., truncated spheres, shown in configuration a (assuming all interfacial energies are isotropic). For configurations a, b and c, the fiber is unconstrained and free to move. For this case, as the grain size increases, e.g., by every other grain consuming its neighbor from configuration a to b to c, the fiber will shrink to maintain a constant ratio between its surface area and its grain boundary area, which depends on the energy ratio of these two interfaces (Kellett and Lange, 1989). In this case, the grains maintain their shape as truncated spheres. It is obvious that the unconstrained fiber continuously decreases its free energy during grain growth until it becomes a single, spherical grain.

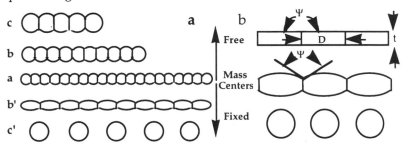

Figure 1 a) Schematic of a polycrystalline fiber that increases its grain size as its mass centers are either free to move (a, b, c) or fixed (a, b', c'). b) Configurational change as a function of the configuration angle, y.

For configurations a, b' and c', the fiber is constrained from shrinking, e.g., by a surrounding matrix.   For this case, the ratio of the surface area to grain boundary area decreases during grain growth, the grains can not maintain their equilibrium shape as truncated spheres, and eventual, the fiber breaks into a number of isolated, spherical grains as observed in experiments (Miller and Lange, 1989).

To determine the conditions where the fiber breaks into isolated grains, Miller and Lange (1989) devised a simple model, shown in Fig. 1b, that allows the free energy change to be calculated and the determination of the equilibrium configuration of a polycrystalline fiber.   The fiber is embedded within a homogeneous matrix material, which has an interfacial energy with the fiber, γ_s.   The initial "bamboo" structure of the fiber is modeled by identical cylindrical grains of length D and diameter t.   The aspect ratio of the grain, 'a' = D/t, is used in subsequent calculations.   The angle, ψ, defined by the surface normals at the grain boundary, is initially 180°.   It was assumed that the grain centers are fixed at a distance L and that the grains develop a barrel shape during grain boundary grooving and second, each grain retains its initial mass.   With these assumptions, the surface and grain boundary area of each grain can be expressed as a function of ψ, which describes the deepening of the grain boundary groove.

The total free energy of each grain is given by:

$$E = A_s \gamma_s + A_b \gamma_b \qquad (1)$$

where A_s is the grain surface area, A_b is the grain boundary area, γ_s is the surface energy, and γ_b is the grain boundary energy.   The surface energies are related through Young's relation:

$$\frac{\gamma_b}{\gamma_s} = 2 \cos \frac{\psi_e}{2}$$

(2)

where ψ_e is the equilibrium dihedral angle. The determination of A_s and A_b as a function of 'a' and ψ allows the energy of a grain, normalized by the energy of the initial, cylindrical grain to be expressed as a function of 'a' and ψ.

Figure 2a shows the normalized energy plotted as a function of the configurational angle ψ for an equilibrium dihedral angle, $\psi_e = 150°$. The free energy is plotted for three grain aspect ratios; for a fixed fiber diameter, the aspect ratio will increase as the grain size increases. For each case, the free energy, E_f, is plotted as ψ decreases from 180° to the value of ψ where the grain boundary disappears. Beyond this point, the free energy is assumed to decrease continuously until it coincides with the normalized free energy of the sphere, E_s, when $\psi = 0$.

a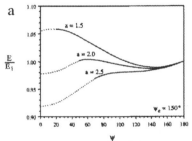
b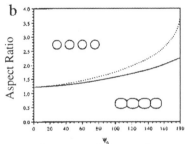

Figure 2 a) Normalized free energy of grain as a function of the configuration angle, for 3 different aspect ratios. b) Equilibrium Configurational Diagram for Fibers with different aspect ratios.

As shown in Fig. 2a, the free energy of the fiber with the smallest aspect ratio, 'a' = 1.5, decreases to a minimum while the fiber is still continuous; for this minimum ψ is near ψ_e. Since any further decrease in ψ will result in a higher free energy, the continuous fiber is the equilibrium configuration. For an increased aspect ratio of 'a' = 2, the normalized free energy again decreases to a minimum corresponding to a continuous fiber. In this case, further decreasing ψ would first increase and then decrease the free energy of the fiber, reaching another minimum for a row of spherical grains. The minimum free energy of the connected fiber is higher than that of the unconnected spheres. Thus for this case, the intact configuration is metastable; to achieve the equilibrium configuration, an energy barrier must be overcome. When the aspect ratio is further increased to 'a' = 3, the row of spherical grains again has the lowest free energy. In this case the free energy continuously decreases with ψ. Such fibers will always break into a row of spheres, provided that mass transport can occur.

The lowest energy configuration for any given set of initial conditions can be represented with an equilibrium configuration diagram. For polycrystalline fibers, the minimum energy configuration can be displayed as regions in aspect ratio vs. equilibrium dihedral angle ('a' vs. ψ_e) space. The boundary between these regions is determined by the condition that both configurations have identical free energies as shown by the solid line in Figure 2b. Whenever the initial conditions of 'a' and ψ_e fall below this line, the lowest free energy configuration is a continuous fiber. Above this line, the lowest free energy configuration is a row of identical spheres. The region of continuous fiber metastability can also be illustrated on the equilibrium configuration diagram. In this region, which is illustrated by the middle curve 'a' = 2 in Fig. 2a, an energy barrier must be overcome. The lower bound of this region is the solid line, whereas the upper bound is the dashed line.

As shown, the instability is a natural phenomena when grain growth occurs in a polycrystalline fiber constrained from shrinking. With sufficient grain growth, the fiber will break into isolated spheres to reduce its free energy, i.e., its surface to volume ratio. If these same isolated spheres were touching one another, they would sinter together to further lower their free energy and produce one of the configurations shown in Fig. 1a for the unconstrained fiber. The ratio of free energy for the unconstrained fiber (E_u) and the constrained fiber that has broken into isolated spheres (E_c) is the energy reduction due to sintering can be expressed as [6]

$$\frac{E_u}{E_c} = \left[\frac{1}{2}\cos\frac{\psi_e}{2}\left(3 - \cos^2\frac{\psi_e}{2}\right)\right]^{\frac{1}{3}} \tag{3}$$

When the dihedral angle, ψ_e, is 150 °, it can be seen from Fig. 2a that when grain growth occurs such that the aspect ratio 'a' becomes equal to 2.5, the constrained fiber breaks into isolated spheres to reduce its free energy by ≈ 7 % (free energy reduction for grain growth between 'a' = 1.5 to 2.5), whereas eq (3) shows that if the same fiber were unconstrained, it would be able to reduce its free energy by another 28 %. Thus, the free energy reduction produced when the constrained fiber breaks into isolated spherical grains is significant (7 %), it is only 1/5 the free energy reduction (0.07/[0.07+0.28]) that could have been realized if the fiber were free to shrink.

3. Morphological Instability of Polycrystalline Films

Whereas Nichols and Mullins (1965) had shown that holes would not develop in either amorphous or liquid thin films, Srolovitz and Safrin (1986) were the first to show that the growth of pin holes in a polycrystalline film was possible by deepening of the groove at a three grain junction during grain growth. Miller, Lange and Marshall (1990) confirmed this idea through experiments and further developed the thermodynamics of this break-up process in the same manner described above for the polycrystalline fiber.

The break-up of a polycrystalline film into isolated grains due to grain growth is, an important problem. Generally this break-up phenomenon is unwanted, e.g., the film may be expect to either protect the underlying substrate, or carry an electric current. On the other hand, use can be made of the film after it breaks into isolated island, e.g., the isolated islands can be 'seeds' for grain growth during subsequent deposition of material (Miller and Lange, 1991).

The instability phenomena associated with polycrystalline thin films is directly analogous to that discussed above for fibers. Namely, once the grains within the film develop a columnar microstructure (each grain spans the film thickness), further grain growth will cause a free standing film to shrink. When shrinkage is constrained by the adherence to a substrate, the film will lower its free energy by eventually breaking into isolated islands. The free energy change associated with this instability can be model similar to the fiber. The initial film is assumed to be flat with a thickness t, composed of identical grains with the shape of hexagonal prisms with a center to center distance (grain size), D. The configuration angle, ψ initiates with a value of 180° and decreases during grain boundary grooving and further decreases to -180 ° as the isolated grain 'dewets' the substrate. (A wetting angle is defined, $\theta = (\pi - \psi)/2$, which physically explains the negative values of ψ once the grain boundaries disappear.) In the case of the polycrystalline film, the free energy per grain is given by:

$$E = A_s \gamma_s + A_b \gamma_b + A_i \gamma_i + A_{sub} \gamma_{sub} \qquad (4)$$

where subscripts s, b, i, and sub stand for surface, grain boundary, interface and substrate, respectively. Analogous to the fiber case, the different interfacial areas can be expressed in terms of the aspect ratio of the grain, (L/t) and the configuration angle, ψ. The only different between the free energy functions for the fiber and film is the larger number variables. Similar to the fiber case the free energy (eq 4) can be expressed as a function of the configuration angle, for a specific set of interfacial energies, expressed by ratios, and thus, by angles (analogous to the dihedral angle). Like-wise, similar to the fiber story, all information can be summarized with multi-variable, equilibrium configuration diagram were the boundaries in this diagram are conditions where two configurations (e.g., the covered and 'island' configuration) have the same free energy.

4. Constrained Densification of Powders

The densification of powder compacts containing reinforcements for ceramic matrix composites introduced the subject of constrained densification to the powder densification community. As detailed elsewhere (Lam and Lange, 1994) different models were put forth to explain the incomplete (or absence) of composite shrinkage. Figure 3a, which shows a periodic distribution of spherical, dense inclusions distributed within a powder compact can be used to explain why the reinforcements prevent the composite from achieving full density.

Consider two slices removed from the composite shown in Fig. 3a, one containing the inclusions, and one taken from between the inclusions, that only contains the powder matrix. Fig. 3b shows that if each slice were heated to allow the powder to densify, the slice containing the inclusions were shrink less because it contains a smaller fraction of the powder, which is the only phase that shrinks. In the composite, this differential shrinkage can not be tolerated unless the slices containing the inclusions were subjected to a sufficient compressive stress to cause the densifying matrix to deform into the regions were shrinkage is constrained. When the powder matrix is a crystalline material that undergoes growth as it densifies, and thus becomes very resistant to deformation by the small 'sintering stresses', high density and low density regions develop within the matrix phase during its 'densification' (Sudre and Lange, 1992).

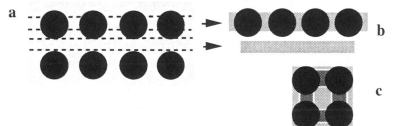

Figure 3 a) Powder matrix containing inclusions. b) Slices (dotted in composites) shrink after matrix densification; matrix slice shrinks more. c) Unit cell showing dense matrix between closer inclusions, and partially dense matrix within cell.

For the periodic array (Lam and Lange, 1994), the matrix between the inclusions with the smallest spacing (cell edges), becomes dense, whereas the matrix at the center of the cell is constrained from shrinking as shown in Fig. 4c. When the inclusions do not initially touch one another, then the composite shrinks until the matrix between the inclusions becomes fully dense. If the inclusions already form a touching network, then the composite does not shrink as the matrix undergoes densification by the development of large crack-like voids as discussed below.

Unlike the fibers and films that start off as dense bodies, the de-sintering phenomena within the powder matrix of the composite is concurrent with its densification. This seemingly contradictory statement occurs because powder regions within the matrix do fully densify, but voids within the matrix concurrently grow. As illustrated in Fig. 3c, shrinkage within the powder matrix is not uniform, i.e., dense and low density regions arise due to different constrains. In addition, cracks present in the matrix suffer large opening displacements (without growth) as opposing regions across the crack shrink in opposing directions. Observations show that two- and three-grain bridges shown in Fig. 4a,b exist between denser regions as schematically illustrated in Fig. 4c (Sudre and Lange, 1992). De-sintering is sequentially observed to occur at these bridges, causing the opposing regions to be less constrained (no longer connected to an opposing region) and allowing these regions to increase their density. When the grain bridges de-sinter, the voids appear to grow bigger. Thus, although the composite may not shrink at all, the matrix will undergo densification, where some voids within the powder matrix shrink and disappear, while others link together and grow larger.

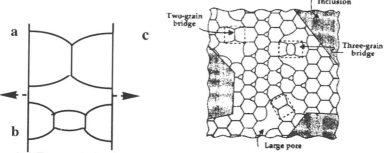

Figure 4 a) Two grain bridge and b) three grain bridge between denser regions in powder matrix phase. c) Schematic of void distribution within matrix phase, where regions are linked by two- and three-grain bridges (Sudre and Lange, 1992).

De-sintering can occur in an identical manner as illustrated above for the fiber, viz., by grain coarsening, which would be best illustrated by Fig. 4b, i.e., the disappearance of the smaller, central grain. Concurrent with grain coarsening, opposing regions will exert a tensile stress on the connecting grain pairs (arrows in Fig. 4), which, in effect, also increases their aspect ratio. Calculations that involve either coarsening, or tensile extension or both, lead to similar free energy functions as discussed above for the grain pairs in the fiber analysis, and similar configurational equilibrium diagrams. As grain become larger, new grain pairs now link other regions, and these grain pairs undergo de-sinter, to further unconstrained the densifying regions. These events are certainly not isolated, but occurs simultaneously throughout the matrix, and continue until a dense, polycrystalline matrix is formed that contain reinforcements and voids. Since the constrain to shrinkage within one regions is directional, depending on the location of neighboring inclusions, the voids tend to develop a directional character, and appear crack-like-not a good omen for a structural ceramic.

5. Concluding Remarks

As reviewed above, the microstructural instabilities associated with grain growth in fiber and film, constrained from shrinking is caused by a phenomenon where previously formed grain pairs, de-sinter to reduce their free energy. A greater free energy reduction would have been achieved if the grain pairs remained intact and their mass centers could move together during grain growth. Similar phenomena occur within the partially dense matrix of ceramic composites. Here, either non-touching or touching reinforcement networks give rise to the constraint to shrinkage. Microstructural observations of the constrained network are very similar to the micrographs shown on pages 482 and 483 of Introduction to Ceramics by Kingery et al. (1976) Since these micrographs are taken for a ceramic powder that does not contain reinforcements, the question arises: What gives rise to the constrain to shrinkage within a powder compact. The answer to this must be the non-uniformity of the powder compact itself, leading to connective network that densify before others, thus constraining the shrinkage of more porous regions. These denser networks are easily observed in the micrographs on pages 482 and 483.

Before this review is closed, it should be pointed out that de-sintering also occurs in two phase systems. One type of two phase system, is where the second phase is non-connective, i.e., an inclusion phase. When the 'inclusion' phase becomes mobile, i.e., diffuses as rapidly as grain boundaries move in the major phase, it collects at 4-

grain junctions (Lange and Hirlinger, 1984). For this case, the ratio of the inclusion to grain size remains constant. Since this subject would require several more pages of this review, the reader is asked to see how this is connected to the phenomenon of de-sintering. When the volume fraction of the second phase is large enough, it becomes a connective, polycrystalline, interpenetrating phase. Growth of grains within the connective, interpenetrating phase also requires de-sintering which the reader might also ponder. In fact, the second phase might be considered a void phase that does not disappear. De-sintering also occurs in powder compacts that sinter and coarsen by evaporation-condensation; in this case the size of the void phase increases with the grain size to produce larger, self-similar microstructures. Thus, as stated above, de-sintering is a pervasive phenomena.

Acknowledgments
This work was supported by AFOSR under Contract No. 91-0125.

References
W. D. Kingery, H.K. Bowen and D.R. Uhlmann, *Introduction to Cermaics, Sec. Ed.,* Wiley, NY, 1976.

K. T. Miller and F. F. Lange, "The Morphological Stability of Polycrystalline Fibers," Acta Met. 37 [5] 1343-7 (1989).

J. Plateau, "Statique Expérimentale et Théorique des Liquides Soumis aux Seules Forces Moléculaires," Paris, 1873.

J. S. W. Rayleigh, *Theory of Sound,* Vol. II, Dover, New York, 1945.

F. A. Nichols and W. W. Mullins, "Morphological Changes of a Surface of Revolution to Capillarity-Induced Surface Diffusion," J. Appl. Phys., 36 1826 (1965).

B. J. Kellett and F. F. Lange, "Thermodynamics of Densification: Part I, Sintering of Simple Particle Arrays, Equilibrium Configurations, Pore Stability, and Shrinkage," J. Am. Ceram. Soc. 72 [5] 725-34 (1989).

D. J. Srolovitz and S. A. Safran, "Capillary Instabilities in Thin Films. I. Energetics," J. Appl. Phys. 60 [1] 247-54 (1986).

D. J. Srolovitz and S. A. Safran, "Capillary Instabilities in Thin Films. II. Kinetics," J. Appl. Phys. 60 [1] 255-60 (1986).

K. T. Miller, F. F. Lange, and D. B. Marshall, "The Instability of Polycrystalline Thin Films: Experiment and Theory," J. Mat. Res. 1 [5] 151-60 (1990).

K. T. Miller and F. F. Lange, "Highly Oriented Thin Films of Cubic Zirconia on Sapphire Through Grain Growth Seeding," J. Mat. Res. 6 [11] 2387-92 (1991).

D.C. Lam and F. F. Lange, "Microstructural Observations on Constrained Densification of alumina Powder Containing a Periodic Array of Sapphire Fibers," J. Am. Ceram. Soc. 77 [7] 1976-8 (1994).

O. Sudre and F. F. Lange, "The Effect of Inclusions on Densification. Part I: Microstructural Development in an Al_2O_3 Matrix Containing a High Volume Fraction of ZrO_2 Inclusions," J. Am. Ceram. Soc. 75 [3] 519-24 (1992).

F. F. Lange and M. M. Hirlinger, "Hindrance of Grain Growth in Al_2O_3 by ZrO_2 Inclusions," J. Am. Ceram. Soc. 67 [3] 164-8 (1984).

Evolution of Strength Determining Flaws During Sintering

Brian D. Flinn, Rajendra K. Bordia and Jürgen Rödel,*
Department of Material Science and Engineering, University of Washington,
Seattle, WA 98195; *Technische Hochschule Darmstadt, Germany

Abstract
The strength and reliability of brittle materials are controlled by the stress intensity factors associated with defects in the body and the resistance of the material to crack growth. The mechanisms (particle rearrangement, neck formation, densification, grain growth, etc.) which occur while sintering a green ceramic body affect both the evolution of the largest flaw and the fracture resistance, which depends on the average microstructure. Thus, in order to gain a full understanding of how processing parameters influence the strength of ceramics it is critical to combine the evolution of the extreme microstructural features (flaws) with average microstructural features.

In this study, we report on the evolution of elastic modulus, crack tip toughness and strength as a function of density. In addition, we have also studied the evolution of the size of artificially induced strength limiting flaws. Knowing the fracture toughness, strength and size of the flaw, the shape factor has been estimated as a function of density.

1. Introduction
The evolution of the elastic modulus during densification has been studied in detail. Two good reviews which compare the various theoretical and experimental results are by Ramakrishnan & Arunachalam (1993) and Rice(1989). Over a wide range of density, a linear relationship has been found between the modulus and density (Coble and Kingery, 1956; Gibson and Ashby, 1988; & Lam, et. al., 1994). In addition, Green, et. al.(1990) have shown that during the early stages of sintering, the modulus increases without significant increase in density due to rapid inter-particle neck growth. In contrast, the evolution of strength determining fracture toughness during sintering has not been well investigated. For materials with no or a shallow R-curve, fracture is determined exclusively by the crack tip toughness, K_o.(Seidel, et. al., 1994). Two recent studies have proposed an experimental technique that can be used to measure K_o for dense or partially sintered bodies (Seidel and Rödel, 1995 & Knechtel, et al., 1995). The crack tip toughness was found to be

proportional to the elastic modulus. Finally the steady state toughness (K_{IC}) has also been shown to be proportional to density (Lam et. al., 1994).

The fracture strength is controlled by both the average and extreme microstructural features. The strength of porous materials has been phenomenologically described by Ryshkewitch(1953) using an exponential function. This expression has found wide use and fair experimental agreement in the range of porosity 0.03-0.6. More recent work by Maiti, et. al.(1984) has started to include the effects of flaw size and fracture resistance of the solid materials on strength. In his review, Rice (1989) has indicated the effect of various pore geometries and pore to crack size ratios on strength. However, the evolution of the size and shape of the strength limiting extreme microstructural features during sintering is not well understood. In this study, we report preliminary results on the evolution of both natural and artificially induced strength limiting flaws.

2. Materials And Measurements

A very fine, high purity Al_2O_3 powder (99.99%) which sinters to full density (1350° C for 0.5 hr.) with a small final grain size (<1 µm) was used (Tamai DAR). Billets for mechanical test specimens were slip cast from an electrostatically dispersed suspension (pH 3.7, 40 vol.% solids) with 2 wt% Carbowax 8000 binder. In order to produce samples with artificial flaws, very low volume fractions (0.1 vol.%) of uniform polystyrene spheres (102 µm ±1µm) were introduced in the suspension (the spheres burn out during subsequent heat treatment leaving behind well defined spherical pores). After mixing and ultrasonication the suspension was coagulated by the addition of NH_4Cl (Velamakanni, 1994) to minimize the segregation of spheres due to buoyant forces. After casting and drying, green density was measured (geometrically) and the billets sintered at 1050-1350 °C for 0.5 hours to produce samples with relative densities varying between 0.60 - 1. The linear intercept grain size and density as a function of sintering temperature are shown in Figure 1. A typical microstructure from a fully dense sample is shown in Figure 2a. From these figures it is evident that a very high density material with small, uniform equiaxed grain size can be produced at low temperatures. An artificially introduced spherical defect is shown in Figure 2b.

Mechanical test specimens (25 mm x 3 mm x 4 mm) were cut from the sintered billets, and then ground, chamfered and polished by standard diamond finishing techniques to a final grit size of 1µm. After final polishing, density was determined by the Archimedes method to determine open and closed porosity. Grain growth was measured by quantitative microscopy on sintered samples. The elastic modulus was measured using a sonic resonance technique, according to ASTM C 1198.

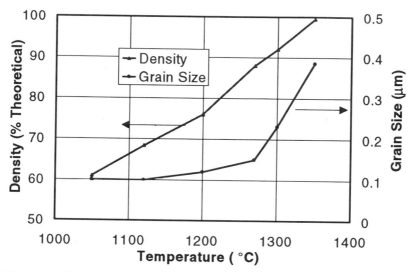

Figure 1. *Sintered density and grain size of Al_2O_3 Powder as a function of isothermal sintering temperature (0.5 hours at temperature)*

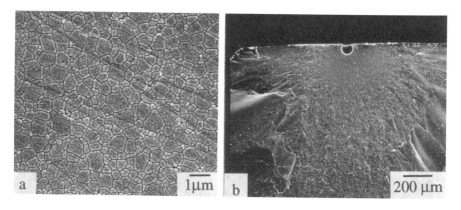

Figure 2. *(a) Microstructure of fully dense sample (1350° C, 0.5 hour). (b) Fracture surface illustrating artificial defect.*

Fracture strength was measured in fully articulated, four point bending. After failure, the fracture surfaces were quantified by stereo and scanning electron microscopy for failure origin, defect type, defect size and position relative to the tensile surface.

Crack tip toughness, K_o, was determined from crack tip opening profiles on indentation induced cracks calculated from the Barenblatt (1962) expression using the technique described by Seidel and Rödel (1995).

$$K_O = uE'\sqrt{\frac{\pi}{8x}} \tag{1}$$

where u is the 1/2 crack opening, x is the distance from the crack tip and E' is the plane strain elastic modulus. Measurements were taken 5 to 100 μm behind the crack tip, and compared with the calculated curve from Eqn. 1. K_o was taken from the average value of specific measurements over the range which fit Eqn 1, generally from 5 to 40 μm.

3. Results

The modulus of elasticity scaled linearly with density, ranging from 40-390 GPa for ρ/ρ_{th} 0.60-.99. The crack tip toughness increased from 0.2 to 1.8 MPa-m$^{1/2}$ over the same density range. Linear regression of the data gave the following empirical relationships:

$$E = 903\left(\frac{\rho}{\rho_{th}}\right) - 510 \tag{2}$$

and

$$K_o = 4.13\left(\frac{\rho}{\rho_{th}}\right) - 2.26 \tag{3}$$

In Eqn's. (2) and (3) the units for E are GPa and MPa-m$^{1/2}$ for K_o. As previously reported, these properties increase linearly with density in agreement with other published results (Lam, et. al., 1994; Green, et. al, 1990 & Knechtel, et. al., 1995).

During sintering, the polymer spheres burned out leaving spherical pores. The size of the spherical pores, C was found to scale with the linear shrinkage of the specimens.

$$C = C_o[1 - \frac{1}{3\rho}(\rho - \rho_o)] \tag{4}$$

where C_o is the initial size of the sphere, ρ and ρ_o are the current and initial theoretical density.

Fractography of modulus of rupture (MOR) specimens containing the spherical pores confirmed that the pores can act as strength limiting defects (see fig. 2b). The position of the strength limiting flaw was characterized with respect to the tensile surface of the test beam as a surface defect or volume defect. The modulus of rupture as a function of density is shown in Figure 3 for

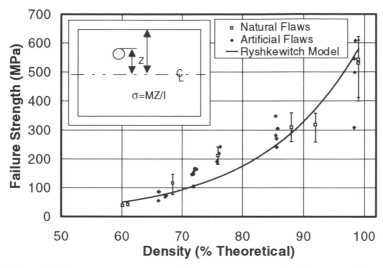

Figure 3. Fracture strength vs. density for natural and artificial flaws.

samples with and without artificial flaws. The average strength of fully dense specimens without artificial defects was over 550 MPa. When the strength limiting flaw was an artificial defect positioned below the tensile surface (volume defect), the failure stress was corrected to reflect the stress experienced by the defect by scaling with the distance from the neutral axis as shown schematically in the insert in Figure 3. The strength vs density data for samples which failed from natural flaws fits reasonably well with the Ryshkewitch model.

4. Discussion

The failure strength, σ_f, of brittle ceramics is given by

$$\sigma_f = K_{crit} / Y\sqrt{C_{crit}} \tag{5}$$

where K_{crit} is the critical fracture toughness, C_{crit} is the critical crack size and Y is a geometric parameter which describes the crack shape. Using crack tip toughness (K_o) data and fracture strength, the equivalent critical flaw size, Y^2C, was determined for all specimens from $Y^2C_{crit}=(K_{crit} /\sigma_f)^2$ by setting $K_{crit}= K_o$. As can be seen in Figure 4 the equivalent flaw size for natural flaws decreases rapidly in the early stages of densification. The equivalent flaw size for the artificial flaws follows the same trend. Also shown in Figure 4 is the measured and actual diameter of the spheres which only decreased approximately 10% over the same range. The solid line through these data points is Eqn. 4, indicating a good fit between the measured and calculated flaw size.

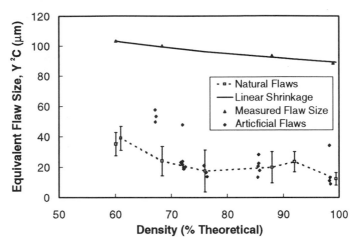

Figure 4. Equivalent flaw size vs density. Actual size of artificial defects also shown on figure for comparison.

The shape factor Y can be obtained for the samples in which the failure origin were artificially induced pores, since their size is known as a function of density. The average calculated shape factors at different densities are given in Table I. Notice the large decrease in the early stages of densification. By comparing the decrease in size of the flaw (10%) with the decrease in the shape factor (factor of 2.2), it can be seen that the change in the shape factor with densification is more significant than the flaw size.

Density (% Theoretical)	67	72	76	85	98
Shape Factor, Y	.76	.5	.41	.45	.34

Table I. Shape factor vs density

Previous work on failure from spherical defects has shown that there is no stress intensity factor for a spherical void, unless a circumferential crack is present (Barrata, 1981; Rice, 1984; Green, 1980) as shown in Figure 5. Using the stress intensity factor estimates by Baratta (1981), crack tip toughness and artificial defect size, R, from our specimens, the circumferential crack length, L, was calculated iteratively. For the fully dense samples, the average calculated circumferential crack length was 0.6 μm which is on the order of the average grain size. This compares very well with models which have suggested that the circumferential crack length should scale with grain size (Evans and Tappin, 1972). However, the calculated circumferentail crack length for the 67% theoretically dense samples increases to almost 6 μm, while the grain size is only 0.2 μm at this density. It is not clear what microstructural characteristic could correlate with dimension of the calculated circumferential crack for the porous samples. It may be necessary to develop other models to account for the stress intensity factor of spherical voids in porous bodies which do act as strength limiting flaws.

Figure 5 Pore with circumferential crack model after Baratta.

5. Conclusions

The linear variation of the elastic modulus and fracture toughness with porosity is relatively well understood and modeled, however the effect of porosity on fracture strength is not as clear. In order to fundamentally investigate the effect of strength on porosity, fracture strength was measured for samples with both natural and artificial flaws of known geometry. For the case of failure from natural flaws, the dependence of strength on density followed the well established Ryshkewitch empirical relation. Using the toughness as a function of density, the equivalent flaw size (Y^2C) was calculated as a function of density. However, since the actual size or shape of the natural flaws was not known their dependence on density could not be obtained. When failure occurred from artificial flaws, it was shown that the functional dependence of strength on density was similar to that for natural flaws. Therefore artificial flaws can be used to fundamentally investigate the evolution of strength during sintering. For failure from artificial flaws, the strength increased by a factor of seven as density increased from 67 to 99 %. This strength increase was due to three factors: increase in toughness (a factor of 3.6), decrease in flaws size (roughly 10 %) and decrease in shape factor (a factor of 2.2). The increase in toughness and the decrease in flaw size is well understood however, the evolution of microstructure and its influence on the shape factor needs to be further investigated to gain a full understanding of the dependence of strength on density. We have used a circumferential crack around a pore as the geometric model of the failure origin for samples that failed from artificially induced pores. This model explains the results at high density well (i.e. the circumferential crack is of the order of the grain size). However, at low density, the calculated crack size does not correspond to any measured microstructural features.

Acknowledgment

This work was funded in part by NSF grants: DMR 9257027 and MSS9209775.

References

N. Ramakrishnan and V. S. Arunachalam, "Effective Elastic Moduli of Porous Ceramics," J. Am. Ceram. Soc., **76** [11] 2745-52 (1993).

R. W. Rice, "Relation of Tensile Strength-Porosity Effects in Ceramics to Porosity Dependence of Young's Modulus and Fracture Energy, Porosity Character and Grain Size", Mat. Sci. & Engr., A112, 215-24 (1989).

R. L. Coble and W. D. Kingery, "Effect of Porosity on Physical Properties of Sintered Alumina," J. Am. Ceram. Soc. **39**, 377 (1956).

L. J. Gibson and M. F. Ashby, *Cellular Solids,* Pergamon Press, Oxford, 1988.

D. C. C. Lam, F. F. Lange and A. G. Evans, "Mechanical Properties of Partially Dense Alumina Produced from Powder Compacts," *J. Am. Ceram. Soc.*, **77** [8] 2113-2117 (1994).

D. J. Green, C. Nader and R. Brezny, "The Elastic Behavior of Partially-Sintered Alumina," Sintering of Advanced Ceramics, Ceramics Transactions Vol. 7, pp. 345-56, American Ceramic Society (1990).

J. Seidel, N. Claussen, and J. Rödel, "Reliability of Alumina Ceramics I. Effect of Grain Size," J. Eur. Ceram. Soc. **15**, 395-404 (1995).

J. Seidel and J. Rödel, "Crack Tip Toughness of Alumina as a Function of Grain Size," submitted to *J. Am. Ceram. Society* (1995).

M. Knechtel, C. Cloutier, R. K. Bordia, and J. Rödel, "Mechanical Properties of Partially Sintered Alumina," submitted to *J. Am. Ceram. Society* (1995).

E. Ryshkewitch, "Compression Strength of Porous Sintered Alumina and Zirconia" communication *J. Am. Ceram. Soc.*, **36** [2] 65-68 (1953).

S. K. Maiti, M. F. Ashby and L. J. Gibson, "Fracture Toughness of Brittle Cellular Solids," *Scripta Met.* **18**, 213-17, (1984).

B. V. Velamakanni, F. F. Lange, F. W. Zok, and D. S. Pearson, "Influence of Interparticle Forces on the Rheological Behavior of Pressure-Consolidated Alumina Particle Slurries," *J. Am. Ceram. Soc.*, **77**[1] 216-20 (1994).

G. I. Barenblatt, "The Mechanical Theory of Equilibrium Cracks in Brittle Fracture," *Adv. Appl. Mech.*, **7**, pp. 55-127 (1962).

F. I. Baratta, "Refinement of Stress Intensity Factor Estimates for a Peripherally Cracked Spherical Void and a Hemispherical Surface Pit," communications of The American Ceramic Society, C3-C4, January 1981.

R.W. Rice,"Pores as fracture origins in ceramics,"*J. Mat. Sci.*,**19**,895-914(1984).

D. J. Green, "Stress Intensity Factor Estimates for Annular Cracks at Spherical Voids," *J. Am. Ceram. Soc.* **63**,4-5, pp. 342-344 (1980).

A. G. Evans and G. Tappin, "Effects of Microstructure on the Stress to Propagate Inherent Flaws,: *Proc. Br. Ceram. Soc.*, **20**, pp. 275-97(1972).

Problems in Viscoelastic Neck Growth

C. Argento, S. Mazur, and A. Jagota

Central Research & Development Division
E.I. Dupont de Nemours & Co. Inc. - P.O. Box 0356
Wilmington, DE 19880-0356

Abstract

Particle coalescence is central to many industrial fabrication processes for organic polymers. While formally analogous to the sintering of inorganic materials, the mechanics and kinetics which govern polymer coalescence are fundamentally different, in large part because of the viscoelastic rheology of high molecular weight polymers. These issues are reviewed in relation to neck growth between spheres. Our objective is to develop physically rigorous computational models for visco-elastic coalescence of polymer particles based on realistic rheology, particle size, and force fields.

I- Viscoelastic Response of Amorphous Polymers

Coalescence of an aggregate into a uniformly dense body requires deformation of the component particles into space-filling shapes. A sufficient description of the time-dependent response to deviatoric stresses for an isotropic material is provided by the shear compliance $J(t)$ and Poisson's ratio ν (which may also be time-dependent). $J(t)$ equals the ratio of strain at time t to a constant stress imposed on the fully relaxed sample at $t=0$. The stress-strain relation is defined using the Boltzman superposition principle:

$$\mathbf{e}(t) = \int_{-\infty}^{t} J(t-t') \frac{d\,\mathbf{s}(t')}{dt'}\, dt' \tag{1}$$

where $\mathbf{e}$ is the deviatoric strain and $\mathbf{s}$ is the deviatoric stress. An analogous expression applies for the bulk compliance.

For linear high molecular weight polymers, $J(t)$ reflects kinetic processes which span several decades in time (Plazek, 1980) and can be partitioned into recoverable and Newtonian viscous contributions:

21

$$J(t) = J_r(t) + t / \eta_0 \tag{2}$$

The recoverable compliance $J(t)$ represents that part of the time-dependent strain which will spontaneously recover if the stress is removed at time t.

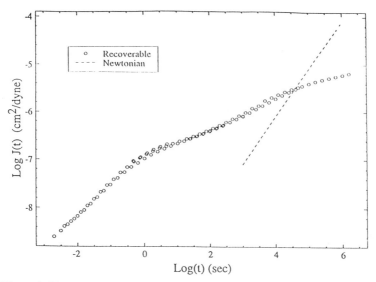

Figure 1: Shear creep compliance for poly(methyl methacylate co ethyl acrylate), molecular weight 101,000. Data at four different temperatures has been reduced to 133°C (Mazur & Plazek, 1994).

Figure 1 illustrates typical behavior for an acrylic copolymer with broad molecular weight distribution at 20°C above its glass temperature T_g. $J(t)$ increases from the glassy value ($J(0) < 1$GPa^{-1}) to about 2 MPa^{-1} in the "plateau" regime. It is noteworthy that viscous flow remains negligible relative to $J_r(t)$ for more than three hours. Since η_0 increases strongly with polymer molecular weight while the growth in $J_r(t)$ through the plateau regime is relatively insensitive, it is not uncommon to find even greater temporal separation of the two deformation mechanisms. Moreover, the principle of time/temperature super-position (Ferry, 1980) (thermo-rheological simplicity) dictates that retardation times for the growth of $J_r(t)$ and t / η_0 respectively will decrease by the same factors with increasing temperature. Thus even at much higher temperatures, recoverable viscoelastic deformation is always several orders of magnitude faster than viscous flow.

For times and temperatures corresponding to the plateau regime, polymer melts resemble an ideal elastic material with time-invariant compliance J_n, ($J_n \sim 2\text{MPa}^{-1}$ in Fig. 1). A logical starting point for analyzing deformation of viscoelastic particles is the corresponding mechanics for elastic materials.

II - Elastic Deformation for Various Forces

Particle deformation and the growth of interparticle contacts may be driven by different kinds of forces depending upon processing conditions. Analytic results for equilibrium elastic contacts under applied loads or surface tension provide some insights relevant to actual processing conditions for polymers.

In compression molding a polymer powder is heated above T_g and subjected to applied load. For a pair of elastic spheres of radius r and compliance J_n pressed together by an axial force F, the radius of the contact x may be calculated at small contacts according to Hertz (Timoshenko & Goodier, 1955):

$$\frac{x}{r} = \left[\frac{3(1-\nu)J_n F}{8r^2}\right]^{\frac{1}{3}} = \left[\frac{3\pi(1-\nu)J_n P}{8}\right]^{\frac{1}{3}} \tag{3}$$

where P is the mean interparticle pressure. Starting from a regular packing of spheres, the extent of densification can be related to growth of the dimensionless contact radius x/r. For realistic packing densities complete densification is achieved at $x/r \cong 0.58$ (Mazur, 1995). According to Eq. 3 this limit corresponds to $P \approx 1/(3J_n)$ and is independent of particle size (r). For the acrylic polymer of Fig. 1, Eq. 3 predicts that full density should be achieved with pressures of order 200 KPa. In the absence of other processes, elastic densification is perfectly reversible so that the sample would return to its initial density upon removing the load. In practice irreversibility is a consequence of the relaxation of the elastic stresses and diffusion of polymer chains across the particle interfaces. These processes require times comparable to the terminal relaxation time $\tau_n = J_n \eta_0$ (Mazur, 1995). Thus compression molding times are primarily governed by internal relaxation which has little to do with densification *per se*.

In the absence of external loads, neck growth and particle deformation may be driven by interfacial tension γ between the polymer and surrounding medium (for example, air). By assuming that the surface forces operate only across the contact, Johnson, Kendall, & Roberts (1971) (JKR theory) extended Hertz's analysis to obtain the following expression:

$$\frac{x}{r} = \left[\frac{9\pi(1-\nu)J_n \gamma}{8r}\right]^{\frac{1}{3}} \tag{4}$$

Unlike the simple Hertz case, here x/r depends explicitly on r. For a given material such that J_n, ν, and γ are constants, the extent of deformation may be increased by decreasing r. Thus to achieve $x/r=0.58$ requires $r \cong 9\gamma J_n$.

These models for elastic materials reveal the importance of $J_r(t)$ for neck growth in the plateau regime, but they are of limited predictive value. Firstly, equations (3) and (4) are derived in the limit of small contact area with uncertain reliability for $x/r \sim 0.5$. Secondly, the relationship between an equilibrium elastic contact and the kinetics of viscoelastic growth of a contact is non-trivial. Substitution of $J_r(t)$ for J_n in eq.(4), which is inconsistent with Boltzmann's superposition principle, predicts neck growth kinetics in qualitative agreement with experiment (Mazur & Plazek, 1994) but results in large quantitative error in the time scale. A more sophisticated approximation (Lee & Radok, 1960) (still limited to small contacts) requires independent knowledge of the stress history and is therefore inapplicable to neck growth driven by surface tension.

III- Interfacial Driving Forces

A rigorous analysis of neck growth requires accurate description of the interfacial forces originating from deviation in molecular interactions near a surface relative to those in the bulk. These forces may be manifest in continuum mechanics in three different ways depending upon the size and structure of volume elements over which the molecular interactions are formally integrated.

The least detailed manifestation of surface forces (the thermodynamic approach) corresponds to the integration of molecular interactions over the entire body. Thus the interfacial force associated with a change in some linear dimension of an object equals the corresponding derivative of the surface energy (surface area times γ). This isotropic average force reveals nothing about the local deviatoric stress field responsible for specific changes of shape. A more localized manifestation of surface forces corresponds to integration of molecular interactions along the "immediate continuous vicinity" of a local volume element. The driving force for deformation of the volume element may be obtained from the Young-Laplace equation (Adamson, 1976), which gives an effective surface traction proportional to the product of surface energy and surface curvature. The tractions acting on a curved surface is then:

$$\mathbf{t} = -\gamma \, \mathrm{tr}(\mathbf{k})\mathbf{n} \qquad (5)$$

where $\mathbf{t}$ is the surface traction from the excess free energy, γ is the surface free energy, $\mathbf{k}$ is the surface curvature and $\mathbf{n}$ is the surface normal. This approach has been successfully used to study fluid deformation, as in viscous sintering (Jagota & Dawson, 1990; Hiram & Nir, 1983). However, preliminary calculations showed that it fails when the material is elastic or viscoelastic. The main contribution of this term is the expansion of the contact area, not the creation of new contact, which is qualitatively different from what is observed in experiments.

None of these manifestations can describe the interactions of molecules operating across a gap of a second medium. In this case, the complete description of the interaction potential is necessary. For non-polar polymer molecules, it is assumed that the interactions are of Van der Waals type and can be described by a Lennard-Jones potential. Continuum mechanics assumes a uniform density distribution which is a valid approximation only for intermolecular distances larger than the minimum in this potential (the nearest neighbor separation). For this purpose only the attractive part of the potential is important and it will be necessary to truncate integrations at some cut-off distance, which is large compared to the minimum in the true molecular potential.

The attractive potential is expressed by:

$$w = -A / \left(\pi^2 d^n \right) \tag{6}$$

where A is the Hamaker constant for the given material configuration and d is the distance between the two molecules . For the case of Van der Waals interactions, in the non retarded regime, $n=6$ (Israelachvili, 1994). This contribution has been widely studied under the scope of elastic adhesion (Derjaguin, Muller & Toporov, 1975) (DMT theory). In the DMT model, the adhesion is presumed to result entirely from Van der Waals attraction outside of the area of contact. In contrast, in the JKR model, the adhesive forces are derived from the thermodynamic approach and assumed to be entirely restricted to within the contact area.

The force acting on a point of body V_1 due to another volume V_2 is given by the volume integration:

$$\mathbf{f} = \int_{V_2} \nabla w \, dV_2 \tag{7}$$

This volume integration is very time consuming and extremely sensitive to the mesh coarseness. Since these forces are concentrated near the surface (Tabor, 1981), it would be useful to express them in terms of a surface force resulting from a surface integration. Therefore, the approach adopted was the one derived by Argento, Jagota & Carter (in preparation). It consists of a new integration method for inverse power law body forces of high order ($n>3$) that allows the reduction of the three dimension volume integration to a two dimension surface integration. The body force field $\mathbf{f}$ is replaced by a surface stress tensor $\mathbf{h}$, of the form:

$$\mathbf{h} = \int_{S_2} \mathbf{G}\, d\mathbf{S}_2 \qquad (8)$$

where S represents the surface of bodies 1 and 2. The kernel $\mathbf{G}$ has the following form:

$$\mathbf{G} = \frac{x_i \mathbf{e}_i}{(3-n)(\mathbf{x} \cdot \mathbf{x})^{\frac{n}{2}}} \qquad (9)$$

where $\mathbf{x}$ is the vector linking the point on the surface of 1 to a point on the surface of 2, x_i is the component of $\mathbf{x}$ on the direction $\mathbf{e}_i$, which is the coordinate system unitary base vector. In the case of near parallel surfaces, like the region near the contact zone between two large spheres, this scheme reduces to the well known "Derjaguin approximation" (Derjaguin, 1934).

IV- Numerical Study

Since no adequate analytical model exists to describe the kinetics of viscoelastic neck growth under the conditions described beforehand, it seems convenient to study this problem with the help of a numerical tool such as the finite element method. Therefore, the method described in the previous section was implemented in the ABAQUS™ finite element code (Hibbitt, Karlsson & Sorensen, Inc.). Calculations were made of contacts of elastic spheres where surface forces were represented by a combination of Young-Laplace and van der Waals contributions.

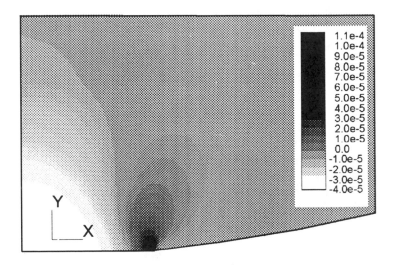

Figure 2: Vertical stresses near the contact between a deformed viscoelastic sphere and a plane.

In preliminary calculations the results were found to be sensitive to the cut-off distance. For an appropriate choice of the cut-off (by trial and error), we obtained a good agreement with the JKR analytical model. Figure 2 shows the vertical stresses in a close-up around the contact between a deformed viscoelastic sphere and a rigid plane. It can be seen that the stress profile is close to the one predicted by the JKR theory, the external ring of the contact is under high tensile stresses while the core of the contact is under compressive stresses.

Some preliminary viscoelastic calculations have also shown that, in agreement with experimental results (Mazur & Plazek, 1994) the time scale of relaxation of the contact radius is orders of magnitude greater than the time scale the time scale of relaxation of the material.

V- Conclusion

A numerical model to describe elastic and viscoelastic neck growth under interfacial forces was implemented in a finite element code and some preliminary calculations were performed. So far, the following can be concluded:

- the curvature formulation based on the surface energy alone is not sufficient to describe the kinetics observed.
- the choice of the cut-off distance for the integration of the attractive potential is very important, a rational criterion for this choice is being worked out.
- qualitative agreement between numerical calculations and experiments has been obtained for viscoelastic sintering, showing the difference of many orders of magnitude between the relaxation time of the radius of contact and the relaxation time of the material.

Acknowledgment
This work has been supported by the Air Force Office of Scientific Research under Grant F49620-95-C-0008

References
D.J. Plazek, in "Methods of Experimental Physics", Vol. 16c, Chap. 11, Academic Press, (1980).

J.D. Ferry, "Viscoelastic Properties of Polymers", 3rd Ed., John Wiley & Sons, (1980).

H. Hertz, as summarized in S.P. Timoshenko & J.N. Goodier, "Theory of Elasticity", 3rd Ed., McGraw-Hill, London, (1955).

S. Mazur, Chap. 8 in "Polymer Powder Technology", M. Narkis & N. Rosenzweig, Ed's., John Wiley & Sons, (1995).

K.L. Johnson, K. Kendall, & A.D. Roberts, "Surface Energy and the Contact of Elastic Solids", *Proc. Roy. Soc. Lond., A.*, **324** 301-313 (1971).

S. Mazur & D.J. Plazek, "Viscoelastic Effects in the Coalescccence of Polymer Partciles", *Prog. Org. Coat.*, **24** 225-236 (1994).

E.H. Lee & J.R.M. Radok, "The Contact Problem for Viscoelastic Bodies", *Trans. ASME, J. Appl. Mech.*, 438-444 (1960).

C. Herring, "Effect of Change of Scale on Sintering Phenomena", *J. Appl. Phys.*, **21** 301-303 (1950).

A.W. Adamson, "Physical Chemistry of Surfaces", pp 4-6, 3rd Ed., John Wiley & Sons, (1976).

A. Jagota and P.R. Dawson, "Simulation of viscous sintering of two particles", *J.Am.Cerm.Soc.*, **73** [1] 173-177 (1990).

Y. Hiran & A. Nir, "A Simulation of Surface Tension Drive Coalescence", *J. Colloid Interface Sci.*, **95** [2] 462-470 (1983).

J. Israelachvili, "Intermolecular & Surface Forces", pp. 155-159, 2nd edition, Academic Press, New York, (1994).

B.V. Derjaguin, V.M. Muller & Y.P. Toporov, "Effect of Contact Deformations on the Adhesion of Particles", *J. Colloid Interface Sci.*, **53** [2] 314-326 (1975).

D. Tabor, "Surface Forces and Surface Interactions", *J. Colloid Interface Sci.*, **58** [1] 2-13 (1977).

C. Argento, A. Jagota & C. Carter, "Surface Formulation for Body Forces", in progress.

B.V. Derjaguin, "Untersuchungen uber die Reibung und Adhasion, IV", *Kolloid Z.*, **69** 155-164 (1934).

Hibbit, Karlsson & Sorensen, Inc., Tel.: (401) 727 4200.

Elasto-Creep Sintering Analysis by Macro-Micro Modeling

F.Tsumori,   T.Aizawa   and   J.Kihara

Department of Metallurgy, University of Tokyo, 7-3-1
Hongo, Bunkyo-ku Tokyo 113, Japan

ABSTRACT

Multi-level modeling is proposed as a new methodology of sintering analysis for quantitative evaluation of geometric changes of green and brown materials together with precise prediction of their microscopic shrinkage in sintering. Variational principles in the elasto-creep frame are utilized to make formulations both for macro and micro models and to simulate time history of geometric configuration, creep strains and stresses by a time incremental procedure. Several numerical examples are employed to verify the validity and effectiveness of the present method.

1 INTRODUCTION

High quality in powder metallurgy requires for quantitative description of mechanical behavior and precise prediction of porosity structure in powder compaction and sintering. For example, since open porosity is still left in the brown body after debinding the injected green compact, significant shrinkage in geometry and dimension is often observed during sintering in the MIM(metal injection molding) process. In this case, reliable sintering analysis is necessary to predict the final shape of sintered products.

In the literature, sintering analysis is generally classified into two approaches : continuum mechanics[1,2] and granular / powder mechanics[3,4] In the former, the elasto-plastic or the rigid-plastic finite element methods are utilized with phenomenological constitutive equations. Hence, they can describe macroscopic shrinkage in geometry, but they have no means to predict local changes of porosity connectivity and to trace evolution of local densification process. In the latter, the microscopic sintering process is directly modeled by time integration of the temporal arrangement of powder particles. The local driving mechanism of sintering can therefore be incorporated into a model, but there is no way to describe macroscopic shrinkage.

In the present paper, a multi-level model is proposed to link the macroscopic model for description of geometric and dimensional shrinkage with the microscopic model for prediction of porosity structure change and local densification. Assuming that constitutive behaviors of materials can be modeled by an elasto-creep equation, the homogenization theory is employed to make theoretical formulation of this multi-level modeling. The time incremental procedure is used to trace both the macroscopic geometric change and the microscopic shrinkage. Several numerical examples are employed to demonstrate the validity and effectiveness of the present model.

2   CONSTITUTIVE MODEL

In this study, viscous sintering behavior is modeled by an elasto-creep constitutive relation for matrix materials. Hence, a material deforms elastically at $t = 0$, and is followed by an elasto-creep deformation for $t > 0$. Among various representations for creep behavior, the following uniaxial creep constitutive equation is selected:

$$\frac{d\varepsilon_c}{dt} = \frac{16\alpha\beta\Omega D_S}{\pi l^2 kT}\sigma = \kappa\sigma \tag{1}$$

where α and β are materials constants nearly equal to unity, Ω the atomic volume, D_S the self-diffusion coefficient, l the edge length of a grain, K the Boltzmann coefficient, and T the temperature.

Owing to the normality rule for creep, Equation(1) is extended to a multiaxial equation:

$$(\dot{\varepsilon}^c)_{ij} = \frac{24\alpha\beta\Omega D_S}{\pi l^2 kT}S_{ij} = \frac{3}{2}\kappa S_{ij} \tag{2}$$

where S_{ij} is the deviatoric stress tensor.

The above elasto-creep behavior is theoretically formulated by using an incremental variational equation. Assuming that all the physical quantities $\{u_0, \varepsilon_0, \sigma_0\}$ are known at $t = t_0$, a true incremental displacement Δu must make the following variational equation stationary during the prescribed time increment Δt :

$$0 = \int_V \delta\Delta\varepsilon^t D\Delta\varepsilon dV - \int_V \delta\Delta\varepsilon^t \Delta\sigma_c dV - \int_S \delta\Delta u\Delta P dS \ , \tag{3}$$
$$\text{or} \qquad 0 = \delta F(\Delta u) \ .$$

From the obtained Δu, both strain and stress increments are calculated by

$$\Delta\varepsilon_{ij} = \frac{1}{2}(\Delta u_{i,j} + \Delta u_{j,i}) \tag{4}$$

and

$$\Delta\sigma_{ij} = D_{ijkl}\Delta\varepsilon_{kl} - \Delta\sigma_{ij}^c \tag{5}$$

where $\Delta\sigma_{ij}^c$ is the creep stress.

To be noted, porosity has no influence on the above elasto-creep behavior of materials; in the following multi-level modeling, both D_{ijkl}^H and $\Delta\sigma_{ij}^H$ in the macroscopic model change with porosity, and creep / recovery process in the microscopic model can be accelerated or decelerated by the transferred stress gradients from the macroscopic model.

3 MULTI-LEVEL MODELING

Macroscopic shrinkage in geometry, dimension and microscopic evolution of porosity connectivity, and powder particle rearrangement during sintering are formulated into multi-level modeling (macro-micro modeling). As depicted in Fig. 1, the macro-model is a porous continuum to be sintered under the prescribed temperature and loading. The micro-model is constructed by a periodic structure with use of a unit cell. Each subdivision of the macro model can have its own unique periodic structure with a different unit cell structure and periodicity. Theoretical combination of macro and micro models is based on the homogenization theory.

Using the asymptotic expansion, true displacement of a solid body can be represented by

$$u(x, y) = u^0(x) + \varepsilon u^1(x, y) \quad \text{for} \quad y = \frac{x}{\varepsilon} \tag{6}$$

where $\{x, y\}$ and $\{u^0, u^1\}$ are, respectively, the coordinate systems and the displacements of macro and micro models, and ε is a periodicity.

Considering that a size of microscopic structure is far smaller than a macroscopic dimension, the periodicity ε must be negligibly small. Hence, $F(u)$ can be modeled by

$$F(u) \simeq F(u^0, u^1). \tag{7}$$

Owing to the periodicity, the original variational equation can be reduced to a system of two variational equations for macroscopic and microscopic domains:

$$\int_V \delta\Delta\varepsilon_x^t(u^0)D^H\Delta\varepsilon_x(u^0)dV = \int_V \delta\Delta\varepsilon_x^t(u^0)\Delta\sigma^H dV + \int_{S_\sigma\cap V}\delta\Delta u_x^t\Delta PdS_\sigma \ , \tag{8}$$

and

$$\int_{V_e} \delta\Delta\varepsilon_y^t D\Delta\varepsilon_y dV_e = \int_{V_e} \delta\Delta\varepsilon_y^t \Delta\sigma_c dV_e + \int_{S_\sigma\cap V_e}\delta\Delta u_y^t\Delta\gamma dS_\sigma \ . \tag{9}$$

As shown in Fig. 1, both physical quantities and properties are mutually transferred between macro and micro models. First, both models are initialized by their prescribed conditions;

- Work material to be sintered is discretized into macroscopic finite element models $\{V_x\}$ with the prescribed loading conditions $P(t)$ but their elastic stiffness $[D^H]$ and creep stress $\{\Delta\sigma^H\}$ are still unknown,

- The matrix in the unit cell is also subdivided into finite elements $\{V_y\}$ with the prescribed surface traction. From the calculated microscopic strains and stresses, $[D^H]$ and $\{\Delta\sigma^H\}$ can be determined for each macro-element V_x.

At $t = 0$, only elastic analysis is made first for each microscopic model to obtain microscopic strain and stress distributions and to determine $[D^H]$ in elasticity. Using the obtained $[D^H]$ for each macro element, another elastic analysis is done for a macroscopic model, which determines an initial macroscopic stress distribution $\{\sigma^0\}$. For $t > 0$ elasto-creep analysis is executed for both models. Using an average stress distribution $\sigma^0(x)$, true displacements, strains and stresses are calculated for local shrinkage. From the calculated stress and stiffness by the above elasto-creep analysis, both $[D^H]$ and $\{\Delta\sigma^C\}$ are updated to be corresponding to current local porosity and creep.

Macroscopic elasto-creep analysis updates current stress distribution by $\{\sigma^0\} = \{\sigma^0\}+\{\Delta\sigma\}$ and predicts macroscopic shrinkage under the prescribed local increment $\Delta P \ (\equiv P(t+\Delta t) - P(t))$.

The above computation is terminated when the time reaches the prescribed holding time t_{hold}. At $t = t_{hold}$, the updated geometry for each unit cell provides local sintering behavior with local stress distribution. Final shape of a sintered product is given by macroscopic model configuration.

4   VERIFICATION ANALYSIS

Copper powders are employed as model materials, the sintering behavior of which is governed by volume diffusion. Necessary data for the following simulations are listed in Table 1. Three examples are taken for verification of the present creep model and the multi-level modeling.

4.1   Uniaxial Creep and Recovery Test

Since the present approach is based on the elasto-creep model, a microscopic model must represent uniaxial creep and recovery behaviors as its spe-

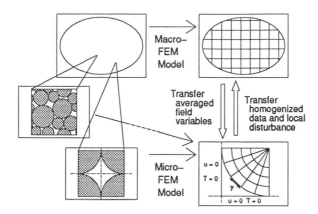

Fig.1: Multi-level modeling

Table1: Computational conditions

α	$\simeq$	1		β	$\simeq$	1	
D_s	$=$	1.45×10^{-13}	m^2/s	l	$=$	1.0×10^{-4}	m
Ω	$=$	1.18×10^{-29}	m^3	k	$=$	1.3807×10^{-23}	J/K
D	$=$	1.298×10^{11}	Pa	ν	$=$	0.343	
T	$=$	1273	K				

cial cases when uniform stress or strain is prescribed. Fig. 2 depicts the time history of microscopic creep strain and stress. As theoretically predicted, creep strains are accumulated linearly with time when a uniform stress is applied and kept constant, while creep stress is exponentially relaxed with time when a uniform strain is applied and kept constant.

4.2 Local Shrinkage by Multiaxial Elasto-Creep

To demonstrate the validity of the present approach to describe local stress redistribution and local shrinkage of a pore, consider an isotropically pressed unit cell with a pore at the center also subjected to surface tension. Due to geometric symmetry, only a quarter of unit cell is discretized as shown in Fig. 3.

For reference, an analytical solution can be calculated by one dimensional analysis for a thick-wall cylinder in stationary creep subjected to constant pressure. As depicted in Fig. 3, in the case when a pore is isolated and far from the interfacial boundary, the calculated stress distribution $\sigma(x, y = 0)$

in the stationary creep region is found to be in good agreement with the analytical solution. Note that there is a change in the sign of the stress from tension to compression along the x-axis. As summarized in Table 2, the local shrinkage rate can be directly evaluated for a unit cell.

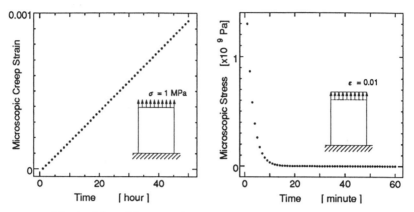

Fig.2: Time history of microscopic strain and stress.

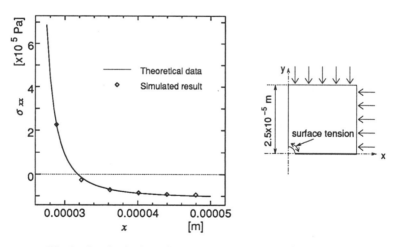

Fig.3: Analytical and simulated data of $\sigma(x, y = 0)$.

4.3   Macro-Micro Interactions in Compression

Uniaxial compression of porous medium is employed to consider stress transfer between macro and micro models. Theoretically, compressibility

Table2: Shrinkage rate of porosity

condition	analytical data	simulated result
only surface tension applied	$-1.10 \times 10^{-14} m/s$	$-1.69 \times 10^{-14} m/s$
$1 \times 10^5 Pa$ iso-static pressed	$-1.50 \times 10^{-14} m/s$	$-2.02 \times 10^{-14} m/s$

and sinterability depend on the density distribution; hence, even when externally applied stresses are uniform, local stress gradients are expected to be different for each portion. In Fig. 4, a material to be sintered has three different porosity structures A to C with different porosity density f.

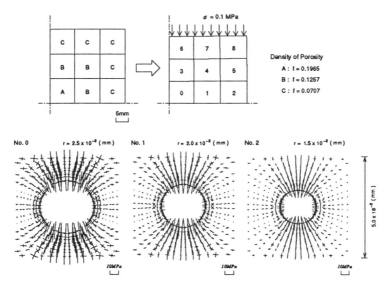

Fig.4: Stress distribution in micro models under uniaxial compressing in macro model.

In this case, the whole body is descretized into nine elements $V_e^x (0 \le e \le 8)$, and a porosity structure A is allocated to V_0^x, B to V_1^x, V_3^x and V_4^x, and C to other elements. First, equivalent elastic stiffness D^H is determined for each V_e^x by taking account of a porosity, and then, macroscopic analysis is done by Equation(8) to determine the average stress distribution σ_e^0. Through microscopic stress analysis by Equation(9), local stress distribution for each element can be obtained by adding the local stress disturbance to σ_e^0.

5   CONCLUSION

Multi-level modeling with the homogenization theory is developed not only to give a macroscopic description of geometric configuration changes in sintering but also to predict local shrinkage of pores in creep. The present paper is limited to macro-micro two-stage modeling in the frame of elasto-creep mechanics; our methodology can be extended to multi-stage modeling by using various nonlinear mechanics. Further developments will be discussed in Ref[6].

References

1) A. Nobara, T. Nakagawa and T. Yabu, "Numerical Simulation of Hot IsostaticPressing Process with HIPNAS.", R-D Kobe Steel Eng. Report, **40**(1990)24.

2) K. Mori, T. Hirano and K. Osakada, "Tool Design Analysis for Net Shaping of Ceramic Parts by Finite Element Method.", Proceeding of Spring Meeting of JSTP. , (1992) 693.

3) K. Shinagawa, K. Mori and K. Osakada, "FEM Simulation of Sintering with Micro Modeling." Proceeding of Spring Meeting of JSTP. , (1992) 689.

4) F. Tsumori, T. Aizaswa and J. Kihara, "Sintering Analysis by Granular Modeling.", Journal of JSPM., **42**(1995)501.

5) N. Kikuchi, "Advanced Computational Methods for Material Modeling", ASME AMD**180**(1993)99-114.

6) F. Tsumori, T. Aizawa and J. Kihara, "Viscous Sintering Analysis by Macro-Micro Modeling.", Abstract of 98th April Meeting of AmCer (1996, Indianapolis).

Estimation of the densification kinetics of particle aggregates through the simulation of the deformation of an average interparticle neck

Didier Bouvard and Robert M. McMeeking

Laboratoire GPM2 / ENSPG,
Institut National Polytechnique de Grenoble,
BP 46, 38401 Saint Martin d'Hères, FRANCE

Mechanical and Environmental Engineering Department,
University of California, Santa Barbara, CA 93106, USA

Abstract

A numerical model has been developed to describe the deformation of an interparticle neck by coupled grain boundary and surface diffusions and to estimate densification kinetics of an aggregate of spheres during free sintering or isostatic pressing. The obtained results contribute to the investigation of constitutive behaviour of metal and ceramic powders during sintering.

1. Introduction

During free or pressure-assisted sintering of metal or ceramic compacts, significant shape changes may occur, mainly due to density gradients, which induce complex, heterogeneous stresses within the compact. Numerical simulation of the sintering process, using for example the finite element method, can allow prediction of the final shape of a sintered part (Riedel and Sun, 1992). The simulation should be based on constitutive equations describing the mechanical behaviour of the material under realistic states of stress. These equations can directly be drawn from experimental data. However, since relevant experiments are difficult and often incomplete or inaccurate, physical modelling can help the choice of constitutive law. Most proposed models describe the constitutive behaviour of aggregates of spheres from the deformation of two-sphere contact either by viscous flow (Jagota et al., 1988) or grain boundary diffusion (McMeeking and Kuhn, 1992; Riedel et al., 1993). Although such models are not able to describe complex microstructure and mechanisms as observed in real materials, it is thought that they can give some basic information on the structure of constitutive equations, which should be next specified with relevant experimental data (Ducamp and Raj, 1989).

This paper aims at contributing to the research of the constitutive law of powder material during sintering through the development of a numerical model for the deformation of interparticle necks by diffusion creep. As already explained by various authors, modelling of grain boundary diffusion, which is often the preponderant densification mechanism (Arzt et al., 1983), should take into account surface diffusion. Although this coupling can be analytically investigated (Swinkels and Ashby, 1983), it is thought that numerical models can more precisely describe the evolution of particle geometry and account for the interaction between neighbouring contacts (Nichols and Mullins, 1965; German and Lathrop, 1978; Bross and Exner, 1979; Svododa and Riedel, 1994). The model presented in this paper, which uses the finite-difference method, simulates the deformation of a neck between two spherical particles during free sintering or when an external force is applied to squeeze the particles together.

In previous simulations, (Bouvard and McMeeking, 1995) the applied force had been kept constant and relations between the shrinkage, the neck growth and the applied force had been derived, from which constitutive law can be drawn through adequate micro-macro process. In the present paper this model is used to directly calculate densification kinetics of powder aggregates during free sintering and hydrostatic pressing.

2. Basic equations and methods
2.1. Two-sphere problem

Two identical spheres of radius R are in contact through a plane circulair join of radius x, which is treated as a grain boundary. Since the system is symmetrical about the axis perpendicular to the join, cylindrical coordinates (r, z) are used. A curvilinear coordinate s is also defined along the intersection of the particles free surface with the meridian plane. The dihedral angle at the neck is ψ. The gradient of chemical potential in the join results in a radial diffusion flux at the edge of the neck, j_r, defined as the volume of material passing out of the grain boundary through unit area in unit time. This radial flux is classically calculated as :

$$(1) \qquad j_r \; = \; \frac{4 \, D_g \, \Omega}{x \, k \, T} \left(\sigma + \gamma_s \, K(x) + 2 \frac{\gamma_s}{x} \sin\!\left(\frac{\psi}{2}\right) \right)$$

where D_g is the diffusivity in the grain boundary, Ω is the atomic volume, k is Boltzmann's constant, γ_s is the surface energy, T is the absolute temperature, σ is the average compressive stress on the neck and K(x) is the curvature at the edge of the neck (Johnson, 1969). Curvature is defined to be positive when the center of curvature is outside of the particle. The total volume of material delivered to the edge of the neck in unit time is then $2\pi x \delta_g j_r(x)$, where δ_g is grain boundary thickness. The shrinkage rate, i.e. half of the rate of approach of one particle towards the other, is :

$$(2) \qquad \dot{w} \; = \; \frac{2 \, \delta_g \, j_r}{x}$$

The matter flowing out of the grain boundary is redistributed on the free surface of the particle by surface diffusion. This diffusion flux j_s, which is entirely in the meridional direction, is expressed as :

$$(3) \qquad j_s \; = \; \frac{\delta_s \, D_s \, \Omega \gamma_s}{k \, T} \frac{dK}{ds}$$

where δ_s is the thickness of the surface layer in which diffusion takes place, D_s is the diffusion coefficient on the surface, $K(s)$ is the curvature of the free surface. The surface flux results in the deposition or removal of material leading to a displacement rate normal to the surface :

$$(4) \qquad v = -\frac{1}{r}\frac{d(r j_s)}{ds}$$

From this equation new particle boundary created by diffusion during any time increment can be calculated. The grain boundary diffusion analysis provides the value of j_s at the edge of the neck. Other boundary conditions are obtained from the dihedral angle at the neck and from an assumed symmetry of the problem. Two symmetry conditions have been investigated. To represent the sintering of a row of particles, the equator has been used as a symmetry plane and the particle surface there has been kept orthogonal to the plane of the equator. A three-dimensional packing of particles has also been schematized by assuming a symmetry condition where the angle between the tangent to the surface and r axis is equal to $\pi/4$ and by forcing this angle to remain $\pi/4$ for that point, as if there was a band of contact all around the equator of the particle. In the present paper only calculations with the "packing" condition are presented. This boundary condition is supposed to represent, on average, the geometry of a contact in a random packing of spheres through densification from 0.64 to 0.85 of relative density, which corresponds to a coordination number between 7 and 9.5 (Arzt, 1982). For higher densities, the interactions between neighbouring contacts become important and it is thought that this model is not relevant.

This problem of neck deformation has been solved with a one-dimensional finite difference calculation using a curvilinear array of points representing the current shape of the free surface particle. Details of the numerical analysis can be found elsewhere (Bouvard and McMeeking, 1995). For free sintering simulations Equation (1) was used with $\sigma = 0$, whereas for isostatic pressing, σ was chosen as explained below.

2.2. Micro-macro model

Consider an isotropic homogeneous aggregate of monosized spherical particles with radius R, which is sintering and being subject to an isotropic external pressure p. It is assumed that every interparticle contact within the aggregate experiences the same contact force and deform with the same neck growth and shrinkage rates. The densification rate can thus be calculated as :

$$(5) \qquad \frac{\dot{D}}{D} = 3\frac{\dot{w}}{R - w}$$

which is integrated as :

$$(6) \qquad D = D_0 \Big/ \left(1 - \frac{w}{R}\right)^3$$

where D_0 is the initial relative density, about 0.64 for random close packing. The average axial stress transmitted to a neck is classically expressed as :

$$(7) \qquad \sigma = \frac{4R^2}{ZDx^2}p$$

where Z is the number of necks per particle, which has been calculated as a function of D by Arzt (1982). The relation found by Arzt in diffusion case can be approximated as $Z = 11D$.

These equations, which relate neck parameters to macroscopic variables, can be used to deduce densification kinetics during isostatic pressing from the numerical model of neck deformation. Consider actual neck geometry, characterized in particular by the neck radius x and the shrinkage w, which corresponds to a relative density of the aggregate through Relation (6). The external stress acting on the neck can be deduced from Relation (7). Then the numerical model allows estimate of the instantaneous shrinkage rate, from which the densification rate is deduced through Relation (5). This process can be iterated so that the densification rate under constant isotropic pressure is obtained as a function of relative density.

2.3. Constitutive equations

Most constitutive laws developed for powder materials during sintering have a linear viscous form, which, for isostatic pressing, reduces to :

$$(8) \qquad \frac{\dot{D}}{D} = \left(\frac{\dot{D}}{D}\right)_s + \frac{p}{K}$$

where $(\dot{D}/D)_s$ is the densification rate during free sintering and K is the volumetric viscosity. $(\dot{D}/D)_s$ and K are functions of relative density and temperature. The sintering stress, defined as the opposite of the negative pressure which would cause zero shrinkage, is expressed as $p_s = K(\dot{D}/D)_s$.

The model described in the previous section should allow estimate of the free sintering densification rate, the sintering stress and the volumetric viscosity, if a linear relation as Equation (8) is applicable, as functions of relative density and material parameters.

3. Results
3.1. Effect of material parameters on neck growth and shrinkage

Free sintering has first been simulated to study the effects of material parameters and to compare the results with the prediction of anaytical models. Complete results are presented in (Bouvard and McMeeking, 1995). To compare grain boundary diffusion and surface diffusion the following parameter has been defined :

$$(9) \qquad \xi = \delta_g D_g / \delta_s D_s$$

Values of ξ between 0.02 and 6 and dihedral angles between 90 and 170° have been investigated. It has been found that the dihedral angle had little effect on the evolution of neck size and shrinkage and that neck growth was little influenced by ξ, whereas the effect of ξ was more significant on the shrinkage rate, which decreases with decreasing ξ. The analytical relations proposed by Coblenz et al.

(1980) to describe sintering by grain boundary diffusion correctly represent the simulated neck growth whatever ξ and shrinkage only for the highest values of ξ.

Modelling of sintering with constant average stress σ has been also achieved with a dihedral angle equal to 130° and various values of ξ between 0.02 and 2. Relations for shrinkage and neck growth rates as functions of neck radius and stress have been fitted.

3.2. Densification kinetics

To estimate the densification kinetics during isostatic pressing, simulations of neck deformation were performed as explained in Section 2 for a constant dihedral angle equal to 130° and various values of ξ and of normalized pressure p^*, defined as pR/γ_s. The calculations have been stopped at a relative density of 0.85. The densification rate obtained in each condition, which has been normalized with regard to $\delta_g D_g \Omega \gamma_s / kTR^4$, is plotted in Figures 1 to 3 as function of relative density.

Two stages are observed in Figures 2 and 3. For low density the densification rate continuously decreases as density increases. From a certain value of density, which is higher when ξ and p are higher, the densification rate gets about constant with respect to density. (Actually it slightly decreases). This second stage corresponds to quasi-equilibrium sintering, when curvature is about constant at every moment along the free surface of the particles, as described in particular by Svoboda and Riedel (1994). This curvature continuously decreases - absolute value - as the densification proceeds. This stage occurs very early when the surface diffusion is much faster than the grain boundary diffusion - $\xi = 0.02$ - and when the applied pressure is low. For the highest values of ξ and p only the first stage is observed within the investigated range of relative density.

A simple expression has been fitted to first stage data :

$$(10) \qquad \left(\frac{\dot{D}}{D}\right)^* = \frac{\alpha(\xi, p^*)}{(D - D_0)^2}$$

The star superscript denotes normalized parameters. α is written :

$$(11) \qquad \alpha(\xi, p^*) = \alpha_0(\xi) + \alpha_1(\xi, p^*)$$

where α_0 describes the densification rate during free sintering. α_0 is equal to 3.2, 0.8, 0.03 for $\xi = 2, 0.2, 0.0$, respectively, which proves that, for an equal grain boundary diffusion coefficient, the densification is faster when the surface diffusion is slower. Figure 4 shows the values of α_1 vs. p^*. For high and intermediate values of ξ, α_1 is independant of ξ and varies linearly with respect to p^*. Then a linear expression of the densification rate as in Equation (8) is convenient and the volumetric viscosity can be estimated as :

$$(12) \qquad K = 0.4 (D - D_0)^2 \frac{\delta_g D_g \Omega}{kTR^3}$$

For the lowest value of ξ, which corresponds to very fast surface diffusion a linear variation is not appropriate and the densification rate is better expressed as

$$(13) \qquad \left(\frac{\dot{D}}{D}\right)^* = \left(\frac{\dot{D}}{D}\right)^*_s + \frac{0.25\, p^{*3/2}}{\left(D - D_0\right)^2}$$

Notice that for the highest pressure the densification rate is independent of surface diffusion coefficient, in the investigated range. The sintering stress is deduced :

$$(14) \qquad p_s = \beta \frac{\gamma_s}{R}$$

with $\beta = 1.28$, 0.32 and 0.215 for $\xi = 2$, 0.2 and 0.02 respectively. p_s is found to be independent of relative density. The above relations are strictly valid before quasi-equilibrium stage, i.e. when the relative density is lower than 0.85, 0.71 and 0.65 for $\xi = 2$, 0.2, 0.02, respectively.

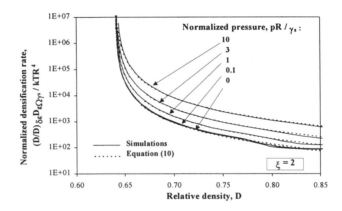

Fig.1: Densification rate deduced from simulations for $\xi = 2$ and analytical fitting.

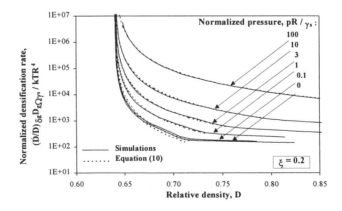

Fig. 2: Densification rate deduced from simulations for $\xi = 0.2$ and analytical fitting.

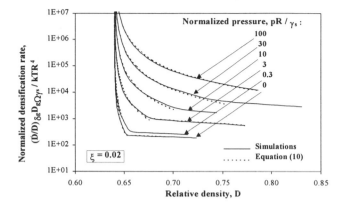

Fig. 3: Densification rate deduced from simulations for $\xi = 0.02$ and analytical fitting.

During quasi-equilibrium stage the densification rate can be written:

$$(15) \quad \left(\frac{\dot{D}}{D}\right)^* = \lambda(\xi, p^*)\, e^{-\mu(\xi, p^*)D}$$

It has not been possible to find simple expressions for λ and μ. However it has been verified that a linear constitutive law as Equation (8) is not valid.

I should be pointed out that the last relation leads to a logarithmic variation of the density with respect to time, whereas Equation (10) gives about a power variation. This is consistent with classical models which predict both variations, respectively during the second stage - pore shrinkage - and the first stage of sintering - neck growth (Exner and Petzow, 1980).

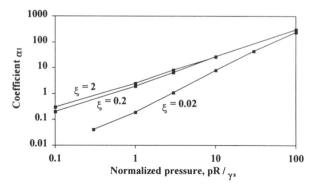

Fig. 4: Influence of the pressure on the densification rate during the first stage.

4. Conclusion

Densification kinetics of an aggregate of spheres during free sintering or isostatic pressing have been obtained from the simulation of the deformation of an average particle contact. A linear constitutive relation has been found in the case when surface diffusion is slow with regard to grain boundary diffusion and simple expressions have been proposed for the sintering stress and the volumetric viscosity. When surface diffusion is fast, two stages have been found, corresponding respectively to neck growth with large curvature gradients, and pore shrinkage with particle surface at quasi-equilibrium. The transition density gets higher when the pressure increases. The relations proposed for both stages result in non linear constitutive laws.

References

E. Arzt, "The influence of an increasing particle coordination on the densification of spherical powders", *Acta Metall.* **30** (1982).

P. Bross and H.E. Exner, "Computer simulation of sintering processes", *Acta Metall.* **27** (1979).

D. Bouvard and R.M. McMeeking, "The deformation of interparticle necks by diffusion controlled creep, *J.Am.Ceram.Soc.*, to be published.

W.S. Coblenz, J.M. Dynys, R.M. Cannon and R.L. Coble, "Initial stage solid state sintering models. A critical analysis and assessment",

V.C. Ducamp and R. Raj, "Shear and densification of glass powder compacts", *J.Am.Ceram.Soc.* **72** (1989).

H.E. Exner and G. Petzow, "A critical evaluation of shrinkage equations", *Proceedings of the Fifth International Conference on Sintering and Related Phenomena*, G.C. Kuczynski (ed.), Plenum Press, New-York, (1980).

R.M. German and J.F. Lathrop, "Simulation of spherical powder sintering by surface diffusion", *J.Appl.Phys.* **36** (1978).

A. Jagota, P.R. Dawson and J.T. Jenkins, "An anisotropic continuum model for the sintering and compaction of powder packings", *Mech.Mater.* **7** (1988).

D.L. Johnson, "New method of obtaining volume, grain boundary and surface diffusion coefficients from sintering data, *J.Appl.Phys.* **40** (1969).

R.M. McMeeking and L.T. Kuhn, "A diffusional creep law for powder compacts, *Acta Metall.* **40** (1992).

F.A. Nichols and W.W. Mullins, "Morphological changes of a surface of revolution due to capillarity-induced surface diffusion, *J.Appl.Phys.* **36** (1965).

H. Riedel, D. Meyer, J. Svoboda and Zipse, Numerical simulation of die pressing and sintering; development of constitutive equations, *Int.J. of Refractory Metals & Hard Materials.* **12** (1993-94).

H. Riedel and D.Z. Sun, Simulation of die pressing and sintering of powder metal, hard metals and ceramics, *Numerical Methods in Industrial Forming Processes*, Chenot, Wood and Zienkiewicz (eds), Balkema, Rotterdam, (1992).

J. Svoboda and H. Riedel, "New solutions describing the formation of interparticle necks in solid-state sintering", *Acta Metall.Mater.* **43** (1995).

F.B. Swinkels and M.F. Ashby, "a second report on sintering diagrams, *Acta Metall.* **29** (1983).

NUMERICAL DETERMINATION OF CRITICAL STRAIN RATE FOR NECK RUPTURE FOR EVAPORATION-CONDENSATION SINTERING OF ISOTROPIC PARTICLES

Jeffrey W. Bullard[*][1] and W. Craig Carter[2]
[1]Building Materials Division and [2]Ceramics Division
National Institute of Standards and Technology
Gaithersburg, MD

ABSTRACT: A numerical technique which was designed to simulate the evolution of a sintering body due to both reduction in surface and in elastic energy in general powder compacts is described and applied to a particular problem which is fundamental to sintering behavior. The critical strain rate for which a neck between two isolated circles neither grows nor shrinks is evaluated in terms of kinetic coefficients appropriate to evaporation-condensation where transport through the vapor phase is rapid compared to surface attachment. As expected, high strain rates and/or slow kinetics favor neck attenuation and eventual rupture while low strain rates and/or fast kinetics favor neck growth.

1 Introduction

A recently developed numerical method (Bullard et. al., 1995a,b) has been used to simulate sintering of complicated many-particle arrangements, such as those found in real powder compacts. This method includes the physics for the two primary driving forces of sintering: capillarity (reduction of surface energy) and stress (reduction of elastic energy). Algorithms and working computer codes exist for surface diffusion and surface attachment limited kinetics (SALK, or evaporation-condensation where diffusion through the vapor phase is rapid compared to surface (de-)attachment processes). Implementation of grain boundary diffusion is in progress. This general method offers unique opportunities for studying complex sintering phenomena. However, the model can also be applied to address fundamental questions about the behavior of sintering for simple particle interactions where analytic treatment would be intractable.

One such fundamental problem is the determination of the critical applied strain rate for which two spheres (which contact each other at a small

---

[*]Current address is the Department of Materials Science and Engineering, University of Illinois at Urbana-Champaign, Urbana, IL.

neck) will either undergo neck growth or neck rupture (de-sinter). The critical issue is the competition between the two driving forces. A tensile strain creates stress which drives the neck between the two particles to shrink and rupture, while driving forces due to the difference in curvature between the neck and the rest of the particle surface cause the incipient neck to grow. As a consequence, large strain rates and/or slow kinetics would tend to lead to rupture; small strain rates and/or rapid kinetics would tend to lead to neck growth. The problem is similar to the analytic solution for stability of an elastic surface evolving by surface diffusion (Rice and Chuang, 1981).

Many numerical studies of shape change due to capillarity for the two-sphere problem have been made. Exner and Bross (1979) used a finite difference technique to study evolution due to surface diffusion for two spheres. Jagota and Dawson (1990) and Herrera and Derby (1995) used finite element techniques to study viscous sintering. With the exception of the viscous sintering studies, the contribution of applied stress to shape changes has not been included in previous modeling studies.

Model sintering experiments have also been performed. Kuczynski (1949) and Kingery and Berg (1955) conducted the first quantitative studies of neck growth rates of spheres sintering by different mass transport mechanisms. Recently, Weiser and DeJonghe (1986) observed local rearrangement of 2D arrays of monosized copper spheres; the rearrangement was supposed to have resulted from inhomogeneous sintering stresses. However, we know of no such experiments for which the strain rate was also a controlled experimental parameter.

Theoretical and experimental studies that can examine the effects of applied strain rates are not merely of academic interest. Stresses develop naturally during the course of sintering of a powder compact and it follows that large tensile stresses which develop near a neck could cause its rupture. Such neck ruptures could produce large stable pores in a finished product, especially if the rupture occurs after significant densification and at large tensile stresses.

Below we present a method for calculating the evolution of an arrangement of discretized particles in two dimensions when the surfaces are isotropic and there are no grain boundaries which are elastically stressed. Results are presented for two circles which are initially in contact across a small neck and subject to various fixed strain rates. The dominant sintering process is taken to be SALK. Material constants were picked that are typical of common metal oxides; however the results could also be presented in non-dimensional forms. Extending the method to three-dimensions is straightforward and will be presented elsewhere.

2 Method

The surface potential, $\mu(s)$ to move a small volume of material from a region of surface located at s is:

$$\mu(s) = \mu_\circ + \vec{n}(s) \cdot \tilde{\sigma}(s) \cdot \vec{n}(s) + \frac{1}{2}\tilde{\sigma}(s) : \tilde{\epsilon}(s) \tag{1}$$

where $\mu_\circ$ is the potential of a stress-free surface, $\vec{n}(s)$ is the local normal to the surface, $\tilde{\sigma}(s)$ is the local stress tensor at s and $\tilde{\sigma}(s) : \tilde{\epsilon}(s)$ is the tensor product for strain energy density.

When the material is elastic, Eq. (1) can be written as:

$$\mu(s) = \mu_\circ + \vec{n} \cdot \tilde{\sigma} \cdot \vec{n} + \frac{1}{2}\tilde{\sigma} : \mathbf{C} : \tilde{\sigma} + \mathcal{O}(\tilde{\sigma}^3) \tag{2}$$

where $\mathbf{C}$ is the fourth-rank compliance tensor and the quantities are understood to depend on s.

When no tractions are applied to the surface at s, then the term which is linear in stress is equal to the pressure at s. If the system is in local equilibrium, then the Gibbs-Thomson equation ($\Delta P = \gamma\kappa$, where γ is the surface tension and κ is the mean curvature) can be applied; then, to second order

$$\mu(s) = \mu_\circ + \gamma\kappa + \frac{1}{2}\tilde{\sigma} : \mathbf{C} : \tilde{\sigma} \tag{3}$$

The second term in Eq. (3) is the increase in potential due to capillarity and depends on geometry alone. Its effect is to move mass from regions of high curvature to regions of low (and possibly negative) curvature. The third term in Eq. (3) is the increase in potential due to elasticity and its effect is to move mass to regions of low stress concentration. Note that the elastic term is independent of the sign of stress—compression raises the potential the same amount as tensile stress of equivalent magnitude.

Methods for rapidly approximating the curvature at the surface of a discretized image of a particle have been developed and are described in (Bullard et. al., 1995c). It is essentially a counting scheme where a template circle (see Fig. 1) is placed at every position on a pixelated[†] surface. It can be shown that the number of pixels in the template circle (or a template sphere in three dimensions), which are not like the pixels at or below the surface, grows linearly with the curvature (Bullard et. al., 1995c).

The same discretization which is used for approximating the curvature is used as a finite element grid to calculate the distribution of stress during the

---

[†]We use the word pixelated to refer to a surface which has discretized with equal-sized squares, or pixels. Pixels from a scanned image are often used as initial data, see (Garboczi and Bentz, 1993).

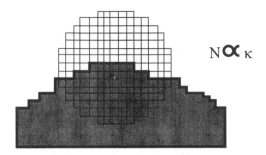

Figure 1: Template method for computing curvature of a surface. A template circle is centered on a surface pixel of interest, and the number N of template pixels lying outside the solid (shown in grey) is proportional to the curvature.

course of evolution. This is performed by setting the stiffness of the pixels representing the material to the appropriate finite value (in the current simulations, we used a Young's modulus of 3×10^{11} Pa and a Poisson ratio of $1/3$ to represent the stiffness of an oxide like alumina, and the stiffness of pixels representing vapor was set to zero. Each time the grid is updated (by a method described below) or the applied strain is changed, a new elastic solution is calculated. Computationally, this is the most time-consuming step.

The net rate of evaporation or condensation of volume at the surface is assumed to be linear in the difference between the potential at the surface, $\mu(s)$, and the potential in the adjacent vapor phase, $\mu_{ambient}$.

$$J(s) = k_{SALK}(\mu(s) - \mu_{ambient}) \qquad (4)$$

where J is the local net flux of material away from a unit area of the surface and k_{SALK} is a mobility term which is related to the activation barrier for (de-)attachment; larger values of k_{SALK} imply faster kinetics. Nonlinear laws, or mobilities which depend on the sign of $(\mu(s) - \mu_{ambient})$, could be easily incorporated into the model.

Since diffusion in the vapor is assumed to be fast compared to surface attachment, $\mu_{ambient}$ is independent of position. However, if volume is to be conserved then $\mu_{ambient}$ must be a function of time. It can be shown that volume conservation requires that $\mu_{ambient}$ be equal to the surface-area-weighted average of the surface potential. During each iteration this average potential is computed and set equal to $\mu_{ambient}$.

SALK transport was simulated by calculating $J(i)$ for each surface pixel i, using Eq. (4), and the change in the solid volume fraction ν within that pixel was updated according to $\Delta\nu(i) = J(i)A(i)\Delta t$, where the surface area

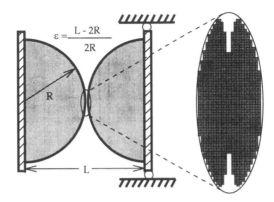

Figure 2: Schematic illustration of the simulated two-circle arrangement, with a detail of the pixel arrangement near the contact. Tensile strain was applied at a fixed rate parallel to the line connecting the circle centers.

$A(i)$ for pixel i is estimated from ν and its unit surface normal (Bullard et. al., 1995c), and Δt is the constant time increment for each iteration. The volume fraction ν lies between 0 and 1, and for the purposes of the FE algorithm, a pixel is assumed to be solid only if $\nu \geq 0.5$.

3   Results and Discussion

The initial conditions for the two-circle problem are illustrated in Fig. 2. The system is discretized into a 159×161 square grid array. Incremental tensile strains are applied to the system simultaneous to the simulated particle evolution. The system is initially stress-free. Examples of system evolution can be found in (Bullard et. al., 1995b).

The neck width (i.e., the number of pixels at the center line) is plotted in Fig. 3 as a function of time for $k_{SALK} = 3.4 \times 10^{-6} \text{ m}^3\text{J}^{-1}\text{s}^{-1}$ and for five different values of strain rate. At low strain rates ($\epsilon_t < 2.0 \times 10^{-5} \text{ s}^{-1}$), the neck grows monotonically with a monotonically decreasing rate. At large strain rates ($\epsilon_t > 1.0 \times 10^{-4} \text{ s}^{-1}$), the neck initially grows as the sharp corner at the particle-particle boundary blunts, but then shrinks and eventually ruptures. At intermediate strain rates, the neck very nearly reaches a steady state width and exhibits a change in growth behavior. At these intermediate strain rates, the blunting due to initial neck growth reduces the stress concentration there, and therefore increasing strains have the effect of counteracting the driving force for further neck growth. Of course if the simulation were continued to very high strains it is expected that the neck would eventually attenuate and rupture since the capillary driving force decreases monotonically with neck width.

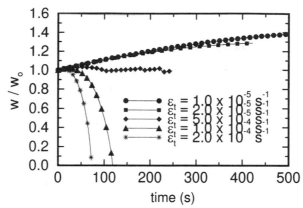

Figure 3: Plot of neck width, normalized by the initial neck width w_o, versus time for varying strain rates. The mobility $k_{SALK} = 3.4 \times 10^{-6}$ m^{3}J^{-1}s^{-1}.

When these intermediate strain rates are investigated as a function of k_{SALK}, a map describing the fundamental behavior of a two particle neck can be obtained as in Fig. 4. In Fig. 4, indications of whether the neck is growing (G) or attenuating (A) are plotted for 12 different combinations of ϵ_t and k_{SALK}.

There are two distinctive regions in Fig. 4: at large strain rates and slow kinetics, necks attenuate, which enhances rupture; at low strain rates and rapid kinetics, necks tend to grow, which stabilizes the compact geometry. The sketched line in Fig. 4 which separates the two regions is a numerical approximation to the critical strain rate for the special case of SALK. The critical strain rate is a quantity which underlies fundamental sintering behavior— neck rupture not only leads to incipient flaws but gives rise to localized reduction in the sintering driving force.

Another interpretation of Fig. 4 is that of a map which delineates the value of kinetic factors which would be favorable in a precoarsening step prior to sintering. Chu et. al. (1991) have found that a pre-sintering step, in which vapor transport was enhanced significantly, increased subsequent densification and improved the homogeneity of the final microstructure. One interpretation is that the absence of densification strain rates during precoarsening allows substantial neck growth. Subsequent sintering could therefore exhibit less neck rupture since 1) the necks are stronger and better able to resist densification strains, and 2) the precoarsening step reduces the capillary driving force which corresponds to lower densification strain rates.

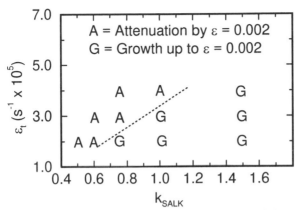

Figure 4: Diagram of regions of neck attenuation (A) and neck growth (G) for different combinations of strain rate ϵ_t and mobility k_{SALK}, where k_{SALK} is the dimensionless mobility scaled to the value used in Fig. 3, $(= 3.4 \times 10^{-6} \text{ m}^3\text{J}^{-1}\text{s}^{-1})$. Attenuation/growth behavior was probed at $\epsilon = 0.002$.

4   Summary

A new numerical technique has been developed that allows simulation of the sintering behavior of complex multi-particle systems when both capillary and elastic driving forces exist. The technique has been applied here to assess the critical strain rate for two-particle neck rupture. The results show that high applied strain rates and/or slow transport kinetics tend to lead to neck rupture, while low strain rates and/or fast transport kinetics favor neck growth and, therefore, more stable microstructures.

Acknowledgment

JWB gratefully acknowledges the National Research Council for postdoctoral fellowship support.

References

[1] J.W. Bullard, W.C. Carter, and E.J. Garboczi, "Digital-Image-Based Simulations of Sintering: I, Influence of Surface and Vapor-Phase Diffusion on 2-D Microstructural Development," in preparation, (1995a).

[2] J.W. Bullard, E.J. Garboczi, and W.C. Carter, "Digital-Image-Based Simulations of Sintering: II, Competition Between Elastic and Capillary Driving Forces," in preparation, (1995b).

[3] J.R. Rice and T. Chuang, "Energy Variations in Diffusive Cavity Growth," *J. Am. Ceram. Soc.*, **64** [1] 40–53 (1981).

[4] H.E. Exner and P. Bross, "Material Transport Rate and Stress Distribution During Grain Boundary Diffusion Driven by Surface Tension," *Acta Metall.*, **27** [6] 1007–12 (1979).

[5] A. Jagota and P.R. Dawson, "Simulation of Viscous Sintering of Two Particles," *J. Am. Ceram. Soc.*, **73** [1] 173–77 (1990).

[6] J.I. Martinez-Herrera and J.J. Derby, "Viscous Sintering of Spherical Particles via Finite Element Analysis," *J. Am. Ceram. Soc.*, **78** [3] 645–49 (1995).

[7] G.C. Kuczynski, "Self-Diffusion in Sintering of Metallic Particles," *Trans. AIME*, **185** 169–78 (1949).

[8] W.D. Kingery and M. Berg, "Study of the Initial Stages of Sintering Solids by Viscous Flow, Evaporation-Condensation, and Self-Diffusion," *J. Appl. Phys.*, **26** [10] 1205–12 (1955).

[9] M.W. Weiser and L.C. De Jonghe, "Rearrangement During Sintering in Two-Dimensional Arrays," *J. Am. Ceram. Soc.*, **69** [11] 822–26 (1986).

[10] J.W. Bullard, E.J. Garboczi, W.C. Carter, and E.R. Fuller, Jr., "Numerical Methods for Computing Interfacial Mean Curvature," *Comput. Mater. Sci.*, **4** 103–16 (1995c).

[11] E.J. Garboczi and D.P. Bentz, "Computational Materials Science of Cement-Based Materials," *Mater. Res. Soc. Bull.*, **18** [3] 50–54 (1993).

[12] M.-Y. Chu, L.C. De Jonghe, M.K.F. Lin and F.J.T. Lin, "Precoarsening to Improve Microstructure and Sintering of Powder Compacts," *J. Am. Ceram. Soc.*, **74** [11] 2902–11 (1991).

Vapor Transport and Sintering

D. W. Readey, D. J. Aldrich, and M. A. Ritland[*]

Colorado Center for Advanced Ceramics
Colorado School of Mines
Golden, CO 80401

Abstract

Vapor transport promotes neck growth and particle coarsening during solid state sintering of ceramics and metals and inhibit, or eliminate, densification. Vapor transport is important for materials that have a high vapor pressures or in atmospheres that produce high vapor pressures. Vapor transport-enhanced sintering is important not only for the unique microstructures that can be produced, but also for what it tells about the sintering process itself.

1 Introduction

Sintering models [1 ,2], predict that both vapor transport and surface diffusion lead to surface morphology changes in a powder compact that inhibit densification. Many ceramics and metals lose weight by volatilization during sintering implying significant pressures of some vapor species in the powder compact. A rough estimate of the rate of weight loss from an approximately spherically-shaped powder compact can be made from:

$$\frac{1}{m}\frac{dm}{dt} = \frac{3D\Omega}{\delta rRT}p_o \tag{1}$$

where m = sample mass, D = gas diffusion coefficient, r = sample radius, Ω = molar volume, δ = gas boundary layer thickness, R = gas constant, T = temperature, and p_o = pressure of the vaporizing species. Weight loss on the order of one percent per hour from a centimeter-size sample implies a pressure on the order of 10^2 Pa (10^{-3} atm). Such weight loss rates are not uncommon during sintering and pressures of this magnitude imply significant vapor transport that can have a significant effect on microstructure development.

Neither vapor transport nor surface diffusion contribute to densification and are to be minimized during the early stages of sintering in order to achieve

---

[*] Now with Golden Technologies Company, Inc., Golden, CO

high densitics. Currcnt thcory and available data provide no guidance on how to control surface diffusion. Yet surface diffusion is frequently invoked to explain microstructural changes and lack of shrinkage during sintering when no other mechanism or model seems to fit. This is particularly true for ceramics. In contrast, the role of vapor transport on the kinetics of densification and microstructure development has been largely ignored. Nevertheless, vapor transport is an ideal process to study since vapor pressures of the transporting gases can be controlled and transport coefficients are available in the literature or easily calculated. As a result, quantitative comparison between models and experiment is possible. In addition, vapor transport only changes surface area without shrinkage as does surface diffusion; so from microstructural changes produced by vapor transport, similar changes can be predicted for surface diffusion.

On the other hand, from a technological standpoint, reactive vapor phase sintering can be controlled to produce porous ceramics and metals with well-defined microstructures that have a number of interesting applications ranging from filters to new structural materials. Depending on the material of interest, various reactive atmospheres can increase the pressures of transport gases and enhance vapor transport as illustrated by the following examples:

$$Ni(s) + Cl_2(g) = NiCl_2(g)$$

$$Al_2O_3(s) + H_2(g) = Al_2O(g) + H_2O(g) \tag{2}$$

$$MgO(s) + H_2O(g) = Mg(OH)_2(g)$$

Some materials have sufficiently high vapor pressures themselves so that vapor transport will dominate microstructure development; for example, ZnS, CdS, and ice. Systematic investigations on the effect of vapor transport on microstructure development in ceramics have been carried out on a number of different ceramic systems [3 -6]. Selected results will serve to illustrate the effects of vapor transport on sintering in single phase materials. The data demonstrate that enhanced vapor transport indeed has a dramatic effect on densification and microstructure development. However, the results also demonstrate that, although some of the model predictions for the effects of vapor transport do apparently occur, there are other observations that do not fit the simple models.

2 Vapor Transport and Sintering

The initial sintering model of two spheres in contact [1,2] Figure 1, is sufficient to illustrate the potential effects of vapor transport. For mass transfer by diffusion through the gas

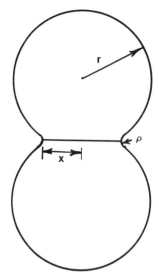

Figure 1. Two-sphere initial
  sintering model [1,2].

phase only, the model predicts that the interparticle neck size, x, should grow with time as:

$$\overset{-3}{x} = \frac{12\,D\,\gamma}{\pi}\left(\frac{\Omega}{RT}\right)^2 P_o\, t \tag{3}$$

where D is the gas diffusion coefficient and γ is the surface energy. Vapor transport only changes the surface geometry, which reduces the driving force for sintering, and does not produce shrinkage and densification; it produces the same microstructural changes as surface diffusion.

Another way to estimate the importance vapor transport plays during sintering can be obtained from the sintering models. The ratio for the rate of neck growth by vapor transport, $\dot{x}_V$, to that for solid state volume diffusion, $\dot{x}_D$, can be shown to be [3]:

$$\frac{\dot{x}_V}{\dot{x}_D} = \frac{2 D_g \Omega}{\pi D_V RT}\frac{\rho}{r} P_o \tag{4}$$

where D_g is the gas diffusion coefficient and D_v is the solid state volume diffusion coefficient that controls the rate of densification. Of course, ρ/r is the shrinkage in the models [1,2] and the actual value of the numerical constant depends on the model of choice but Eqn. [4] predicts the relative transport contributions of vapor and solid diffusion. With the approximate values of $\rho/r \approx 0.01$ (i.e. a shrinkage of about 1 percent), $\Omega \approx 10$ cc/mole, $D_g \approx 1$ cm^2/s, and $T = 1500$ K, Eqn.(4) gives,

$$\frac{\dot{x}_v}{\dot{x}_D} \cong 5x10^{-12}\frac{p_o}{D_v} \tag{5}$$

where p_o is in Pa and D_v in cm^2/s. For gas pressures on the order of 10^2 Pa (10^{-3} atm) (see Eqn. (1) and the discussion following it), vapor transport will be an important process if the solid state diffusion coefficient is less than $D_v \approx 10^{-10}$ cm^2/s, near the value of diffusion coefficient necessary to observe reasonable shrinkage rates.

Vapor transport also enhances grain growth (i.e. particle coarsening or Ostwald ripening) during sintering. For Ostwald ripening controlled by diffusion in the surrounding gas phase [7 ,8]:

$$\bar{r}^3 - \bar{r}_o^3 = \frac{8}{9}D\gamma\left(\frac{\Omega}{RT}\right)^2 p_o t \tag{6}$$

where, $\bar{r}$ is the mean particle size; $\bar{r}_o$ the mean initial particle size; and t, time. A result similar to Eqn. (6) holds for a surface reaction-controlled coarsening process [7]. If vapor transport produces particle coarsening, the rate of densification also decreases due to the reduction of the surface area driving force.

Both Eqns. (3) and (6) assume that diffusion through the gas phase is rate-limiting. However, in reality, the following sequential steps must all occur:

1. the forward reaction;
2. gas diffusion;
3. the back or reverse reaction;
4. the grain boundaries must be mobile.

For gas diffusion to be rate controlling, steps 1,3, and 4 must be rapid compared to step 2. The fourth step may not be so obvious, but can be seen intuitively in that each grain is constrained by its nearest neighbors in each of the three dimensions. Therefore, it cannot grow unless it consumes its neighbors by grain boundary motion. This last requirement implies that, for coarsening to occur, temperatures must be near the sintering temperature where grain boundary diffusion is sufficiently fast so that grain boundaries can act as sources and sinks for the solid-state diffusion that leads to densification. Several reactive vapor phase sintering studies clearly show the importance of mobile grain boundaries [4,6]. Of course, the Ostwald ripening models for coarsening were developed for dilute solutions and really do not apply to powder compacts with particles in contact. Nevertheless, the models still predict the correct parametric dependence with only the constants and predicted grain size distributions changing significantly.

It has been found in all systems studied to date, that sintering in a reactive atmosphere such as HCl [5] produces vapor species through reactions such as

$$TiO_2(s) + 4\ HCl(g) = TiCl_4(g) + 2\ H_2O(g) \qquad (7)$$

or similarly in hydrogen [4]

$$ZnO(g) + H_2(g) = Zn(g) + H_2O(g) \qquad (8)$$

that will prevent shrinkage from occurring if the product gas pressures are in excess of about 10^2 Pa. The dominant microstructure change is significant particle or grain growth as predicted by Ostwald ripening models, Eqn. (6). Therefore, by controlling the initial density, the sintering temperature, and the sintering time, materials with varying porosity and pore (grain) size can be produced.

3 Example of Vapor Phase Sintering

Thermodynamic calculations [9] show that vapor pressures are sufficiently high so that vapor transport should be the dominant transport process for Fe_2O_3 in an HCl atmosphere at all temperatures of interest [10], as shown in Figure 2. The main iron species are $FeCl_3$ and $FeCl_2$. As expected from the discussion above, with such high pressures of iron and oxygen transporting species, little shrinkage would be expected to occur and the particle size would increase. This is the case as shown in Figures 3 and 4.

The experimental results can be quantitatively compared with model, Eqn. [6]. With $\gamma \approx 500$ mJ/m^2, and $D \approx 10^{-4}$ m^2/s, the calculated particle size at 1200 °C in pure HCl after 10 minutes, beginning with a particle size of 0.2 μm, is 11 μm and that measured, 6 μm. This is good agreement considering the approximate values of γ and D used.

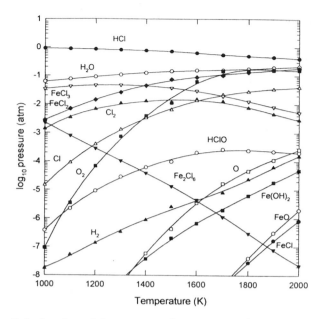

Figure 2. Calculated partial pressures of gaseous species over Fe_2O_3 in HCl versus temperature.

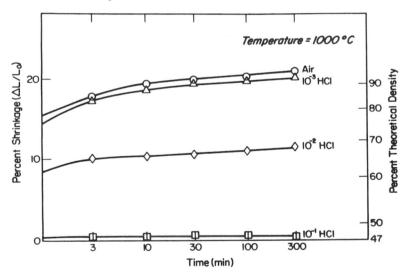

Figure 3. The effect of the reactive HCl atmosphere on the densification of Fe_2O_3 [4]. The numbers on the HCl curves refer to the fraction of HCl in a total pressure of one atmosphere of argon.

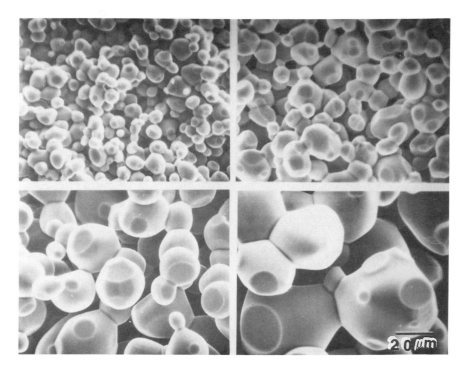

Figure 4. Grain growth of Fe_2O_3 in 10% HCl at 1200 °C for 10, 30, 100, and 300 minutes clockwise from the upper left [4].

4 Conclusions

Enhanced vapor transport during sintering reduces or eliminates shrinkage and produces exaggerated grain growth as predicted. The results of vapor phase sintering experiments also point out things that are not taken into consideration by sintering models. First, significant grain growth can occur at low densities. Second, some interparticle necks must be broken if grains grow without compact shrinkage. Finally, for a multi-component system, several phenomena are possible and are currently under investigation. On the other hand, vapor phase sintering provides a means to produce porous ceramics with independently controlled pore (grain) size and density that have potential for use in a number of applications.

5 Acknowledgments

This work has been supported by ACX Technologies, Inc., ONR, and NASA Microgravity Science and Applications Division.

6 References

1 . G. C. Kuczynski, "Self-Diffusion in Sintering of Metallic Particles," Trans. AIME 185 (2) (1949), 169-78.

2 . W. D. Kingery and M. Berg, "Study of Initial Stage of Sintering Solids by Viscous Flow, Evaporation-Condensation, and Self-Diffusion," J. Appl. Phys. 26 (10) (1955), 1205-12.

3 . D. W. Readey, T. Quadir, and J. Lee, "Effects of Vapor Transport on Microstructure Development," Ceramic Microstructures '86, Role of Interfaces, Materials Science Research Vol. 10, J. A. Pask and A. G. Evans, eds, (Plenum, N.Y., 1987), 485-496.

4 . D. W. Readey, "Vapor Transport and Sintering," pp. in Sintering of Advanced Ceramics, Ceramic Transactions, Vol. 7, C. A. Handwerker, J. E. Blendell, and W. A. Kaysser, eds., (The Am. Ceramic Society, Columbus, 1990), 86-110.

5 . Michael J. Readey and Dennis W. Readey, "Sintering of TiO_2 in HCl Atmospheres," J. Am. Ceram. Soc. 70 (12) (1987), C-358-C361.

6 . Marc A. Ritland and Dennis W. Readey, "Alumina-Copper Composites by Vapor Phase Sintering," Ceramic Engineering and Science Proceedings 14 (9-10) (1993) 896-907.

7 . I. M. Lifschitz and V. V. Slyozov, "The Kinetics of Precipitation from Supersaturated Solid Solutions," J. Phy. Chem. Solids 19 (1/2) (1961), 35-50.

8 . C. Wagner, "The Aging of Precipitates by Dissolution," (in Ger.), Z. Elektrochem. 65 (718) (1961), 581-91.

9 . T. Bessman, "A Computer Program To Calculate Relationships in Complex Chemical Systems," (Rept. ORNL/RM-5775, Oak Ridge National Laboratory, Oak Ridge, TN, 1977).

10 . D. R. Stull and H. Prophet, JANAF Thermochemical Tables, 2nd Ed., NSRDS-NBS 37, (U.S. Government Printing Office, Washington, D.C., 1971).

A Stereological Examination of Porous Alumina Microstructures During Densification

Mitchell D. Lehigh and Ian Nettleship

Department of Materials Science and Engineering
The University of Pittsburgh, Pittsburgh, PA 15261

Abstract

The evolution of porous ceramic microstructures subject to thermal treatment was examined. Global stereological parameters such as surface area density of boundaries were measured as a function of pore volume. The form of the resulting relationships were described and compared with those previously observed for powdered metals and ceramics. The effect of temperature on the evolution of the microstructure was examined.

1. Introduction

In comparison with other complex microstructures such as those of steels, the microstructure of porous materials, and porous ceramics in particular, has received much less attention. The effect of porosity on materials properties tends to be evaluated with a series of empirical equations that include pore volume but little other microstructural information. More sophisticated and systematic approaches are usually inhibited by the wide range of model geometries applied to porous solids over the full range of pore volume. The microstructure of a porous ceramic has also been shown to control sintering and the resulting microstructure of the dense material. Consequently, sintering is the field in which most attention has been paid to porous ceramic microstructures. Microstructure maps based on grain growth trajectories as a function of density have been developed by Yan (1981) and used by Harmer (1984), to examine the effect of dopants. Such maps are useful as long as one can establish the rate-controlling mechanisms of coarsening and densification. Detailed evaluations of porous ceramic microstructures are somewhat rare. This persists despite the existence of appropriate global stereological parameters that have unambiguous meaning throughout the sintering process and the radical changes in microstructure therein. This study is part of a wider effort to characterize the effects of processing on the stability of porous ceramic microstructures. Previous work by Lehigh and Nettleship (1995) has shown that pore volume gradients, controlled by solid loading during casting, have a pronounced effect on the development of the microstructure during thermal treatment.

2. Experimental Procedure

Two alumina powders which will be refered to as powder 1 (Premalox, Alcoa, Pittsburgh PA) and powder 2 (Nanophase Alumina, Alcoa, Pittsburgh PA) where used in this study of thermal stability. Powder 1 was dispersed in deionized water at pH 7 with a polyelectrolyte (Darvan C, Vanderbilt Co) and milled to a median particle size of 0.4μm, while powder 2 was dispersed in deionized water at pH 4 and milled and classified to a median particle size of 0.17μm. Powder 1 had a surface area of $11m^2g^{-1}$ and powder 2 had a surface area of $100m^2g^{-1}$. Both dispersions were cast into plaster molds, at 30vol% solids, and fired in air in the temperature range 1300°C to 1400°C. Sections were then prepared by cutting the samples perpendicular to the casting interface and polishing prior to thermal etching at the firing temperature for a further 30 minutes. Pore volume and surface area density of solid-vapor, S_v^{sv} and solid-solid boundaries, S_v^{ss} were evaluated by point counting techniques, described by Underwood (1970), on SEM micrographs. The magnifications used were calibrated with latex spheres at the same working distance and accelerating voltage used in the measurements. The precision of the experiments are reflected in the error bars which represent the standard deviations for a 95% confidence interval. The accuracy was evaluated by comparing the pore volume with that determined by the water immersion technique. If the two measurements were consistent the micrographs were considered to be representative of the microstructure.

3. Results

The green densities of the compacts were estimated from their measured volume and weight. The compacts derived from powder 1 were in the range 57% to 60% dense. However, the compacts of powder 2 were only 45% to 48% dense. After firing the measurements of pore volume by stereology and water immersion were found to be consistent, all but a few of the immersion densities were within the precision of the stereological result.

Fig.1 shows the relationship between surface area density of pore boundaries and relative density for compacts of both powders. Linear fits to the data were made in all cases. Fig.1(a) shows that for powder 1 fired at 1325°C, the pore surface area density decreased linearly by 50%, from 2.8μm^{-1} at 70% dense to approximately 1.6μm^{-1} at 89%. A similar plot for powder 1 fired at 1375°C gave a higher gradient. Fig.1(b) shows similar plots for powder 2 fired at 1325°C and 1300°C. For this material the results ranged from values as high as 5.3μm^{-1} at 76% to 1.6μm^{-1} at 94% on firing at 1325°C. The pore surface area density in compacts made from powder 2 was noticeably higher than for powder 1 at the same density and firing temperature. Comparing the results for powder 2 fired at 1325°C and 1300°C again suggests that a higher gradient is achieved for the lower firing temperature if a linear relationship is fitted.

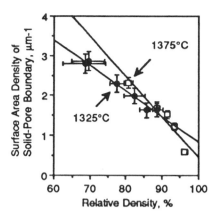

Fig. 1(a) shows the surface area
density of solid-vapor boundaries as
a function of relative density for
powder 1 fired at 1375°C and
1325°C.

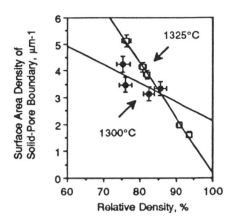

Fig. 1(b) This is a plot of surface
area density of solid-vapor
boundaries as a function of relative
density for powder 2 fired at 1325°C
and 1300°C.

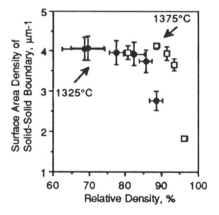

Fig. 2(a) shows the surface area
density of solid-solid boundaries as
a function of relative density for
powder 1 fired at 1375°C and
1325°C.

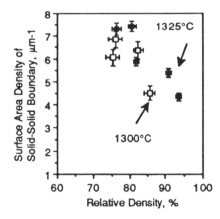

Fig. 2(b) This is a plot of surface
area density of solid-solid
boundaries as a function of relative
density for powder 2 fired at 1325°C
and 1300°C.

The relationship between pore volume and surface area density of solid-solid boundaries is presented in Fig.2. Powder 1 shows a similar relationship for both 1325°C and 1375°C in Fig.2(a). At both temperatures the surface area density of solid-solid boundaries remained at approximately $4\mu m^{-1}$ between 70% and 85%. At higher density the surface area of solid-solid contacts decreased rapidly. The density at which the rapid decrease takes place is lower for the lower temperature (1325°C). Fig.2(b) shows the results for powder 2 fired at 1325°C and 1300°C. The relationship is not as clear for this material, with a general decrease over the density range 70% and 90%. Close to 75% the materials had surface area densities above $7\mu m^{-1}$ and the values are still above $4\mu m^{-1}$ at densities above 90%. Notice this is higher than the maximum surface area density measured for powder 1 over the same density range.

The utility of the surface area density of boundary measurements is the ability to define a variety of length scales in the microstructure. These include grain size, pore size and pore spacing as outlined by Exner and Hougardy (1988). For example the mean grain intercept, λ_g is given by:

$$\lambda_g = 4\rho / (S_v{}^{sp} + 2S_v{}^{ss}) \tag{1}$$

where ρ is the relative density, $S_v{}^{sp}$ is the surface area density of solid-vapor boundary and $S_v{}^{ss}$ is surface area density of solid-solid boundary. Notice that no constant, is used in the definition to avoid an assumption regarding grain shape. Similarly the mean pore intercept λ_p can be defined as:

$$\lambda_p = 4(1-\rho)/S_v{}^{sp} \tag{2}$$

and the pore spacing λ_{ps} as:

$$\lambda_{ps} = 4\rho/S_v{}^{sp} \tag{3}$$

Using these definitions, the relationships between grain size and relative density was determined, as shown in Fig.3. The results for powder 1 are given in Fig.3(a) and mirror the results for $S_v{}^{ss}$ in Fig.2(a). At densities below 85% the grain size at both temperatures remains relatively constant in the range $0.25\mu m$ to $0.4\mu m$. Above this density the grain size increases as the surface area density of solid-solid boundaries decreases. This increase appears to begins at lower density for the sample fired at the lower temperature. The results for powder 2 are shown in Fig.3(b). In this case the grain size is in the range $0.15\mu m$ to $0.3\mu m$ at densities below 85% and increased at higher densities. There is little difference in grain size for materials made with powder 2 and fired to the same density at different temperatures.

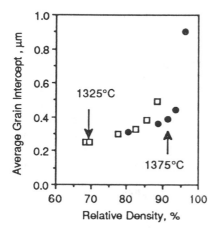

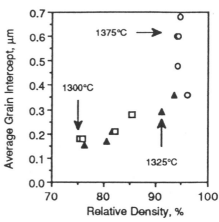

Fig3(a) A plot of average grain intercept against relative density for powder 1 fired at 1375°C and 1325°C.

Fig3(b) A plot of average grain intercept against relative density for powder 2 fired at 1375°C, 1325°C and 1300°C.

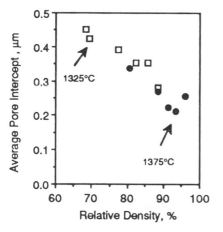

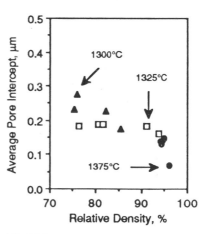

Fig.4(a) Plot of average pore intercept against relative density for powder 1 fired at 1375°C and 1325°C.

Fig4(b) Plot of average pore intercept against relative density for powder 2 fired at 1375°C, 1325°C and 1300°C.

Fig.4(a) shows the relationship between relative density and pore size for compacts of powder 1 fired at 1325°C and 1375°C. Both temperatures show a general reduction in pore size as the relative density is increased. The values range from 0.45μm at 70% dense to approximately 0.2μm at 94% dense. There appears to be little difference in the pore size for materials fired to the same density at different temperatures. Similar data for powder 2 is given in Fig.4(b). The pore sizes in the compacts made from powder 2 are noticeably smaller over the same density range than for powder 1. The results range from approximately 0.25μm at 76% dense to 68nm at 96% dense. Again, there is little difference in pore size for compacts fired to the same density at different temperatures. The results for the pore spacing are not shown, but they closely resemble the results for the grain size, suggesting that the grain size and the pore spacing are the same in the samples over most of the density range studied.

4. Discussion

The form of the relationship between surface area density of pore boundaries and relative density has been more closely studied for metal powders. Work by Aigeltinger and Exner (1977), on the sintering of 45μm and 120μm copper spheres showed the typical linear relationship between density and surface area density of solid-pore boundary, implying a constant pore size in the density range 80% to 100%. The linear fits in Fig.1(a) and Fig.1(b) of this study would also imply a constant pore size but they do not extrapolate to zero surface area at 100% dense. The positive surface area density at 100% dense suggests that the pore size must decrease with increasing density. The calculation of pore size for individual data points is given in Fig. 4. This shows a gradual decrease in pore size throughout the density range studied and not just the density range corresponding to final stage (>90%). This is especially true for powder 1. Further examination of Fig. 1(a) and Fig.1(b) show that the assumption of constant pore size could lead one to conclude that firing at lower temperature would lead to larger pore sizes for both powders because of the lower gradients. However, the calculation of pore size for each data point reveals little difference in the pore sizes of samples fired to the same density at different temperatures, as shown in Fig.4(a) and Fig.4(b). The linear relationships in Fig.1 are only valid if the intercept at 100% dense is zero. Similar stereological experiments on both tungsten and copper powders by Liu and Patterson (1994), have also shown a change in pore size during densification. In this case the pore size was observed to increase during the final 15% of densification and a consequent reduction in apparent pore mobility was observed.

The work on metal powders has also established a commonly observed shape to the relationship between the surface area density of solid-solid contacts and relative density. When the density first increases one would expect the surface area density of solid-solid contacts to increase as the necks between the grains grow and the number of necks increase. Then as the compact approaches the final stage of sintering the grains begin to grow more rapidly and the surface area density of solid-solid decreases. Thus the relationship contains a peak somewhere in the transition from intermediate to final stage sintering. An example of such a peak at 85% dense is given by Aigeltinger and Exner (1977), for the sintering of copper spheres. A peak in the surface area density of solid-solid contacts has also been observed in ceramics, such as the MgO-doped alumina studied by Shaw and Brook (1986). However, in the same study a clear peak could not be defined for undoped material. Lehigh and Nettleship (1995), have also observed a peak in surface area density of solid-solid contacts for alumina samples containing a pore volume gradient developed by differential sedimentation in casting of low solid loading slips. A peak in S_v^{ss} was only present in the part of the compact made up of the fines in the slip. The relationship for powder 1, shown in Fig. 2(a), does not show a clear peak in the density range shown. There is instead a plateau region extending up to approximately 85% dense, with roughly the same value for compacts fired at both 1325°C and 1375°C. Then, at higher densities, there is a rapid decrease in solid-solid contact for both sintering temperatures, probably due to accelerated grain growth. The decrease in solid-solid contacts occurs at lower density for the sample fired at the lower temperature. It could be concluded that the grains are coarsening earlier in the microstructure evolution at lower temperature. The relationship for powder 2 was very different. In this case the surface area of solid-solid contacts is higher than in powder 1 and the decrease is more gradual with no pronounced peak or stable plateau region. It must be remembered that this material had a lower green density and different powder characteristics, in particular a very high surface area. It is thought that this may have prevented the powder from reaching a high packing density during forming. These results suggest that packing efficiencies and other powder processing variables can have a pronounced effect on the evolution of the surface area of solid-solid contact. This in turn suggests that forming methods could be studied as a means of controlling the stability of a porous ceramic microstructure during thermal treatment.

The results for the average grain intercept are a reflection of the results for the surface area density of solid-solid contacts. Fig.3(a) show the results for powder 1, for which the grain size is relatively stable for densities up to 85%. At higher densities the grain size increases as one would expect. The relationship for powder 2 is shown in Fig.3(b). Again the rapid increase in grain size takes place above 90% dense. The average pore spacing intercept gives very similar values to the average grain intercept, which suggests that most of the pores remain on the grain boundaries throughout the densification observed in this study.

5. Summary

The effect of thermal treatment on the microstructure of two porous alumina ceramics has been investigated. It was found that the average pore intercept decreased throughout the density interval studied for both materials studied. The variation of surface area density of solid-solid contacts did not show a distinct peak. It remained relatively stable for powder 1 until the beginning of of final stage sintering. The surface area density of solid-solid contacts for powder 2 decreased gradually over the same density range. It is thought that the powder processing variables may have a considerable effect on the form of this relationship for ceramics.

6. Acknowledgments

This project is funded by NSF grant No CMS-9409456 and also by The Alcoa Foundation. The authors would like to thank Alcoa for donating the powders used in this project.

7. References

E.H. Aigeltinger and H.E. Exner, "Stereological Characterization of the Interaction Between Interfaces and its Application to the Sintering Process," *Met Trans A.*, **8A**, 421-424 (1977).

H.E. Exner and H.P. Hougardy, *Quantitative Image Analysis of Microstructures*, DGM Informationsgesellschaft, Germany, p22, (1988).

M.P. Harmer, "Use of Solid-Solution Additives in Ceramics Processing," pp. 679-696 in *Advances in Ceramics*, **vol.. 10**, ed. W.D. Kingery, The American Ceramics Society, Columbus, Ohio (1984).

M.D. Lehigh and I. Nettleship, "Microstructural Evolution of Porous Ceramics," in *Advances in Porous Materials*, Proceedings of the Materials Research Society Symposium, Vol 371, Materials Research Society, Pittsburgh, (1995).

Y. Lui and B.R. Patterson, "Determination of Pore Mobility During Sintering," *Met. & Mat. Trans.*, **25A**, 81-87, (1994).

N.J. Shaw and R.J. Brook, "Structure and Grain Coarsening During the Sintering of Alumina," *J. Am. Ceram. Soc.*, **69** [2], 107-110, (1986).

E.E. Underwood, *Quantitative Stereology*, Addison-Wesley, Reading MA, p.5 (1970).

M.F. Yan, "Microstructure Control in the Processing of Electronic Ceramics," *Mater. Sci and Eng.*, **48**, 53-72 (1981).

Anomalous Densification of Complex Ceramics in the Initial Sintering Stage

B. Malič, D. Kolar, and M. Kosec

Jožef Stefan Institute, University of Ljubljana, Ljubljana, Slovenia

Abstract

In recent years great advances have been achieved in the preparation of homogeneous precursors to complex ceramics, chemically mixed at the molecular level. However, the different diffusion rates of the various diffusing species during sintering of binary and ternary compounds and solid solutions frequently cause dehomogenization and segregation. These phenomena are particularly pronounced with active fine grained powders with large surface areas. Surface diffusion and segregation in the necks between the powder particles play an important role. Dehomogenization gives rise to an anomalous course of densification in the early sintering stage and influences the later sintering stages, as well as microstructural development. The dehomogenization phenomenon was experimentally demonstrated in the initial sintering stage of $Pb(Zr_{0.5}Ti_{0.5})O_3$ (PZT).

1. Introduction

In recent years, considerable progress in the processing of ceramics has been achieved by improving the quality of powders. Attention is paid to wet-chemical methods of powder preparation, which assure fine particle size, controlled morphology, high purity and high homogeneity. Improved homogeneity is particularly desirable in complex multicomponent ceramics.

It is frequently stressed that wet chemical methods, such as coprecipitation or sol-gel methods, assure chemical homogeneity "on a molecular" level. However, this homogeneity may be temporarily lost during the sintering operation due to the very nature of the sintering process.

Kuczynski et al (1960) pointed out that the vacancy gradient set up between sintered particles by the sharp curvature of the neck can, under favorable conditions, produce considerable segregation in a completely homogenized solid solution. When the diffusion coefficients of constituent atoms in solid solution are different (which is frequently the case), the neck area becomes enriched in the faster diffusing atoms, at least at the early stage of sintering when the vacancy gradient due to the smallness of the radius of neck curvature is large. Such segregation must be a transient phenomenon, since segregation gives rise to a chemical potential gradient arising from the concentration gradient between the neck and regions adjacent to the neck, and acts in a direction opposite to the chemical potential gradient due to the neck curvature. When the radius of curvature of the neck increases, the chemical potential which causes the dehomogenization decreases. With accumulation of faster diffusing atoms in the neck, the concentration gradient also increases and the maximum segregation is reached. After this, the chemical potential gradient due to the concentration gradient predominates and back diffusion from the neck to other regions occurs. As a result, homogeneity is re-established.

Kuczynski et al (1960) demonstrated segregation in Cu-In and Cu-Ag alloys. Mishra et al (1975) demonstrated dehomogenization of Au-Ag alloy and concluded that the initial neck growth as well as segregation takes place by surface diffusion, whereas back diffusion occurs by a combination of surface and volume diffusion.

It may be expected, due to capillary forces in the early sintering stage, that dehomogenization effect occurs in complex ceramic systems as well. It should be particularly pronounced in sintering of nanosized powders, where surface diffusion plays an important role. The segregation-homogenization phenomenon should be reflected in densification curves.

To demonstrate the effect, we examined the sintering of fine sol-gel prepared powders of lead zirconate - lead titanate solid solution, $Pb(Zr_{0.5}Ti_{0.5})O_3$ (PZT).

2. Experimental

Stoichiometric $Pb(Zr_{0.5}Ti_{0.5})O_3$ powder was prepared by alkoxide sol-gel synthesis. Metal content in precursors was determined gravimetrically. The experimental procedure has been described in more detail elsewhere (Malič, Kosec and Kolar, 1995). The PZT powder was calcined at 650°C for 5 hours in a flowing oxygen atmosphere, then milled for 2 hours in a planetary ball mill in n-butanol and dried at 100°C.

The powder was isostatically pressed at 500 MPa into pellets which were then fired at 800 to 1000°C at a heating rate of 10°C/minute in oxygen. Other samples were sintered in $PbZrO_3$ packing powder. The weight losses were not observed after sintering up to 8 h at 1000°C.

The powder was analyzed by SEM, granulometry and BET. The phase composition was determined by XRD (CuK_α radiation, 0.02°/step). Densities of the compacts were determined by measuring the sample mass and dimensions, and densities of ceramics by the Archimedes technique. The theoretical density, used in calculation of relative density, was 8.00 g/cm^3. Sintering curves were recorded by a heating microscope. Densities were calculated from shrinkage data and the initial density. For isothermal heating runs (750 to 1000°C) the samples were pushed into the hot furnace and rapidly cooled.

The samples were polished and thermally etched. Microstructural analysis was performed by a scanning electron microscope equipped with an EDX/WDX analyzer (JEOL JXA 840 A), and an analytical electron microscope operating at 200 kV (JEOL 2000 FX EM). The chemical composition of phases was determined using a Link AN-10000 EDXS system (Energy Dispersive X-ray Spectroscopy) with an Ultra Thin Window Si(Li) detector, connected to the TEM. The Cliff-Lorimer method and absorption corrections were used for quantitative analysis. The accuracy of the concentration measurements for the elements Pb, Zr,

T_1 were better or equal to 5 %. The concentrations of oxygen were calculated from stoichiometry.

3. Results and discussion

$Pb(Zr_{0.5}Ti_{0.5})O_3$ powder crystallized in the pseudo-cubic perovskite structure. The particle size distribution ranged from 0.5 to 2 μm with a median value of 0.7 μm as determined by granulometry. The size of aggregates calculated from the BET specific surface area of 6.6 m^2/g was 0.11 μm.

The density of the green compact pressed at 500 MPa was 4.5 gcm^{-3} (56 % of theoretical density). When pressed at 100 MPa, the green density achieved 49 % T.D.

Fig. 1 shows the densification curve as a function of temperature at a heating rate of 10°C/min. The very rapid densification observed above ≈ 950°C does not support the solid state sintering mechanisms. Dense samples were separately prepared by isothermal heating for 1 hour at temperatures of 850-1000°C. Data are summarized in Table 1. High densities, achieved at low sintering temperature and fine grain size are common features of chemically prepared, fine grained (nanometer size) powders with a narrow particle size distribution. Tomandl, Stiegelschmitt and Böhner (1991) reported that with sol-gel process a relative density of 99 % can already be obtained at 850°C.

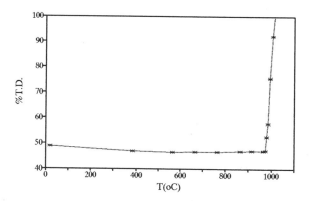

Fig. 1: Sintering curve of $Pb(Zr_{0.5}Ti_{0.5})O_3$ compact in air, pressed at 100 MPa. Heating rate: 10°C/minute.

Table 1: Density and grain size of $Pb(Zr_{0.5}Ti_{0.5})O_3$ ceramics, fired at 850 to 1000°C for 1 hour in oxygen at a heating rate of 10°C/minute in $PbZrO_3$ packing powder. T.D.: theoretical density.

T (°C)	% T.D.	d (μm)
850	89	0.9
900	98.6	1.0
950	99.8	1.2
1000	99.1	1.3

The sudden and steep increase in sintering rate at ≈ 900°C and the well crystallized grains are indicative of liquid phase sintering. The liquid phase may be the PbO-PZT eutectic above ≈ 840°C; however, the compact composition was stoichiometric PZT. To clarify this point, the initial sintering stage was analyzed in more detail.

The isothermal densification curves, presented in Fig. 2, show anomalous behavior at 750°C and 800°C. At 750°C, no densification was registered in the first 210 minutes. After this induction period, densification commenced and the compact achieved 80 % of theoretical density in the next 6.5 hours. Isotherms recorded at higher temperatures exhibited progressively shorter induction periods: 30 minutes at 800°C and 5 minutes at 850°C. At still higher temperatures, the induction period could not be registered.

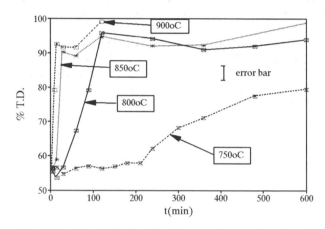

Fig. 2: Density of $Pb(Zr_{0.5}Ti_{0.5})O_3$ ceramics as a function of temperature and time of isothermal heating runs in an air atmosphere

The reason for the anomalous behavior of PZT compacts submitted to low temperature sintering can be ascribed to preferential diffusion, as described in the introduction. Accumulation of faster diffusing species in the necks between the particles, triggered by surface curvature, causes an increased tendency for backward diffusion, sustained by the concentration gradient. Formation of the thermodynamically nonequilibrium phase and its subsequent annihilation interferes with the normal densification process, being reflected as an induction period. Further densification commences only after neck curvature decreases and the material homogenizes again.

At low sintering temperatures, material transport takes place predominantly by surface diffusion and, when possible, by the vapor transport. PZT is known for the high vapor pressure of PbO (Härdtl and Rau, 1969). To the authors' knowledge, surface diffusivities of Pb, Zr and Ti ions in PZT have not been reported. Recently, Slinkina and Doncov (1992), using radioactive tracers, measured the effective self-diffusion coefficients in polycrystalline 99 % dense $Pb(Zr_{0.5}Ti_{0.5})O_3$ ceramics. The effective diffusion coefficient of Pb^{2+} was 5 - 40 times higher than that of O^{2-} and almost two orders of magnitude higher than D_{Ti} and D_{Zr}, which were close to each other. Nakamura, Chandratreya and Fulrath (1980) and Kosec and Kolar (1983) reported that during the formation of PZT from $PbTiO_3$ and $PbZrO_3$, titanium ions diffuse much faster than zirconium ions. The vapor pressure transport of PbO into the necks is also possible. To maintain electrical neutrality, diffusion of cations is accompanied by simultaneous flow of oxygen ions, or through gas-phase transport. Faster diffusion of Pb and Ti, caused by high neck curvature in the initial sintering stage of fine-grained PZT compacts, causes accumulation of Pb and Ti or Zr in the necks, and corresponding depletion of both species in the other regions of PZT grains. According to the phase diagram (Fushimi and Ikeda, 1967), PZT may be Pb deficient up to 2 mol % PbO; however, PbO does not dissolve in PZT. The simplified equation derived by Kuczynski et al (1960), and later verified by Anthony (1975), makes it possible to estimate the maximal excess concentration of the faster diffusing species in the neck area. With 100 nm size particles and a neck diameter 10 nm, the calculated excess concentration is 8 mol %.

Accumulation of PbO in contacts among PZT grains in the initial sintering stage have been already assumed by Prisedskii, Gusakova and Klimov (1976), however without experimental evidence.

Dehomogenization of PZT solid solution in the early sintering stage was confirmed in this work by XRD, SEM and TEM analysis. X-ray diffraction is a sensitive method for examining the compositional homogeneity of PZT powders. $Pb(Zr_xTi_{1-x})O_3$ solid solution exhibits a phase transition from the rhombohedral to tetragonal structure at $x \approx 0.53$. In practice, the boundary is found to have a finite range of x in which the rhombohedral and tetragonal phases coexist. The width of the boundary is linked to the extent of compositional fluctuations: it is broad for conventionally calcined and milled PZT powders, and sharp for coprecipitated, sol-gel or spray-dried powders (Kakegawa et al, 1977, Kakegawa et al, 1984, Ramji Lal, Krishnan and Ramakrishnan, 1988, Saha and Agrawal, 1992). The width of the coexistence range decreases with sintering temperature and firing time (Mabud et al, 1980, Lucuta, Constantinescu and Barb, 1985).

Fig. 3 presents XRD patterns of PZT loose powder and a powdered sintered sample after various soaking times at 800°C. Material sintered for 1 hour exhibits the presence of tetragonal and rhombohedral phases (triplet at $2\theta = 45$ deg), and homogenizes on further heating to pure tetragonal phase. The presence of rhombohedral phase in the loose powder, heat treated under identical conditions, is much less pronounced.

The amount of PbO was not sufficient to be detected by XRD. Its presence in the sample sintered for 90 minutes at 850°C was detected on SEM micrographs as bright inclusions, Fig. 4 a. The size of the inclusions was too small to be analyzed by EDX; however the light color indicates the higher atomic number as compared with the matrix phase. On further heating, the inclusions redissolved in the structure, Fig. 4 b. The inclusions were verified as PbO by EDX analysis in the transmission electron microscope. Further point analysis was performed in neck areas and in grain interiors. The results are presented in Table 2. The difference between the Pb concentration in the neck and the grain interior exceeds the standard deviation, thus confirming the enrichment with Pb in the neck regions. The spread of results for Ti and Zr analysis was too high for reliable conclusions.

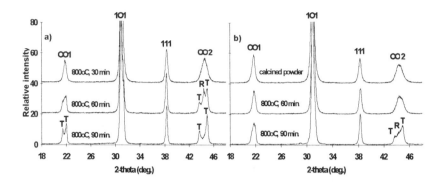

Fig. 3: XRD patterns of a) crushed pellet and b) loose powder Pb(Zr$_{0.5}$Ti$_{0.5}$)O$_3$ after heating at 800°C for various times. The XRD pattern of calcined powder is also included. T: tetragonal, R: rhombohedral phase.

Table 2: Element analysis in the neck region and grain interiors. Data are the averages of three analyses of neck or grain.

	Pb (mol %)	Zr (mol %)	Ti (mol %)	O (mol %)
average*	20.8	10.2	9.3	59.7
σ (%)	1.6	3.5	7.4	2.2
neck	21.1	10.3	9.0	59.9
grain	20.2	10.1	9.8	59.6
theoretical	20.0	10.0	10.0	60.0

*: average of all measurements

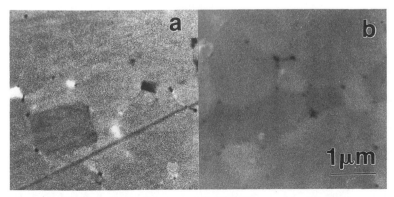

Fig. 4: Micrographs of the polished surfaces of Pb(Zr$_{0.5}$Ti$_{0.5}$)O$_3$ ceramics, fired at 850°C for a) 90 and b) 360 minutes in an air atmosphere.

The findings of the present work open several questions. Sintering of stoichiometric PZT has been analyzed in the past in accordance with the solid state sintering model. The transient presence of PbO, which is likely to form a eutectic at the sintering temperature of PZT, although of transient nature, casts doubt on the correctness of solid state sintering models, at least in the initial sintering stage. Secondly, the microstructural development in PZT should be sensitive to the heating schedule, which is to be proven. And finally, the dehomogenization phenomenon with the transient existence of metastable phases is likely to be a frequent phenomenon in the initial stage of sintering of multicomponent ceramics. It is particularly to be expected in sintering of fine powders with a high driving force for sintering.

4. Conclusions

1. $Pb(Zr_{0.5}Ti_{0.5})O_3$ compacts made of nanosized sol-gel prepared powders sinter at 900°C to above 99 % of theoretical density with an average grain size of ≈ 1 μm.

2. In the initial sintering stage, dehomogenization takes place due to unequal diffusional fluxes of constituting ions, driven by the high neck curvature.

3. In sintering of PZT ceramics with composition near the morphotropic phase boundary, the dehomogenization results in segregation of PbO, which is likely to form a transient liquid phase which in turn influences the densification and microstructural development.

4. Dehomogenization in the initial sintering phase of compacts made of nanosized multicomponent powders is likely to be frequently encountered phenomenon.

Acknowledgment

This work was supported by the Ministry of Science and Technology of the Republic of Slovenia within the National Research program, and by the Commission of the European Communities within the Community's Action for Cooperation in Science and Technology with the Central and Eastern European Countries, 1992. The authors thank Dr. Goran Dražič for performing TEM/EDX analyses.

5. References

T. R. Anthony, Solute segregation in vacancy gradients generated by sintering and temperature changes, *Acta Metallurgica* <u>17</u>, 603-609 (1969)

S. Fushimi, T. Ikeda, "Phase equilibrium in the system $PbO-TiO_2-ZrO_2$", *J. Am. Ceram. Soc.* <u>50</u>, 129-132 (1967).

K. H. Härdtl, H. Rau, "PbO vapour pressure in the $Pb(Zr_xTi_{1-x})O_3$ system", *Solid State Comm.* <u>7</u>, 41-45 (1969).

K. Kakegawa, J. Mohri, S. Shirasaki and K. Takahashi, "Preparation of $Pb(Zr,Ti)O_3$ through the use of cupferron", *J. Am. Ceram. Soc.* <u>67</u> [1] C2-C3 (1984).

K. Kakegawa, J. Mohri, K. Takahashi, H. Yamamura and S. Shirasaki, "Composition fluctuation and properties of Pb(Zr,Ti)O$_3$", *Solid State Comm.* **24**, 769-72 (1977).

M. Kosec, D. Kolar, PZT solid solution formation from PbZrO$_3$ and PbTiO$_3$, *Mater. Sci. Monographs* **16**: *Ceramic Powders*, P. Vinzenzini (ed.) Elsevier, Amsterdam, pp. 421-427 (1983).

G. C. Kuczynski, G. Matsumura and B. D. Cullity, "Segregation in homogeneous alloys during sintering", *Acta Metallurgica* **8**, 209-15 (1960).

Ramji Lal, R. Krishnan and P. Ramakrishnan, "Transition between tetragonal and rhombohedral phases of PZT ceramics prepared from spray-dried powders", *Br. Ceram. Trans. J.* **87**, 99-102 (1988).

P. Gr. Lucuta, Fl. Constantinescu and D. Barb, "Structural dependence on sintering temperature of lead zirconate-titanate solid solutions", *J. Am. Ceram. Soc.* **68** [10], 553-37 (1985).

S. A. Mabud, "The morphotropic phase boundary in PZT solid solutions", *J. Appl. Crystallogr.* **13** [3], 211-16 (1980).

B. Malič, M. Kosec and D. Kolar, "Morphology and cold compaction behaviour of alkoxide-derived Pb(Zr$_{0.5}$Ti$_{0.5}$)O$_3$ powders", pp. 291-295, *Ceramic Transactions* **51** ("Ceramic Processing Science and Technology") H. Hausner, G. L. Messing and S. Hirano (eds.), The American Ceramic Soc., Westerville, OH (1995).

A. Mishra, F. V. Lenel and G. S. Ansell, "The contribution of diffusional flow mechanisms to micro segregation in silver-gold alloys", *Mat. Sci. Research* **10**, 339-347, Plenum, New York, 1975.

Y. Nakamura, S. S. Chandratreya, R. M. Fulrath, "Expansion during the reaction sintering of PZT", *Ceramurgia Int.* **6**, 57-60 (1980).

V. V. Prisedskii, L. G. Gusakova and V. V. Klimov, "The kinetics of the initial stage of sintering of lead zirconate-titanate ceramics", Izvest. Acad. Nauk. USSR, *Neorgan. mat.* **12**, 1995-99 (1976).

S. H. Saha and D. C. Agrawal, "Compositional fluctuations and their influence on the properties of lead zirconate titanate ceramics", *Am. Ceram. Soc. Bull.* **71** [9], 1424-1428 (1992).

G. Tomandl, A. Stiegelschmitt, R. Böhner, "Lowering the sintering temperature of PZT-ceramic by sol-gel processing", *Science of Ceramics* **14**, D. Taylor (ed.), The Institute of Ceramics, Shelton, Stoke-on-Trent, Staffs. UK, 305-308 (1991)

Sintering Kinetics of ZnO During Initial Stages

Leinig Perazolli, Elson Longo and José A. Varela

Instituto de Química - Universidade Estadual Paulista - UNESP, C. P.: 355, CEP: 14800-900, Araraquara, SP, Brazil.

Abstract

Sintering kinetics of ZnO during initial stages were studied using a dilatometer considering isothermal and constant heating rate (CHR). Using Woolfrey-Bannister, Wang-Raj and Chu-Rahaman-De Jonghe-Brook approaches two dominant mechanisms for linear shrinkage of up to 6% were observed. Structural rearrangement is dominant for shrinkage up to 1.5% with apparent activation energy of 290kJ/mol. Grain boundary diffusion takes over in the range of 3 to 6% with apparent activation energy of 220kJ/mol. Isothermal and CHR results are in agreement for the range of 3 to 6%.

I - Introduction

Empiricism was the base for processing ceramics materials until 1945 when Frenkel[1] first proposed a model for the sintering of viscous materials such as glasses. Since that first scientific paper attempting to describe the sintering mechanism, many other models have been proposed based on atomistic mass transport for crystalline materials[2-5], all based on simple geometrical model of adjacent spheres. Considering that the sintering of compacted powder starts with high porosity and small particle size and ends up with small porosity and large grain size, the evaluation of microstructure should be modeled.

Several authors proposed models that considered the entire sintering process. Ashby[6] considered several mechanisms of mass transport and calculated flux equations for all of them and attempted to integrate them in sintering maps. DeHoff[7] used a cell geometry to describe microstructural evolution during sintering. Wang and Raj[8] used the following generalized sintering equation to calculate activation energies for several systems.

$$\frac{d\rho}{dt} = \frac{A\,f(\rho)}{T\,G^m}\exp\left(\frac{-Q}{RT}\right) \tag{1}$$

where ρ is relative density, A is constant, G is the mean grain size, Q is the activation energy, R is the gas constant, T is the temperature in Kelvin, m is the grain growth coefficient and $f(\rho)$ is a density dependent function. However they did not discuss the derivation of this equation and its implication.

Johnson et al[9] have developed a combined-stage sintering model and represented by a single equation that quantifies sintering as a continuous process from beginning to end. In their model the microstructure is characterized by two separate parameters representing geometry and scale.

Chu et al[10] proposed a general sintering model in which the densification and microstructure evolution have been considered. They used the concept of volumetric strain rate, defined as $(1/\rho)(d\rho/dt)$, and the interpore distance (x) to characterize the densification and microstructure during sintering. Their equation for linear strain rate is:

$$\varepsilon'd = \frac{1}{3\rho}\frac{d\rho}{dt} = KD(T)\frac{\phi^{(n+1)/2}}{x(T)^n kT} \tag{2}$$

where K is a constant, k is the Boltzman constant, x is the interpore distance. n the sintering coefficient, D(T) is the diffusion coefficient, and ϕ is the stress sintering factor.

Many of the sintering experiments were carried out using isotherms to evaluate the sintering kinetics. However this technique has several problems including the heat-up period in which considerable microstructural development and densification can occur. Another problem in determining the sintering kinetics is the determination of zero time for the isotherm.

The first work in applying constant heating rate was conducted by Young and Cutler[11] in 1970 and their data were analyzed in terms of an equation developed by Johnson[5] for the sintering of two spherical particles.

Woolfrey and Bannister[12] formulated a methodology to determine all sintering parameters during the initial stage, using constant heating rate sintering. They assumed constant grain size and the general sintering equation:

$$T^2\frac{dY}{dT} = \frac{QY}{(n^\# + 1)R} \tag{3}$$

where $Y = \Delta L/L_0$ is the linear shrinkage, $n^\# = n-1$ and Q is the apparent activation energy determined by the Dorn Method[12].

Lange[13] studied the sintering of Al_2O_3 powder compacts using constant heating rate up to 1500°C. He noticed that the densification strain rate increased with relative density up to a maximum value and then decreased for higher densities. For all heating rate considered the maximum strain rate occurred for the same relative density of 0.77. Lange proposed that the densification process is dominated up to $\rho = 0.77$, after which the coarsening is dominant.

Several studies on sintering of ZnO are reported in the literature[14-20]. Submicron size ZnO powders have been used to study the initial stage[14-16]. Densification and grain growth have been investigated in the intermediate and final stages of sintering[17,18]. The entire sintering process has been investigated using constant heating rate[10-13]. Whittemore and Varela[14] studied the early stage of sintering where no densification occurs. By measuring surface area loss they proposed a surface diffusion controlling mechanism and determined an activation energy of 184kJ/mol. Komatsu et al[15] studied the sintering of ZnO in the range of 600-700°C by measuring shrinkage and found an apparent activation energy of 220kJ/mol. Gupta and Coble[17] studied the sintering of ZnO during the intermediate stage of sintering (temperatures from 900 to 1300°C) using isotherms. They proposed that the lattice diffusion of zinc by interstitial mechanism is controlling the densification of ZnO in this range of temperatures. They obtained an apparent activation energy of 276kJ/mol for shrinkage and 253kJ/mol for grain growth. Grain growth for pure ZnO has been studied by Dutta and Spriggs[18] who found an apparent activation energy of 212kJ/mol. The sintering of ZnO by constant heating rate has been studied by Senos et al[19]. They proposed a rearrangement mechanism for linear shrinkage smaller than 2.0% with an activation energy of 210kJ/mol. In the intermediate stage the sintering is controlled by grain boundary diffusion. Chu et al[10] used constant heating rate to study the densification of ZnO for relative densities ranging from 0.5 to 0.98. Considering that there are contributions for densifing and nondensifying processes for coarsening, the apparent activation energy for nondensifying coarsening (Q_{nd} = 63kJ/mol) is smaller than the apparent activation energy for densifying coarsening (Q_d = 210kJ/mol). By applying equation 2 they found that the mean pore separation data satisfy the model for lattice diffusion controlled sintering.

Considering that there is no agreement about the sintering parameters obtained by different studies, especially for the sintering mechanism and activation energies obtained using different models, the aim of this work is to clarify these issues by comparing the constant heating rate data existing models.

II - Experimental Procedure

99.9% pure zinc oxide powder with surface area of $5.0m^2$/g and mean particle diameter of 0.21 μm was isostaticaly pressed with 210MPa to form small rods reaching a relative green density of 0.60 ± 0.05. The rods were sintered in a dilatometer (NETZSCH, Germany) using constant heating rates of 1, 2, 5 and 10°C/min in atmosphere of dry synthetic air. The compacts were sintered up to 1000°C and then rapidly cooled down to room temperature. Isotherms were obtained for some samples after fast heating up to a determined, fixed temperature, and kept at this temperature for different times. After sintering, each compact was fractured in three pieces for characterization. One piece was polished and its surface was chemically etched. The microstructure was examined using a Scanning Electron Microscope (JEOL model 305, Japan). Another piece was utilized for measurement of surface area by nitrogen adsorption technique (Micromeretics ASAP 2000, Norcross, GA, USA), and for determination of pore size distribution by mercury porosimetry (Micromeretics model 9200).

III - Results

Figure 1 shows the linear shrinkage ($Y = \Delta L/L_0$) as function of time for temperatures ranging from 700 to 800°C. Considering a fixed value of Y = 0.05, the apparent activation energy was calculated by determining the inverse of time necessary to reach such densification as a function of inverse temperature. The value calculated from this figure is Q = 210kJ/mol.

The densification rate (dρ/dt) as a function of temperature is shown in Figure 2. For each heating rate the linear shrinkage rate and densification increase from 500°C (the temperature where shrinkage starts), up to the maximum point and then decrease for higher temperatures. The point for maximum shrinkage rate occurs for the same relative density of 0.75 ± 0.01, independent of the heating rate. These results agree with results obtained by Lange[13] in sintering of alumina.

Mean pore diameter and surface areas for ZnO samples sintered by heating rate of 10°C/min up to a predetermined, fixed temperature, followed by fast cooling are shown as a function of temperature in Figure 3. Both curves show similar behaviors except for temperatures bellow 700°C. For these temperatures the reduction of surface area rate is smaller than the decrease of mean pore size rate. After 700°C both surface areas and mean pore diameter decrease continuously until the point of the equilibrium of dihedral angle (T = 850°C). Above this temperature the surface area decreases slowly.

Grain sizes were obtained by SEM for samples obtained as above described and Figure 3 also shows the plot of grain size versus temperature. Grain size remains practically constant up to 750°C (Y = 6.0% and ρ = 0.72). For higher temperatures grain size increases with temperature.

IV - Discussions

The constant heating rate method shows that the starting point for ZnO densification is about 500°C. To determine kinetics parameters during sintering of ZnO before coarsening, three approaches were considered.

i) Woolfrey-Bannister Approach

In the Woolfrey and Bannister[12] approach the general equation for constant heating rate (equation 3) was considered. Figure 4 shows the plot of $T^2 d(Y)/dT$ versus Y for all heating rates considered in this study. If grain size remains constant this plot should be a straight line with slope given by Q/R(n + 1). The apparent activation energy Q can be determined by the Dorn Method[12]. By inspecting Figure 2, and 4 three regions can be found before the maximum of shrinkage rate is reached. For the first region corresponding to linear shrinkage from 0 to 1.5% a straight line can be drawn for

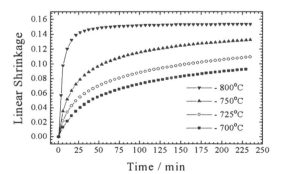

Fig.1-Linear shrinkage as function of time for isothermal sintering.

all heating rates considered. Considering that in this region the activation energy calculated by equation (3) is 290kJ/mol, the determined value for n is nearly zero. This value is consistent for the mechanism of mass transport of viscous flow type. This mechanism is not probable since the temperature is too low and pure ZnO does not form liquid for this range of temperatures. As proposed by Varela et al[20] and by Senos and Vieira[19] this thermally activated process should be rearrangement of particles. Senos and Vieira proposed that this rearrangement can be explained by grain boundary sliding to accommodate the sintering stresses that are generated during sintering for this packed particle system. They found that the apparent activation energy for this early stage of sintering is 260±24kJ/mol which is smaller than the value measured in our work (290kJ/mol).

We propose that in this early stage of sintering (for temperatures bellow 650°C) the nondensifying mechanism in surface diffusion should control neck growth. The asymmetry of these necks due to particle size distribution and the stresses generated by capillary forces in the neck region are sufficient to break down necks with increasing temperature. In this way the activation energy should be compatible with the energy necessary to break bonds in the ZnO system. Martins et al[21] using the quantum mechanical method AM1 obtained a value for the binding energy for the ZnO molecules in the crystalline structure of 3.0 eV (290.7kJ/mol). This value agrees with the value calculated in the early stage of sintering using the Dorn equation (290kJ/mol). The neck growth by surface diffusion[14] is followed by rearrangement of particles after the fracture of these necks leading to densification. The unstable neck formation during the early stage of sintering of MgO was experimentally observed by Rankin and Boatner[22]. The abrupt breakage of the neck was attributed to existing stress in the neck region. Other experimental results, Figure 3, reinforce the dominance of rearrangement in the early stage of sintering of ZnO. This figure shows that grain size remains constant in this early stage and that the rate of pore size decrease is higher than the rate of surface area decrease. Thus rearrangement of particles is responsible for densification in the early stage of sintering and seems to be controlled by the breakdown of necks among particles.

The curve of Figure 4 shows a continuous variation of the linear shrinkage rate from $Y = 1.5\%$ to $Y = 7\%$. In the range of 3.0 to 6.0 a straight line can be drawn with apparent activation energy determined by equation 3 of 210kJ/mol. Using

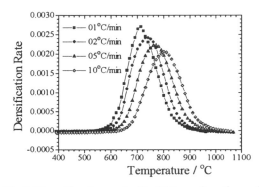

Fig.2-Densification rate ($d\rho/dT$) as function of temperature.

this value for activation energy, the average sintering coefficient calculated from the slope of equation 3 for all heating rate is 2.1± 0.1. This is consistent to with grain boundary diffusion being the controlling mechanism for mass transport and is supported by the fact that grain size remains practically constant up to $\rho = 0.72$ (T = 760°C) as showed in Figure 3. For higher temperatures coarsening becomes more important and equation 3 is no longer valid. Finally in the range of Y = 1.5 to Y = 3.0 most of rearrangement process was concluded, leading to formation of more stable necks with smaller capillary forces. However, due to the particle size distribution, some unstable necks are formed. These necks could be broken leading to additional rearrangement.

ii) Wang-Raj Approach

In the Wang and Raj[8] approach a modified equation 1 was used to determine the apparent activation energy for the sintering of the system. Considering values of $d\rho/dT$ for constant ρ at different heating rates, and that in the initial state grain size (G) is constant, the plot of the first member of equation 3 against the reciprocal of temperature should be a straight line, where the angular coefficient is $-Q/R$.

Plots were considered for constant ρ in the range of $\rho = 0.601$ to $\rho = 0.628$ (early stage) and in the range of $\rho = 0.657$ to $\rho = 0.722$ (first stage). The following average values for activation energies were obtained for both stages: Early stage Q = 290 ± 10kJ/mol, and first stage Q = 220 ± 10kJ/mol.

iii) Chu-Rahaman-De Jonghe-Brook Approach

Considering equation 2 either the activation energy or the sintering coefficient can be determined if one of these values is known. To calculate n values during early and initial stages, a plot of $\ln[(T/\rho)(d\rho/dT)\exp(Q/RT)]$ against ϕ gives a straight line. The slope of this straight line gives $(n + 1)/2$. Considering that $\phi = 5(1-\rho)$[13] for ZnO[13] and activation energies values determined by the Dorn and the Wang and Raj approaches, n was determined in both stages. During the early stage $(0.601 < \rho < 0.628)$ the activation energies obtained by both approaches are in agreement and is 290kJ/mol. Then the value of n determined through the slope of the plot above is smaller than 1 indicating a rearrangement process. For the first

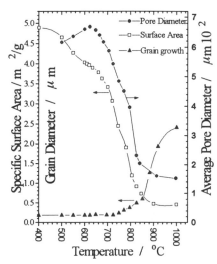

Fig.3-Mean pore diameter, surface area, and grain size as function of temperature for samples sintered at 10°C/min.

stage of sintering (0.657< ρ <0.722) the apparent activation energy determined by Dorn and Wang and Raj approach was 220kJ/mol and values of n determined using the plot described above were around 3 indicating a mechanism of grain boundary diffusion.

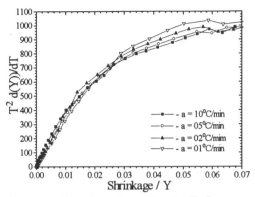

Fig.4-Woolfrey-Bannister plot for ZnO.

V - Conclusions

Using constant heating rate during sintering of ZnO pellets three distinctive regions could be determined before coarsening. Rearrangement seems to be the dominant process in the first region (0.601< ρ <0.628). The second region is characterized by the transition state where more than one mechanism has a significant effect on sintering. The apparent activation energy determined in the first region (290kJ/mol) is in agreement with the calculated energy bond of ZnO. This support the premise that the rearrangement process is governed by breaking these bonds. The apparent activation energy calculated in the third region (220kJ/mol) is in agreement with the grain boundary oxygen diffusion mechanism.

VI - References

1) J. Frenkel, "Viscous Flow of Cristalline Bodies under the Action of Surface Tension", *J. Phys.*, **9** [5] 385-91 (1945).

2) G.C. Kuczynski, "Self-Diffusion in Sintering of Metallic Particles", *Trans. Am.Inst. Mining Met. Eng.*, **185** [2] 169-78 (1949).

3) W.D. Kingery and M. Berg "Study of Initial Stages of Sintering Solids by Viscous Flow, Evaporation-Condensation, and Self-Diffusion", *J. Appl. Phys.*, **26** [10] 1205-12 (1955).

4) R.L. Coble, "Sintering of Crystalline Solids I: Intermediate and Final Stage Diffusion Models", *J. Appl. Phys.* **32** [5] 787-92 (1961).

5) D.L. Johnson, "New method of Obtaining Volume, Grain-Boundary, and Surface Diffusion Coefficients from Sintering Data," *J. Appl. Phys.*, **40** [1] 192-200 (1969).

6) M.F. Ashby, "A First Report on Sintering Diagrams", *Acta Mettall.*, **22** [3] 275-89 (1974).

7) R.T. DeHoff, "A Cell Model for Microstructural Evolution During Sintering", pp. 23-24 In: *Sintering Heterogeneous Catalysis*. Ed. by G.C. Kuczynski, A. E. Miller, and G.A. Sargent. *Plenum Press,* New York, (1984).

8)   J. Wang and R. Raj, "Estimate of the Activation Energies for Boundary Diffusion from Rate-Controlled Sintering of Pure Alumina, and Alumina Doped with Zirconia or Titania", *J. Am. Ceram. Soc.* 73 [5] 1172-75 (1990).

9)   J.D. Hansen, R.P. Rusin, M. Teng, and D. Lynn Johnson, "Combined-Stage Sintering Model", *J. Am. Ceram. Soc.,* 75 [5] 1129-55 (1992).

10) M.Y. Chu, M.N. Rahaman, and L. De Jonghe, "Effect of Heating rate Sintering and Coarsening", *J. Am. Ceram. Soc.,* 74 [6] 1217-25 (1991).

11) W. S. Young and I.B. Cutler, "Initial sintering with Constant Rates of Heating", *J. Am. Ceram. Soc.,* 53 [12] 659-63 (1970).

12) J.L. Woolfrey, and M.J. Bannister, "Nonisothermal Techniques for Studying Initial-Stage Sintering", *J. Am. Ceram. Soc.,* 55 [8] 390-94 (1972).

13) F.F Lange, "Approach to Reliable Powder Processing", pp. 1069-83 In: *Ceramic transactions,* vol. 1, *Ceramic Powder Science IIB.* Edited by G.L. Messing, E.R. Fuller, Jr., and H. Hausner. Ameerican Society, Westerville, OH, (1989).

14) O.J. Whittemore and J.A. Varela, "Initial Sintering of ZnO", *J. Am. Ceram. Soc.* 64 [11] 154-155, (1981).

15) W. Komatsu, Y. Moriyoshi, and N. Sato, "Synergetic Effect of ZnO with Different Dopants on Sintering", *Yogyo Kyokai,* 77 [10] 347-54 (1969).

16) D. Dollimore and P. Spooner, "Sintering Studies on Zinc Oxide", *Trans. Faraday Soc.* 67 [9] 2750-59 (1971).

17) T.K. Gupta and R.L. Coble, "Sintering of ZnO: I, Densification and Grain Growth", *J. Am. Ceram. Soc.,* 51 [9] 521-25 (1968).

18) S.K. Dutta and R.M. Spriggs, "Grain Growth in Fully Dense ZnO", *J. Am. Ceram. Soc.,* 53 [1] 61-62 (1970).

19) A.M.R. Senos and J.M. Vieira, "Pore size Distribution and Particle Rearrangement During Sintering", *Third Euro-Ceramics* vol.1 pp. 821-826, Ed. by P. Durán and J.F. Fernández, Faenza Editrice S.L. (1993).

20) J.A. Varela, E. Longo, V.C. Pandolfelli and C.V. Santilli, "Sinterização de Óxido de Zinco em Atmosferas de Ar Seco, Oxigênio e Dióxido de Carbono", *Cerâmica* 32 [199] 187-190 (1986).

21) J.B.L. Martins, E. Longo and J. Andrès, "ZnO Clusters Models: AM1 and MNDO Study", *Intern. J. Quantum Chem.* 27 643-53 (1993).

22) J. Rankin and L.A. Boatner, "Unstable Neck Formation During Initial-Stage Sintering", *J. Am. Ceram. Soc,* 77 [8] 1987-90 (1994).

VII - Acknowledgment

The authors acknowledge CNPq, FAPESP and FINEP-PADCT for the financial support for this work.

Sintering Processes of Metallic Nano-Particles: A Study by Molecular-Dynamics Simulations

Huilong Zhu[1] and R.S. Averback[1,2]

[1]Materials Research Laboratory and
[2]Department of Materials Science and Engineering
University of Illinois at Urbana-Champaign, Urbana, IL 61801

Abstract

Molecular-dynamics computer simulations were employed to investigate the dynamic processes of sintering and plastic deformation of two ultra-fine single crystal grains of copper at 700 K. Due to ultra-fine size (4.8 nm in diameter) of the grains, the local shear stresses in the necking region exceed the shear strength of copper. This causes plastic deformation, shrinkage, dislocation gliding and neck growth. Glide occurs on the usual glide system in fcc crystals (111):<110>. Such deformation represents a new mechanism for the initial stages of sintering. The relative shrinkage of the center to center distance was about 12%. Shrinkage caused by grain boundary diffusion is negligible over our simulation time scale. The relation between shrinkage and the sliding is discussed. A large relative rotation, $\approx 17°$, and the formation of twist twin grain-boundary between the two particles were also observed in our simulations.

1. Introduction

Nanocrystalline materials have become of great interest in recent years owing to their interesting properties and potential for technical applications (Gleiter 1989, Siegel 1994). A fundamental problem in studying nanocrystalline materials is the sintering processes of two ultra-fine particles as they first come into contact (Kingery and Berg 1955, Nichols 1966, Ashby 1973, Kellett and Lange 1989, Soppe, Janssen, Bonekamp, Correia and Veringa 1994). This problem, however, is complex and includes many superimposed mechanisms. For example, currently available macroscopic models are based on the kinetic and thermodynamic properties of equilibrium solids, which is not the case for nanocrystals. In addition, the local stresses of grain boundary (GB) in nanocrystalline metals become comparable to their strengths when the sizes of the particles are less than a few tens of nanometer (Easterling and Tholen 1972). In this case, all known models no longer can be used. Instead, it is likely that a more atomistic approach will be necessary to describe the kinetic behavior. Molecular dynamics (MD) simulations, in particular, provide a powerful means to probe the atomic motion that takes place during sintering of ultra-fine grains, and because of the small number of atoms in each grain, it becomes a practical means as well. The aim of the present work, therefore, was to test the feasibility of using atomistic simulations of the dynamic processes which occur during the early stages of two-particle sintering.

2. Method

The simulations were performed using a modified version of the DYNAMO code (Daw, Basker and Foiles 1991). Two fcc monocrystal copper spheres, each sphere containing 4688 atoms and having a diameter of 4.8 nm, were used in the simulations. For each simulation, a constant background temperature of 700 K was maintained throughout the sintering process. In order to examine the effect of GB orientation on sintering, two different GB angles were selected. The embedded atom potential of Sabochick and Lam for Cu (1991), which was proven reliable in earlier studies, was employed. The averaged surface energy calculated by the potential is about 1445 ergs/cm² for a sphere of Cu with R = 2.4 nm at temperature of 700 K, where R is the radius of the sphere. It agrees reasonably well with the experimental value of 1780 ergs/cm² (Kimmel and Scheithauer 1974).

3. Results

It is illuminating to examine a projected view of the atoms at various instants of time to gain a qualitative understanding of how atoms move during the sintering of two spheres. The first sample, S_1, consists of two randomly rotated spheres with a center-center distance of (2R + 0.2 nm). Figure 1 shows the projections of all atoms. When the two spheres were placed in contact the attraction between the spherical surfaces drew them together (Johason, Kendall and Roberts 1971). An upper limit, x_{ul}, for the contact radius due to elastic deformation is (Ashby 1973)

$$x_{ul} \approx (\frac{\gamma_{eff}}{RG})^{1/3} R \qquad (1)$$

where R is the radius of the spheres and G is the shear modulus. γ_{eff} is an effective surface energy, $2\gamma_s - \gamma_{gb}$, where γ_s and γ_{gb} are surface and GB energies, respectively. For copper, $\gamma_{gb} = 625$ ergs/cm² (Kimmel and Scheithauer 1974), G = 4.21×10^{11} dyne/cm² (Overton and Gaffney 1955) and so we obtain, $x_{ul}/R \approx 0.31$ from eq. (1) with R = 2.4 nm. By 5 ps, shown in Fig. 1(a), the contact radius reached its maximum, 0.84 nm, due to elastic deformation given by eq. (1). From 5 to 40 ps, the neck radius continued to increase, up to ≈1.6 nm. If we assume that standard models are adequate to describe our simulation, the neck growth after elastic deformation is mainly due to GB diffusion (Berrin and Johnson 1967), surface diffusion and lattice diffusion (Kuczynski, 1949). Only lattice and boundary diffusions allow the centers of the spheres to approach and result in densification and shrinkage, which are important properties of a nanomaterial in the sintering. The relative shrinkage, $\Delta L_{c-c}/2R$, is plotted in Fig. 2, where ΔL_{c-c} is the change in distance between the centers of the spheres. We calculate ΔL_{c-c} by following the locations of two atoms at the centers of the two respective grains. By 5 ps the shrinkage decreased very quickly which is caused by the elastic deformation and soft

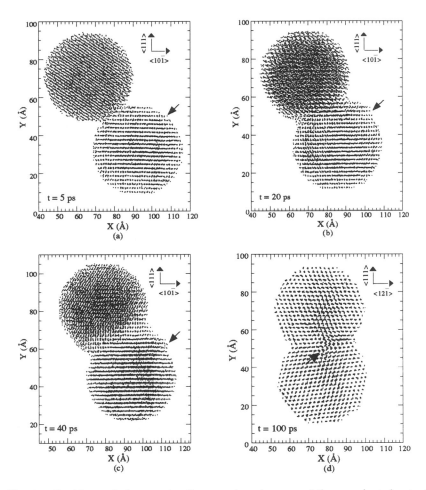

Fig. 1   Positions of all atoms in S_1, at various instants of time, projected onto the $(12\bar{1})$ plane of the sphere on the bottom for Figs. 1(a) - (c) and onto the $(10\bar{1})$ plane for Fig. 1(d).   Each dot represents one Cu atom.   The arrows in Figs. 1(a) - (c) show the sliding plane. In Fig. 1(d) the arrow shows the location of grain boundary dislocation.

collision between the spheres. The final shrinkage due to diffusional, plastic and elastic processes is about 0.58 nm (= 2R×0.12). It is noteworthy that a twin boundary, involving a rotation of 180° around the <111> direction, has formed. This can be seen in Fig. 1(d). Very recent experiments (Zhang, Han, Wang and Wang 1995) have revealed nearly the same twin boundary structure in the contact of two, 5 nm, Au particles as that predicted here.

As mentioned-above, only lattice diffusion and GB diffusion can cause shrinkage. In our simulation no vacancies were introduced inside the spheres, so

GB diffusion remains as the only diffusional mechanism that can be responsible for the shrinkage. The relative shrinkage, $y(= \Delta L_{c\text{-}c}/2R)$, due to GB diffusion can be calculated using (Berrin and Johnson 1967)

$$\dot{y} = -\frac{2CA_{gb}D_{gb}(x_{nk} - r)}{\pi x_{nk}^4 r} \quad \text{and} \quad \dot{x} = \frac{4CA_{gb}D_{gb}(x_{nk} - r)}{A_r x_{nk}^2 r} \tag{2}$$

where $C = \gamma_s \Omega/(kTR^3)$, $A_{gb} = 2\pi x_{nk}\delta/R$ = the area of the neck surface intersected by the GB, $A_r = 4\pi r(\theta(x_{nk} + r) - r\sin\theta)$ = the total surface area of the neck, Ω = atomic volume, δ = thickness, D_{gb} = grain boundary diffusivity, k = Boltzmann constant, T = absolute temperature, x_{nk} and r are radii, normalized by R in eqs. (2), of the neck and of the curvature of the neck surface, $r = (x_{nk}^2 - 2y + y^2)/(2(1 - x_{nk}))$.

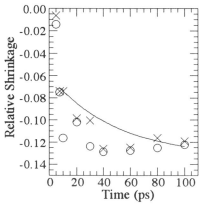

Fig. 2 The relative shrinkage as a function of time. x and o denote the shrinkage for S1 and S2, respectively. The solid line is the curve calculated by Berrin and Johnson model for S1.

We solved eqs. (2) numerically in the time interval from 10 to 100 ps. The solid line plotted in Fig. 2 is the curve of y given by eqs. (2) with $\Omega = 1.18 \times 10^{-23}$ cm^3, $\delta = 0.51$ nm (Ashby 1973), T = 700 K and the fitting parameter $D_{gb}(S_1) = 1.0 \times 10^{-2}$ cm^2/s, where $D_{gb}(S_1)$ denotes GB diffusivity of S_1. It is found that the value of D_{gb} must be greater than Cu liquid diffusivity, $D_{lq} \approx 0.5 \times 10^{-4}$ cm^2/s (Hsieh, Diaz de la Rubia, and Averback, 1989) by a factor of 200. It is far too large for grain boundary diffusion at T = 700 K. We will return to this question later.

We also simulated a rather special case: two crystallographically aligned spheres brought into contact at the start of the event. This sample is called S_2 and all the sintering conditions are the same as S_1 excepting the relative orientation. Figure 3 shows the projections of all S_2 atoms in the (001) plane at times, 10 ps and 100 ps, respectively. It can be seen that by 10 ps the lattice structure in the neck was distorted by the high stress and the collision between the two grains. At 100 ps, the neck region of S_2 formed a single crystal with a vacancy left behind. For both S_1 and S_2, the final shrinkages are about 12% which can yield a 21% increase in the atomic density if the initial configuration in a nanocrystal is close-packed. In this case, the final density is $\approx 90\%$ of the perfect crystal since the initial atomic density of a close-packed configuration is 74%.

Another way to estimate the GB diffusivity is to calculate the relative mean square displacement (Zhu and Averback 1995), $\Delta R^2(t, t')$, of the atoms in the boundary from time t' to t. To find $\Delta R^2(t, t')$ in the boundary, we first

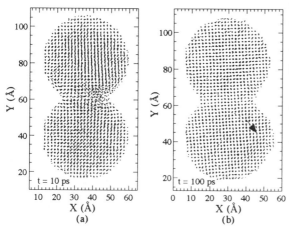

Fig. 3 Positions of all atoms in S_2 projected onto the (001) plane. (a) at 10 ps and (b) at 100 ps. The arrow indicates the gliding direction of the dislocation.

marked all atoms in the GB formed by the necking at time t' and followed these atoms until time t. The thickness of the GB used here was $1.5a_0$ ($a_0 = 0.362$ nm). For S_1 we obtained $\Delta R^2(40, 5)=0.087$ nm^2 or a GB diffusivity $D'_{gb}(S_1) = 6.2 \times 10^{-6}$ cm^2/s which is much less than $D_{gb}(S_1)$. Similarly, for S_2 we calculated $D'_{gb}(S_2) = 4.66 \times 10^{-6}$ cm^2/s. The reason for $D'_{gb}(S_2) < D'_{gb}(S_1)$ is due to the fact that the GB formed in the event S_2 has a higher structural order than S_1. If the values of D'_{gb} are used in eqs. (2), the relative shrinkages are less than 0.1% which is much smaller than the values obtained from our simulations. Consequently, the shrinkage caused by GB diffusion is negligible on our simulation time scale, and some factor other than diffusion must be responsible for the shrinkage.

If shear stress in a material exceeds its shear strength, crystal plane sliding will occur. For two nano-particles' sintering, as indicated by Easterling and Tholen (1972), the local stress in the neck region may be very close to the strength of metals when the particle radii are about 30 nm. Actually, we can roughly estimate the stress in the neck. When two spheres of radii R are pressed together with a normal force F_s, the average pressure, $\bar{P}$, over the neck and the maximum shearing stress, σ_{max}, were calculated by Hertz (1881)

$$\bar{P} = \frac{F_s}{\pi x^2} = 0.41 \left(\frac{F_s E^2}{R^2}\right)^{1/3} \text{ and } \sigma_{max} = 0.465\bar{P} \qquad (3)$$

where E(= 12.6×10^{11} dyne/cm^2 for Cu) is Young's modulus for the material under consideration. In the situation of two spheres in contact without external forces, one can obtain $F_s \approx \pi \gamma_{eff} R$ (Johnson, Kendall and Roberts 1971) and then eq. (3) becomes $\bar{P} \approx 0.6(\gamma_{eff} E^2/R)^{1/3}$. Thus we get $\sigma_{max} \approx 3$ GPa for R = 30 nm and $\sigma_{max} \approx 7.4$ GPa for R = 2.4 nm. The shear strength, σ_c, of single crystal Cu is about 1.4 GPa, roughly estimated by G/30 (Kittel 1986). It can be seen that $\sigma_{max} > \sigma_c$ which means that crystal plane sliding is possible when R < 30 nm. The shear stresses in the glide plane of S_1 at several times are shown in Fig. 4. The glide plane is the region: $[y_s(t) - 0.3$ nm, $y_s(t) + 0.3$ nm], where $y_s(t)$ is the position, at time t, of the 20th plane (indicated by the arrows in Figs. 1(a)-(c)) counting from the bottom plane of the lower sphere. The protrusion on the lower nano-particle at (x, y) = (10.4 nm, 5.5 nm) in Fig. 1(b) (also indicated by an arrow) clearly shows that the non-elastic shrinkage of the two nano-particles is due to slip. The maximum shear stress at 5 ps is about 8 GPa which is almost the same as the value, 7.4 GPa, given by eq. (3). However, the absolute value of the shear stress decreased dramatically from 5 ps to 40 ps. This is attributed to the deformation of the particles. At 40 ps, the averaged shear stress over the neck region is ≈ 0.3 GPa which is less than the shear strength of 1.4 GPa. For S_2 a detailed study (Zhu and Averback 1995) showed that from 10 to 20 ps, the neck recrystalized and a dislocation formed. This dislocation glided from the neck, along the direction indicated by the arrow in Fig. 3(b), to the surface of the lower sphere within 80 ps. As a result, an additional crystal plane, was pushed out in the <100> (or x) direction.

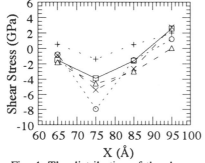

Fig. 4 The distribution of the shear stress in the sliding plane of S_1 at various instants of time: O - 5 ps, ◻ - 10 ps, △ - 20 ps, X - 30 ps and + - 40 ps.

In order to quantitatively find the relative movement of crystal planes, the averaged displacements (ADs) of atoms in the various atomic planes have been calculated. The AD between time t and t' in the region of R_g is defined as

$$\Delta x(R_g) = \frac{1}{N(R_g)} \sum_j (x_j(t) - x_j(t')), \qquad (4)$$

where the summation is over the atoms in the region of R_g. In the real computation we select the atoms at time t in a slab with a constraint $x_j > x_d$ or $x_j < x_d$, and then $N(R_g)$ is the total number of atoms in the slab at time t, where x_j is the x component of the position of the jth atom in the slab and $x = x_d$ defines a plane that parallels the $(10\bar{1})$ plane as well as passes point $(x_d, 0, 0)$ (see Fig. 1(c)). The reason for using the constraints, $x_j > x_d$ and $x_j < x_d$, is to extract some information about grain rotation. Similarly, we can define $\Delta y(R_g)$ and $\Delta z(R_g)$. We calculated Δx, Δy and Δz as a function of y (see Fig. 1(c)) and in the slabs: $[y - \Delta l, y + \Delta l]$ with $x_d = 86$ Å, where $\Delta l = 0.1$ nm. Figures 5(a) and (b) show the ADs between time 5 ps and 40 ps. We have chosen the values of Δx, Δy and Δz at y = 5.2 nm with x>8.6 nm, as the origins of Δx, Δy and Δz, respectively. It is found that the largest displacement (Δx), ≈ 1.2 nm for both cases of $x > x_d$ and $x < x_d$, is in <10$\bar{1}$> direction of the lower sphere and the smallest displacement (Δy), ≈ 0.15 nm, in the <111> direction. The fact that $\Delta x(x>8.6nm) \approx \Delta x(x<8.6nm)$ indicates that the glide direction was along the $<10\bar{1}>$. It is interesting to note that the values of $\Delta z(x>8.6nm)$ and $\Delta z(x<8.6nm)$ are quite different, which implies that a relative rotation occurs about the $<111>$ direction. One can easily estimate the rotation angle, α, by $\alpha \approx \sin^{-1}[(\Delta z(x<8.6nm)|_{y\,=\,7.5\;nm} - \Delta z(x>8.6nm)|_{y\,=\,5.2\;nm})/R]$. Therefore, we obtain $\alpha \approx$ - $15.6°$ at 40 ps, where the negative sign means the rotation is clockwise. The particles rotate with respect to one another as illustrated in Fig. 5(c). Before 20 ps, no obvious rotation occurs. From 20 to 55 ps the two spheres rotate $\approx 17°$ relatively to each other.

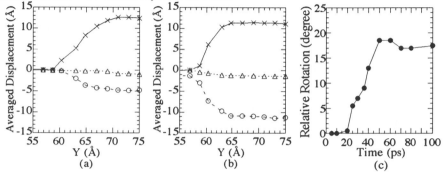

Fig. 5 Crystal plane sliding and relative rotation in S_1. Figures (a) and (b) show the averaged displacements, calculated by eq. (4), for $x \geq 8.6$ nm and $x < 8.6$ nm, respectively and X - Δx (in $<10\bar{1}>$), Δ - Δy (in $<111>$) and O - Δz (in $<12\bar{1}>$). The relative rotation of the spheres as a function of time is plotted out in Fig. (c). The rotation axis is $<111>$ crystallographic axis of the sphere on the bottom.

The exact quantitative relationship between shrinkage and dislocation motion is subject to further detailed study. Here we have provided a preliminary picture based on our simulation data. First, we note that the shrinkage and the sliding took place in the same period of time as can be seen by comparing Figs. 1, 2 and 5. For example, the primary shrinkage occurred before 40 ps. Similarly the protrusion in S_1 appeared prior to 40 ps, as indicated by the arrow in Fig. 1(c). From Fig. 5 we obtain the displacement in the $<10\bar{1}>$ direction, which is ≈ 12 Å. The angle between $<10\bar{1}>$ direction and the line joining the centers of the two sphere is $\approx 25°$. Thus we obtain the shrinkage, $\Delta L_{c-c}/2R = \sin(25) \times 12/48 = 10.6\%$. This value is close to the shrinkage, 12.6% , which has been shown in Fig. 2.

Acknowledgments
One of the authors (HZ) is grateful to Dr. L.C. Wei for helpful discussions. This research was supported by the U.S. Dept. of Energy, Basic Energy Sciences under grant # DEFG02-91ER45439. Grants of computer time from NCSA at the University of Illinois and NERSC at Lawrence Livermore National Laboratory are also gratefully acknowledged.

References

H. Gleiter, "Nanocrystalline materials", Progr. Mater. Sci. **33** 223-315 (1989).

R.W. Siegel, "Mechanical properties of nanophase materials", pp233-261, G.C. Hadjipanayis and R.W. Siegel (eds.), Nanophase Materials, (1994).

W.D. Kingery and M. Berg, "Study of the initial stages of sintering of solids by viscous flow, evaporation-condensation, and self diffusion", J. Appl. Phys. **26** 1205 (1955).

F.A. Nichols, "Coalescence of two spheres by surface diffusion", J. Appl. Phys., **37** 2805 (1966).

M.F. Ashby, "A first report on sintering diagrams", Acta Met. **22** 275 (1973).

B.J. Kellett and F.F. Lange, "Thermodynamics of densification: I, Sintering of simple particle arrays, equilibrium configurations, pore stability and shrinkage", J. Am. Ceram. Soc. **72** 725 (1989).

W.J. Soppe, G.J. Janssen, B.C. Bonekamp, L.A. Correia and H.J. Veringa, "A computer simulation method for sintering in three dimensional powder compacts", J. Mater. Sci. **29** 754 (1994).

K.E. Easterling and A.R. Tholen, "Surface energy and adhesion at metal contacts", Acta Met., **20** 1001 (1972).

M.S. Daw, M.I. Basker and S.M. Foiles, private communication.

M.J. Sabochick and N.Q. Lam, "Radiation-induced amorphization of ordered intermetallic compounds CuTi, $CuTi_2$,, and $CuTi_4$: a molecular dynamics study", Phys. Rev. **B43** 5243 (1991).

E.R. Kimmel and W. Scheithauer, Jr., "Dispersoid growth in $Ni-2ThO_2$ due to self diffusion sintering of particle clusters", Met. Trans., **5** 1495 (1974).

K.L. Johason, K. Kendall and A.D. Roberts, "Surface energy and the contact of elastic solids", Proc. R. Soc. Lond. **A324** 301-313 (1971).

W.C. Overton and J. Gaffney, "Temperature variation of the elastic constants of cubic elements. I. Copper", Phys. Rev. **98** 969 (1955).

L. Berrin and D.L. Johnson, "Precise diffusion sintering models for initial shrinkage and neck growth", pp.369-392, Sintering and Related Phenomena, edited by G.C. Kuczynski, N.A. Hooton and C.F. Gibbon, Gordon and Breach, New York, (1967).

G.C. Kuczynski, "Self-diffusion in sintering of metallic particles", Trans. AIME **185** 169 (1949).

H. Hsieh, T. Diaz de la Rubia, and R.S. Averback, "Effects of temperature on the dynamics of energetic displacement cascades: a molecular dynamics study", Phys. Rev. **B40** 9986-9988 (1989).

H. Zhu and R.S. Averback, "Molecular dynamics simulations of densification processes in nanocrystalline materials", Mater. Sci. and Eng. A, in press.

H. Hertz, see pp.413-414, Theory of Elasticity, Third Edition, edited by S.P. Timoshenko and J.N. Goodier, McGraw-Hill Book Co., Singapore, (1982,).

C. Kittel, Introduction to Solid State Physics, sixth edition, p559, John Wiley & Sons, Inc., New York, (1986).

Zhang, H.Q., Han, M., Wang, Q., and Wang, G.H., "HREM characterization of gold cluster structures", Surf. Rev. and Letts, in press.

H. Zhu and R.S. Averback, unpublished data.

Sintering and Grain Growth of Ultrafine CeO$_2$ Powders

M. N. Rahaman

University of Missouri-Rolla, Department of Ceramic Engineering
Rolla, Missouri 65401

Abstract

The sintering and grain growth of ultrafine ceramic powders (particle size 10-20 nm) synthesized under hydrothermal conditions are considered. Compared to coarse (micron-sized) powders, the densification rate of compacts formed from ultrafine powders increases significantly. A problem is that the rate of the non-densifying mechanisms that lead to coarsening of the microstructure can also be significantly enhanced. The production, by sintering, of dense solids with ultrafine grain size is therefore not guaranteed. However, processing techniques such as the control of the heating schedule and the use of dopants, commonly used for microstructural control in coarse powders, are also effective for ultrafine powders. The use of these techniques is outlined for the sintering of ultrafine powders of CeO$_2$. Bulk solids and thin films with almost full density coupled with ultrafine grain size (<200 nm) can be produced below ≈ 1350 °C.

I. Introduction

Ultrafine (or nanocrystalline) powders, with particle sizes less than 50-100 nm, can have significantly enhanced sintering rates and decreased sintering temperatures compared to coarse-grained (micron-sized) powders (Eastman et al, 1989; Hahn et al, 1990; Zhou and Rahaman, 1993). They may allow the fabrication of novel materials (e.g., solids with high density and ultrafine grain size) that cannot be produced by the sintering of their coarse-grained counterparts.

While the significantly enhanced densification rates for ultrafine powders have been well demonstrated, the control of grain growth during sintering has received relatively little attention. As shown by the work of Komarneni et al (1988) and Hahn et al (1990), the attainment of high density can be accompanied by coarse-grained microstructures. According to Herring (1950), for geometrically similar clusters of particles, the rate of matter transport varies as $1/G^m$, where G is the particle or grain size and m is an exponent that depends on the mechanism, e.g., m = 4 for grain boundary diffusion or for surface diffusion, m = 3 for lattice diffusion, and m = 2 for vapor transport (evaporation/condensation). Herring scaling law predicts that the rates of the densifying mechanisms (e.g., grain boundary and lattice diffusion) as well as the rates of the non-densifying mechanisms (e.g., surface diffusion and vapor transport) increase significantly with decreasing particle size. Therefore, in the sintering of ultrafine powders, the attainment of high density coupled with ultrafine grain size is not guaranteed.

Several fabrication techniques (e.g., the use of dopants, the application of an external pressure and the control of the heating schedule) have been shown to be effective in controlling the microstructure of coarse-grained systems (Brook, 1982). However, ultrafine powders are characterized by a large percentage of molecules on their surfaces, a high density of grain boundaries in the consolidated material and a high volume of grain boundary material. The consequences of these characteristics for microstructural control in ultrafine-grained systems are not clear.

In the present paper, our recent work on the sintering and grain growth characteristics of ultrafine powders powders is described briefly. Data are presented for model powders of CeO_2 synthesized under hydrothermal conditions. The effectiveness of controlled heating schedules and the use of dopants for microstructural control are discussed.

II. Synthesis of Ultrafine Powders

While several methods can be used to synthesize ultrafine ceramic powders, the inert gas condensation technique described by Birringer et al (1984) and by Eastman et al (1989) has received much recent attention. As outlined earlier, the experiments described in the present paper utilized CeO_2 powders synthesized by chemical precipitattion under hydrothermal conditions. The procedure has been described in detail by Rahaman and Zhou (1995). Undoped CeO_2 and CeO_2 doped with up to 20 at% of various divalent and trivalent cations (Mg^{2+}, Ca^{2+}, Sc^{3+}, Y^{3+} and Nd^{3+}) were synthesized. X-ray diffraction (XRD) and high resolution transmission electron microscopy (TEM) revealed that the powders consisted of single crystal particles with an average size in the range of 10 to 15 nm. As an example, Fig. 1 shows a high resolution TEM micrograph of the CeO_2 powder doped with 6 at% Ca. The particles have faceted sides and within the limit of resolution, the lattice fringes cover the whole particle. As shown in Fig. 2 for the undoped powder, the particle size distribution is fairly narrow and can be well described by the Lifshitz-Slyozov-Wagner (LSW) theory of Ostwald ripening controlled by the diffusion step.

III. Sintering of Powder Compacts

Powder compacts (6 mm in diameter by 4 mm) for sintering and grain growth studies were formed by uniaxial pressing in a tungsten carbide die at ≈50 MPa. The compacts were sintered in air in a dilatometer that allowed continuous monitoring of the shrinkage kinetics. Sintering was performed at constant rates of heating (1, 5 and 10 °C/min) to 1350-1400 °C.

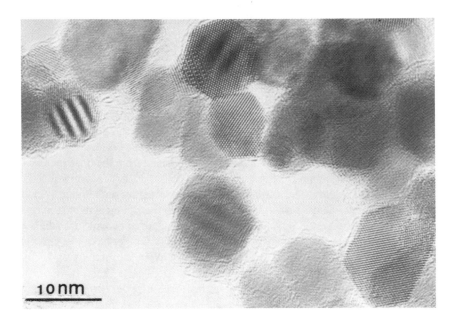

Figure 1 High resolution TEM of CeO$_2$ powder doped with 6 at% Ca^{2+}.

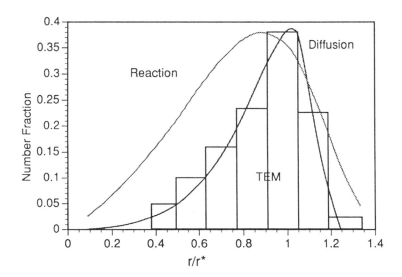

Figure 2 Data for the particle size distribution of undoped CeO$_2$ powder (histogram) compared with the predictions of the LSW theory.

The sinterability of the ultrafine CeO_2 powder during heating at 5 °C/min is shown in Fig. 3, where the relative density, ρ, is plotted as a function of temperature, T. Also shown are the data for a coarser CeO_2 powder (average particle size $\approx$100-200 nm) prepared by precipitation from cerium nitrate solution at room temperature and atmospheric pressure and for a commercial high-purity CeO_2 powder (average particle size $\approx$500 nm; Cerac, Inc., Milwaukee, WI). The data show that the commencement of measurable shrinkage occurs at approximately the same T ($\approx$800-900 °C) for all three powder compacts. However, after the commencement of sintering, the densification rate increases rapidly with decreasing particle size. Figure 3 also shows that the compact formed from the ultrafine CeO_2 powder reaches a maximum relative density of $\approx$0.95 at $\approx$1200 °C, after which the density decreased slightly. The difficulty in achieving densities higher than 0.95 at temperatures greater than $\approx$1200 °C is associated with the chemical reduction of CeO_2 to Ce_2O_3. However, almost full density can be achieved by control of the heating schedule (e.g., isothermal sintering below 1200 °C). For example, sintering for 2 h at 1150 °C leads to almost full density with an average grain size of $\approx$100 nm (Zhou and Rahaman, 1993).

As outlined earlier, dopants have been shown to be very effective for controlling the microstructure of coarse-grained ceramics. It is important to determine the effectiveness of the approach for ultrafine-grained ceramics. For dense

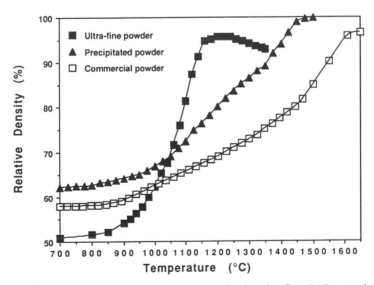

Figure 3 Relative density versus temperature for the ultrafine CeO_2 powder (average particle size 10-15 nm), a coarser CeO_2 powder (100-200 nm) prepared by chemical precipitation and a commercial CeO_2 powder ($\approx$500 nm).

bodies produced from the doped CeO$_2$ powders, the grain growth kinetics showed a minimum at a dopant cation concentration of ≈6 at%. Figure 4 shows the data for the grain size of the Ca-doped CeO$_2$ after sintering at 10 °C/min to 1350 °C. This concentration for the minimum grain growth is approximately the same as that found by Rahaman and Hu (1993) for a coarse-grained CeO$_2$ powder. The data indicate that despite the significantly higher interfacial area associated with ultra-fine powders, control of grain growth can be achieved with approximately the same dopant concentration used in coarse-grained systems of the same composition.

For a fixed dopant cation concentration (6 at%), the effect of the elemental dopant composition on the sintering kinetics of the ultrafine CeO$_2$ powder is shown in Fig. 5. Compared to the undoped powder, all of the dopants (Mg^{2+}, Ca^{2+}, Sc^{3+}, Y^{3+} and Nd^{3+}) lead to a shift of the sintering curve to higher temperatures. How-ever, the samples doped with Mg^{2+} and Sc^{2+} reach a limiting density of only ≈0.95 while those doped with Ca^{2+}, Y^{3+} and Nd^{3+} reach almost full density. Electron microscopy of the sintered compacts revealed that, compared to the undoped CeO$_2$, the grain growth kinetics are enhanced by Mg^{2+} and Sc^{2+} but reduced significantly by Ca^{2+}, Y^{3+} and Nd^{3+}. The effect of the dopants on the sintered microstructures are illustrated in Fig. 6, which shows scanning electron micrographs of the fractured surfaces of the materials doped with Mg^{2+}, Ca^{2+}, Y^{3+} and Nd^{3+}. The grain sizes of the materials are 1 μm (Mg dopant), 30 nm (Ca), 200 nm (Y) and 400 nm (Nd).

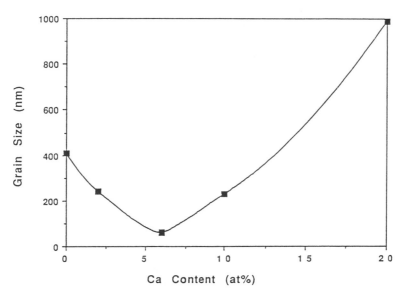

Figure 4 Grain size versus dopant concentration for the Ca-doped CeO$_2$ powder compacts sintered at 10 °C/min to 1350 °C.

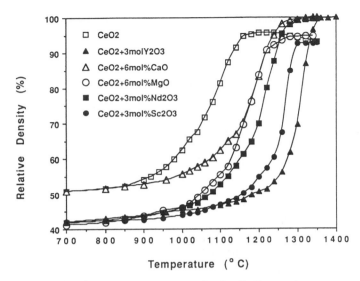

Figure 5 Relative density versus temperature for the CeO_2 powder compactsdoped with 6 at% of divalent and trivalent cations during sintering at 10°C/min to 1350°C.

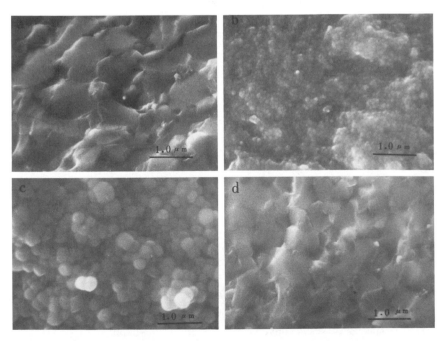

Figure 6 SEM of the fractured surfaces of the CeO_2 doped with 6 at% of (a) Mg^{2+}, (b) Ca^{2+}, (c) Y^{3+} and (d) Nd^{3+} after sintering at 10°C/min to 1350°C.

IV. Sintering of Thin Films

The sintering benefits outlined earlier for ultrafine powders can be used effectively for the production of thin films (Yang, 1995). With adequate control of microstructural coarsening, dense as well as porous films with thicknesses in the range of a few tenths of a micron to a few microns can be produced from concentrated suspensions of ultrafine particles. This range of film thickness is conveniently between those achieved with the commonly used methods employing solutions (less than a few tenths of a micron for a single coating) or tape casting of concentrated suspensions (≈ 10 μm to several hundred microns).

The stability of the suspension of ultrafine particles controls the homogeneity of the deposited film and, hence, the microstructural evolution of the fired film. Electrostostatic or steric repulsion can be used to stabilize the suspensions. Dilute suspensions of CeO₂ powders doped with 6 at% Y^{3+} (particle size 10-20 nm) were stabilized electrostatically at pH values of 3.5 to 4. However, particle concentrations above ≈ 0.2 vol% produced flocculation. Higher particle concentrations (≈ 10 vol%) were achieved without flocculation through the use of sterically stabilized suspensions with polyvinylpyrrolidone (relative molecular weight $\approx 30,000$) as a dispersant. Because of the very high specific surface area of the powder, a fairly high concentration (25 wt%) of the dispersant was required to produce good stability. As described in detail elsewhere (Yang, 1995), adherent thin films on substrates of Al_2O_3 or Y_2O_3-doped ZrO_2 were prepared from the sterically stabilized suspensions by spin coating. Measurements (by SEM) of the thickness of the film at several temperatures during constant heating rate sintering (10 °C/min) gave a shrinkage curve that followed the same trends as those for a powder compact of the same composition. Figure 7 shows the top surface and a fractured surface of a film sintered at 10 °C/min to 1300 °C. The film (thickness

 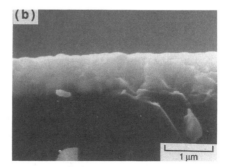

Figure 7 SEM of (a) the top surface and (b) a fractured surface of an Y-doped CeO₂ film (6 at% Y) sintered at 10 °C/min to 1300 °C.

$\approx 1 \mu m$) has a fairly smooth surface. Furthermore, the magnitude of the shrinkage coupled with the absence of porosity in the SEM micrographs indicated that the film was almost fully dense.

V. Conclusions

The enhanced sintering rates and low sintering temperatures associated with ultra-fine (nanocrystalline) powders can be used effectively for the production, by sinter-ing, of bulk solids and thin films with high density and controlled microstructure. However, for the achievement of ultrafine grain size ($<\approx 200$ nm) in the fabricated material, control of the grain growth is important. Techniques such as the use of dopants and control of the heating schedule, commonly used for coarse (micron-sized) powders are also effective for ultrafine powders. With these techniques, bulk solids and thin films of CeO_2 with nearly full density and grain sizes <200 nm were produced by sintering.

References

R. Birringer, H. Gleiter, H.-P. Klein and P. Marquard, "Nanocrystalline Materials: An Approach to a Novel Solid Structure with Gas-like Disorder", Phys. Lett., **102A** 365-69 (1984).

R.J. Brook, "Fabrication Principles for the Production of Ceramics with Superior Mechanical Properties", Proc. Brit. Ceram. Soc., **32** 7-24 (1982).

J.A. Eastman, Y.X. Liao, A. Narayanasamy and R. W. Siegel, "Processing and Properties of Nanophase Oxides", in Processing Science of Advanced Ceramics. Edited by I. A. Aksay, G. McVay and D. Ulrich. Mater. Res. Soc. Symp. Proc., **155** 255-66 (1989).

H. Hahn, J. Logas and R.S. Averback, "Sintering Characteristics of Nanocrystalline TiO_2", J. Mater. Res., **5** [3] 609-14 (1990).

C. Herring, "Effect of Change of Scale on Sintering Phenomena", J. Appl. Phys., **21** 301-303 (1950).

S. Komarneni, E. Fregeau, E. Breval and R. Roy, "Hydrothermal Preparation of Ultrafine Ferrites and Their Sintering", J. Am. Ceram. Soc., **71** [1] C26-C28 (1988).

M. N. Rahaman and C.-L. Hu, "Effect of Dopants on Sintering and Microstructure Development", Ceram. Trans., **32** 309-22 (1993).

M.N. Rahaman and Y.-C. Zhou, "Effect of Solid Solution Additives on the Sintering of Ultra-fine CeO_2 Powders", J. Europ. Ceram. Soc., **15** 939-50 (1995).

X. Yang, "Chemical Precipitation Techniques for the Production of Ceramics with Controlled Microstructure", Ph.D. Thesis, University of Missouri-Rolla (1995).

Y.-C. Zhou and M.N. Rahaman, "Hydrothermal Synthesis and Sintering of CeO_2 Powders", J. Mater. Res., **8** [7] 1680-86 (1993).

Sintering Behavior of Nano-sized Silicon Powders

H. Hofmann and J. Dutta

Laboratoire de Technologie des Poudres, Département des Matériaux
Ecole Polytechnique Fédérale de Lausanne,
CH-1015 Lausanne, SWITZERLAND

Abstract

The sintering behavior of nano-sized silicon powders prepared by plasma-induced dissociation of silane has been studied. It was observed that the powder morphology has a very important influence on the sintering behavior.Si-clusters which are formed during the powder synthesis acts as seeds for the crystallization of polycrystalline particles. Classical sintering models are insufficient to explain the sintering behavior but hard core/sinterable coating or the model of particle sliding can explain the sintering rate of this powder.

1. Introduction

The fabrication of dense parts starting from nanoscaled powder (n-powder) while retaining the grain size on a nanometric scale is one of the most interesting challenges facing the materials engineering community. By now, a lot of techniques have been developed or modified for the production of n-powders. The problem now consists in working out appropriate methods for the consolidation of these powders including the shaping as well as the sintering steps. All consolidation methods which can be cited as typical representatives for powder technology may be used in n-powder processing. However, compacting, hot pressing, sintering, etc., as applied to n-powders share some features which are connected with their activity and with the high levels of inter particle friction(1). The small particle size, or in other words the high surface area of the powder leads to high reactivity and therefore to a low sintering temperature. Also the formation of agglomerates is enhanced since the forces like van der Waals or electrostatic forces are stronger than the gravitational forces. Additionally, most of the powder production methods lead to the formation of a powder with agglomerates or aggregates. This is true for all high temperature synthesis methods (Laser, Plasma) or the wet chemical routes. These agglomerates are the origin of a broad pore size distribution and therefore a broad sintering range. The advantage of the high surface can only be used for the internal sintering of the agglomerates. After this first sintering step, we obtain more or less dense aggregates following which the sintering behavior is determined by the size of these aggregates which are generally 10 to 100 times larger than the original powder. Therefore, to achieve a dense microstructure the sintering temperatures have to be increased which leads to an increase of the rate of grain growth. A dense and nanosized microstructure with a grain size < 100 nm can only be achieved with agglomerate free powder and with green bodies which show a homogeneous pore size distribution. The aim of this work is to investigate the sintering behavior of amorphous nanosized silicon powder, as a simple representative system to form a basic understanding of the sintering behavior of silicon based alloys (Si_3N_4, SiC etc.). Especially the behavior of the powder during the heating up as well as during the first sintering step has been studied in detail.

2. Experimental

The powders were prepared by gas phase reaction of silane in a capacitively coupled radio-frequency Plasma-Enhanced Vapor Deposition (rf-PECVD) system and collected from the electrodes after synthesis. Further details of fabrication of nano-sized silicon powders in the PECVD reactor will be reported elsewhere (2). Samples for transmission electron microscopy (TEM) studies were sprayed on carbon-coated copper grids and studied in a Philips EM 430 microscope operated at 300 kV. For in-situ observation of the sintering process, the sample was mounted on a GATAN 'hot stage'. High temperature in-situ experiments were carried out between 300 and 900 °C during several hours. The microstructure was studied from the video recording carried out during the in-situ heating sequence. Additionally, powders were heated in a furnace at different temperatures between 300 and 900°C during 1 - 8.5 h in forming gas (10%H_2 and 90% N_2). The microstructure as well as the transformation behavior of these powders were examined in a High Resolution-TEM (HRTEM; JEM 4000EX) operated at 400 kV. Real and Fourier space image processing of the micrographs recorded at a magnification of 500'000 times were carried out. Estimation of the grain size was made from the TEM measurements, while the specific surface area was determined by gas adsorption techniques (BET).

3. Results

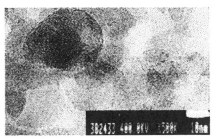

Figure 1: Microstructure of the as-received powder.

The morphology of the as-received powder is shown in fig. 1. The powder is amorphous, slightly agglomerated with particle sizes between 14 and 17 nm. Specific surface area of the powder was measured as 162 m^2/g. The annealing of the powders in a furnace as well as during in-situ heating in the TEM leads to the crystallization of the amorphous powder.

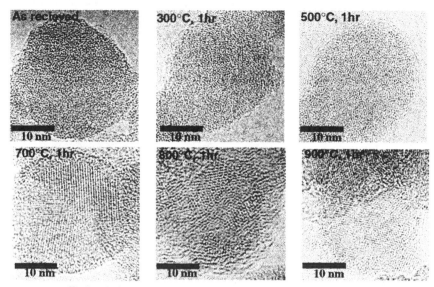

Figure 2: Development of the microstructure during heating.

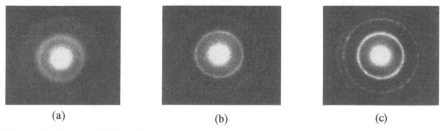

(a) (b) (c)

Figure 3: Electron diffraction pattern of the silicon powder. a) as-received; b) 10 mins at 650°C; c) 1h at 650°C.

Fig.2 shows the development of the microstructure with respect to temperature for a constant annealing time of 1h at each temperature. At temperatures below 500 °C the powders are amorphous while at temperatures > 600 °C crystalline phases of silicon can be observed. It is interesting to note that the crystallite sizes are much smaller compared to the actual particle size as the crystallization process is limited to the primary particle. Quantitative metallography as well as the analysis of the x-ray peaks result in a mean crystallite size of only 4.5 nm (700 °C) and 6 nm (900 °C) respectively in the polycrystalline particles. Twins as well as partially amorphous domains between the crystallites were observed. The kinetics of the crystallization as followed during in-situ annealing in TEM is shown in fig 3. After 10 min annealing at 650 °C, first signs of crystalization could be noticed. Complete crystallization (when the diffraction pattern did not change anymore) is reached after 1h. The grain growth after crystallization at 850 °C is shown in fig 4. During annealing, the grains grow from a mean diameter of 4.5 nm up to 10 nm after 7.5h.

Figure 4: Grain growth of silicon grains at 800°C.

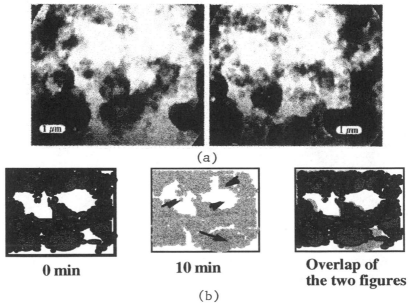

Figure 5: Sintering behavior of Si-powder. a) In-situ observation in the TEM; b) Schematic view of the particle movement.

In Addition to the crystalline grains in the particles, an amorphous surface layer of silicon oxide of 1.5 to 2 nm is observed. The results of the in-situ observation of the sintering behavior of the silicon powder mounted on a GATAN hot stage at 900 °C are shown in fig 5a which is schematically represented in fig 5b. Fig 5a shows the difference in the microstructure after 10 min sintering. The shrinkage range is still very small, but we can observe some particle movement (rearrangement). This particle movement is more clearly shown in fig 5b, the arrows indicating the direction of the movement. A neck formation or a neck growth were however not observed.

4. Discussion

The formation of the powder in rf silane discharge passes through three distinct regions: the initial clustering phase followed by the formation of larger particles and finally the agglomerates. The polymerization occurs via negative ion induced clustering such as (3)

$$Si_nH_x{}^- + SiH_4 \rightarrow (Si_n+H_{x'}{}_-) + (H, H_2 \text{ products})$$

These clusters agglomerate and forms in the second step of the powder formation, aggregates with a high density and a diameter of 14 to 17 nm. Fig.1 shows, that the particles are nearly 100% dense with respect to the amorphous phase, while for the crystalline Si, we can assume a density of 84% (6). The final step of the powder formation consists of the agglomeration of the primary particles with a very low density and a estimated average size of 150 nm. During heating up to the sintering temperature, the amorphous powder crystallize. Our results, which shows a grain size of 4.5 nm supports the theoretical approach of Veprek et al. We believe that the primary Si-clusters in the powder particles acts as seeds for the crystallization process. The crystallization temperature is between 600 to 650 °C (873 - 923 K) 0.5 Tm of the bulk Si. This value is very high as compared to the thermodynamic calculations carried out by Veprek (7) where a stable crystalline size of 3 nm was observed (>1000 Si-atoms per cluster). Therefore, we can conclude that in this system the grain growth is very slow [8]. The relatively large size of the nucleus explains also why no grains smaller than 3 nm could be observed. Additionally, the amorphous oxide-layer around each grains inhibit the grain growth across the particle-particle interface and limit also the maximum grain size. Summarizing the results and the discussions of the as received as well as heated powder, we can describe the system before sintering as following (Fig.6). The powder is loosely packed and slightly agglomerated. The powder particles with a mean diameter of 17 nm are polycrystalline with a crystal size of 4.5 to 6 nm with relatively thick grain boundaries of 1 to 2 nm. Also twins as well as larger amorphous domains in the particles could be observed. All particles are coated with an amorphous oxide layer of 1 - 2 nm thickness which acts as a barrier for a grain growth across the particle interfaces.

The sintering of Si nanoparticles is still not very well-known. Si shows very strong covalent bonds and the sintering temperatures of coarse Si grains is between 0.75 T_m (beginning of the densification)and 0.98 T_m (max. density). It is well-known that the melting temperatures (T_m) decreases with decreasing particle size; particles with a size > 10 nm shows a linear decrease of T_m (9). For a particle size of 17 nm the melting temperature is 95% of the melting temperature of the bulk material. This small difference of T_m can not really influence the sintering behavior of nanosized powders. Also the size of the crystals in the particles leaves a lower T_m. Shi (9) shows that only free nanoparticles have a lower T_m, while nanoparticles in a matrix have T_m's which depends from the surrounding matrix. For example, it was observed that the melting point of Sn in an amorphous oxide matrix is much lower compared to the bulk melting point of Sn (10), whereas in a C matrix T_m increases. Unfortunately, no adequate interpretation has been offered in the literature regarding the observations of superheating as well as melting point suppression for the same nanocrystals in different matrixes (9). Therefore, the melting point of the Si-crystals in the particles is unknown, but because of the similarities between Si and Sn we can assume a depression of T_m and so also a suppression of the sintering temperature. But this general approach shows the suppression of the starting temperature of sintering (T_{ss}) for the silicon nanopowders, lower than the crystallization temperatures (0.5 T_m of bulk Si), as we observe in metals and oxides (0.3 T_m), (11) will not be possible.

Starting from the description of the microstructure of this system (fig. 5) we have to discuses the sintering behavior in more detail with the model of particles with a hard, not "sinterable" core and a soft sinterable layer. Following Jagota (11), for equal-sized, spherical, coated particles with a diameter d, forming a packing with solid fraction ρ, the minimum coating thickness, c, to achieve full density is :

$$c/d = (1/\rho)^{1/3} - 1$$

In agglomerated powders with ρ of 50%, a coating thickness of 2 nm is enough for full densification, whereas in areas between the agglomerates a coating of 5.5 nm is necessary. In this work, we will concentrate on the sintering of the agglomerates where c = 0.25d because only for this area the observed coating of 2 nm is enough for a full densification.

The relative density after 10 min sintering at 900 °C as estimated from fig 4 (in reference 11) using the normalized time τ' which is defined as:

$$\tau' = (3/4\ \pi)1/3\ \tau\gamma/(\eta(d+c)$$

where, τ = sintering time; γ = surface energy; η = viscosity of the coating at the sintering temperature (for SiO_2 at 900 °C: 2 10^{13} Pa s).

For estimating value of the surface energy of amorphous SiO_2 nano particles, the following equation for their estimation was used (12):

$$\gamma' = \gamma\ (\ 1 - \theta_3\ a/l + \theta_4\ (a/l)^2 + ..)$$

where, γ' the surface tension of nano particles; θ = numerical coefficient θ_3, θ_4 = 1, 1 = radius of the particles ; and a lattice constant or here the distance between the next neighbors.

Using 0.2 J/m^2 for γ and 0.06 nm for a typical distance between tetrahedral SiO_4 (13), the calculated γ' for 17 nm amorphous SiO_2 -particles is 0.19 J/m^2. With this value and a sintering time of 10 min, the change in the relative density is only 1%. These results show clearly, that for this mechanism (viscous flow with hard core) in the system Si/SiO_2 even in the nanosized domain at temperatures > 0.5 T_m, only limited sintering occurs (sintering time 1 month). This model thus cannot describe the observed rearrangement.

Therefore a more detailed model based on the work of Gryaznov and Trusov (see for example 12) was used for the explanation of the observed rearrangement. We can assume that internal plastic deformation is inhibited in Si-nanoparticles especially in our case of polycrystalline particles because the critical length for dislocation will be larger than the diameter of the crystallites or the particles. Therefore, the contribution of inter particle sliding to nanopowder shrinkage becomes substantial. In the agglomerates the typical size of the pores is of the same order as the particle size. Such ensemble of nanoparticles allows inter particle sliding where nanoparticles as a whole slip into pores. The driving force for such a process is the surface tension and the shrinkage rate may be described approximately by the equation:

$$d\rho/dt = D_p\gamma\rho/kT$$

where D_p = effective diffusion coefficient of nanoparticles in the agglomerate. The characteristic sintering time can be calculated as $t' = kT(d/a)^{1.5}/D_p$.

D_p depends upon the particle size and on the surface diffusion :

$$D_p = (a/d)^{3/2}D_s$$

Where, D_s = 1.3 10^{-12} m^2/s, for Si-nanoparticles D_p = 8.6 10^{-16} m^2/s and therefore the characteristic sintering time is <1 s. This value is comparable with the characteristic sintering time of Ni-nanoparticles sintered at 600 K. These results show that the sintering of Si nanoparticles coated with SiO_2 in the agglomerates is fast at relatively high temperatures. This leads to a densification of the agglomerates and a complete densification of the sample is impossible or possible only at very high temperatures as observed for submicron or micron sized Si-powder. It is interesting to note, that the use of the classical approach of the sintering theory to explain the sintering of Si-nanoparticles gives much lower characteristic sintering times ($t' << 1s$) (14). The reason for this difference is still unknown.

5. Conclusion

The investigation of the sintering behavior of nanosized Si-powder shows clearly, that the powder morphology has a very important influence on the sintering behavior. The particles produced by a plasma-process are amorphous.

During heating a crystallization at half the melting temperature could be observed. Since during the powder synthesis Si-clusters are formed, which acts as seeds for the crystallization of polycrystalline particles. Additionally a amorphous silica layer on the surface of each particles could be observed. The sintering behavior was determined by this silica layer. The used models, like hard core/sinterable coating or the model of particle sliding can explain the sintering rate of this powder. Even at 900 °C (0.63 Tm) only a sintering of the agglomerates could be observed whereas the sintering of the sample was negligible. The use of a classical model to explain the sintering process was not successful.

6. Acknowledgments

The authors are grateful to Dr. H. Hofmeister and Dr. I. M. Reaney for transmission electron microscopy. This work has been funded by the Fonds National project of the Federal Government of Switzerland under contract no. 2100-039361.93/1.

7. References

1. R.A. Andrievski, "Review Nanocrystalline High Melting Point Compound-based Materials", *J. of Materials Sci.* 29 614-631 (1994).
2. J. Dutta, W. Bacsa and Ch. Hollenstein, "Micro-structural Properties of Silicon Powder Deposited by rf Plasma-induced Dissociation of Silane", *J. Applied Phys.* 77 3729-3733 (1995).
3. Ch. Hollenstein, J.L. Dorier, J. Dutta, L. Sansonnens and A.A. Howling, "Diagnostics of Particle Genesis and Growth in rf Silane Plasmas by Ion Mass Spectrometry and Light Scattering", *Plasma Sources Sci. Technol.* 3 278-285 (1994).
4. M.V. Ramakrishna and J. Pan, "Chemical Reactions of Silicon Clusters", *J. Chem. Phys.* 101 (9) 8108-8118 (1994).
5. H. Hofmeister, *private communication.*
6. C. Bossel, J. Dutta, R. Houriet, J. Hilborn and H. Hofmann, "Processing of nanoscaled silicon powders to prepare slip-cast ceramics", Mat. Sci. & Engg. A (1995), accepted for publication.
7. S. Veprek, Z. Iqbal and F.-A. Sarott, "A Thermodynamic Criterion of the Crystalline-to-Amorphous Transition in Silicon", *Philosophical Magazine B*, 45 [1] 137-145 (1982).
8. K. Lu, "Interfacial Structural Characteristics and Grain-Size Limits in Nanocrystalline Materials Crystallized from Amorphous Solids", *Physical Review B*, 51 [1] 18-27 (1995).
9. F.G. Shi, "Size Dependent Thermal Vibrations and Melting in nanocrystals", *J. Mater. Res.*, 9, [5] 1307-1313 (1994).
10. K.M. Unruh, B.M. Patterson and S.I. Shah, "Melting Behavior of $Sn_x(SiO_2)_{100-x}$ Granular Metal Films", *J. Mater. Res.*, 7 [1] 214-218 (1992).
11. A. Jagota, "Simulation of the Viscous Sintering of Coated Particles", *J. Am. Ceram. Soc.*, 77 [8] 2237-2239 (1994).
12. V.G. Gryaznov and L.I. Trusov, "Size Effects in Micromechanics of Nanocrystals", *Progress in Materials Science*, 37 [4] 289-401 (1993).
13. R.K. Iler, The Chemistry of Silica", edited by John Wiley & Sons, New York, (1979).
14. F.E. Kruis. K.A. Kusters and S.E. Pratsinis, "A Simple Model for the Evolution of the Characteristics of Aggregate Particles Undergoing Coagulation and Sintering", *Aerosol Science and Technology*, 19 514-526 (1993).

SINTERING OF SUPPORTED GOLD CATALYSTS

Suiying Su,[1,2] Charles E. Hamrin, Jr.[2]
and Burtron H. Davis[1]

[1]Center for Applied Energy Research
3572 Iron Works Pike
Lexington, KY 40511

[2]Department of Chemical Engineering
University of Kentucky
Lexington, KY 40506

Abstract

Gold sols with a narrow range of particle sizes have been prepared and added to high surface area supports. At lower temperatures, TEM and XRD measurements indicate that sintering occurs by continuous particle growth but at higher temperatures particle growth is rapid and is followed by "sintering" of polycrystal to single crystal particles.

Introduction

The production of catalysts that are stable for long periods of use continues to be a goal of catalysis research. This is especially true for catalysts that contain expensive metals where, in order to optimize metal usage, a high dispersion of the active component is present on a high surface area support. In addition to coke formed during hydrocarbon reactions, sintering is a major reason for the reduction of activity of supported catalysts. Both empirical and mechanistic models that are based on atomic and/or particle migration have been proposed.

Dispersion is usually the dependent variable in sintering studies and can be calculated from the dimensionless form below:

$$D = (3V/aN_0)\ (1/r_{av})$$ [1]

where a is the area per surface metal atom (nm^2/atom), N_0 is Avagadro's number (atoms/mole), V is the molar volume of the metal (nm^3/mole), and r_{av} is the average particle size (nm). In this discussion, crystal size will be used to define the average dimension of a single crystal of the metal and particle size will be used to designate the size of the agglomerate of the single or polycrystalline metal units. In practice, the crystal size is what is determined by X-ray diffraction measurements whereas the particle size is what is obtained from transmission electron microscopic measurements.

For a spherical shaped Pt particle, Wanke and Flynn (1975) showed that equation [1] can be reduced to:

$$D = 0.509/r_{av}$$ [2]

Because the density, crystal structure and atomic radii of Au and Pt are very similar, and the uncertainties introduced by the assumption of spherical particles, equation [2] can also be utilized for gold.

With non-contacting supported metal entities, two sintering mechanisms are possible: (1) transport of individual metal atoms or small metal "molecules" from smaller to larger particles or (2) surface migration of metal particles that coalesce upon collision.

Wynblatt and Gjostein (1976) modified the atomic migration models developed by Chakraverty (1967). In this model it is assumed that the rate of capture of atoms migrating to a particle is described by Fick's second law and that the boundary condition of an average concentration of atoms on the surface applies at distances from the particle greater than the unit surface migration rate. The average surface concentration is taken to be equal to the concentration predicted by the Kelvin equation for a metal particle with a radius such that the net rate of gain of metal atoms is zero. It is assumed that the average particle radius is the same as the one assumed for the Kelvin equation approximation. It is further assumed that the metal particle size distribution is separable into time and radius dependent functions. Thus, the solution then predicts that

$$r = (r_o^4 + k_1 t)^{1/4}$$ [3]

if surface diffusion of the migrating species is rate controlling and

$$r = (r_o^3 + k_2t)^{1/3} \qquad \qquad [4]$$

if escape of metal atoms from a particle is rate controlling. In order to obtain agreement between theory and experimental results, it was proposed that at some particle size, growth is temporarily inhibited until nucleation of a new layer occurs. Flynn and Wanke (1974) modified this model by assuming that the absolute rate of loss of atoms was independent of particle size and was rate controlling.

A particle migration mechanism for the sintering of supported metal catalysts has been advanced by Ruckenstein and Pulvemacher (1973) and by Wynblatt and Gjostein (1975). In this model, reviewed in detail by Ruckenstein (1987) particles migrate on the surface, colloid and coalesce.

The models predict similar behavior for the rate of change of metal dispersion as a function of time. Experimental data on the influence of the initial particle size distribution as a function of sintering time would be useful for model discrimination. To this end, a rather narrow size range of particles has been prepared and these have then been deposited on a support. The initial size and the change of the size distributions with time have been obtained from X-ray diffraction and transmission electron microscopy measurements.

Experimental

Preparation of Catalysts

Experiments were performed to obtain uniform, highly dispersed gold or platinum suspensions. Sodium citrate was used as a good reducing agent for synthesis of gold colloidal suspensions (Thompson and Collins, 1992; Su, 1995) (Table 1).

Table 1. Colloidal Solutions Used in Catalyst Preparation

No.	Preparations	Color	Au Conc.
A	0.8 1 0.33 mM $HAuCl_4$ mixed with 50 ml Na citrate (1 wt.%). Stir 2 hr., store 4 months.	Purple	52 ppm
A4	14 ml Na citrate (1 wt%) added to boiling 0.25 1 0.31 mM $HAuCl_4$, boil 30 min.	Deep wine red	41 ppm

Support materials were calcined at 500 °C for five hours before use. Properties and source of the supports are given in Table 2. Metal colloids in Table 1 and other dispersions listed in reference 9 were selected to prepare the catalysts and were dialyzed for 6 days before use. Mixtures of 8 to 20 grams of the support with 200-900 mℓ of the colloidal solution were agitated for 5 hours and then placed in wide shallow plates where an air stream was directed over the surface. When all the water had evaporated, the powder was collected and dried at 110 °C in air overnight. The Au catalysts which were prepared are listed in Table 3.

Table 2. Properties of Support Materials

Name	Specific Area, m^2/g	Pore Vol., cm^3/g	Pore Size nm	Source
Al_2O_3	198	0.389	114	UCI
SiO_2	10.7	0.159	130	Davison
ZrO_2	43.5	0.264	134	CAER

Table 3. Properties of Supported Catalysts.

Name	Catalysts	Au, wt.%	$D_0(nm)^{-1}$
SUAU4	Au/Al_2O_3	1.65×10^{-3}	0.051
SUAU3	Au/Al_2O_3	0.839×10^{-3}	0.069
SUAUSi	Au/SiO_2	1.025×10^{-3}	0.068
SUAUZr	Au/ZrO_2	0.861×10^{-3}	0.068

The catalysts [SUAU4 and SUAU3)] were calcined in air at 100°C intervals between 200 and 700 °C. Au/SiO_2 and Au/ZrO_2 were calcined in air at 400 °C or 700 °C. At each temperature, a sample was calcined for successive 5 hour time intervals and the resulting catalyst was analyzed by TEM and X-ray diffraction following each calcination period.

The H-800 NA electron microscope was capable of analyzing elements, microdiffraction (2 nm size) and observing a high resolution image in a nanometer region.

A powder sample was dispersed in alcohol, and a drop of sample was placed on a Ni or Cu grid. The size of at least 200 particles was measured and population fractions of different particle size intervals were calculated. The arithmetic mean diameter was found to match best the XRD diameter for 2 samples and was therefore used for all samples.

The X-ray diffraction spectra were measured in the region of 2θ between 30 to 50 degrees. The peak width, β, at half maximum of the major metal peak [(111) for Au or Pt] was measured. The particle size, D_d, was calculated using the following equation:

$$D_d = 0.94\lambda / \beta \cos\theta \qquad\qquad [5]$$

where $\lambda = 1.5418$ Å. Metal dispersion D was calculated from average crystallite radius data using Equation (2).

Results and Discussion

Sample (Au4) had initially a narrow distribution of particle sizes that fall within the range of 10 to 40 nm TEM (0 hrs, TEM in Figure 1.)

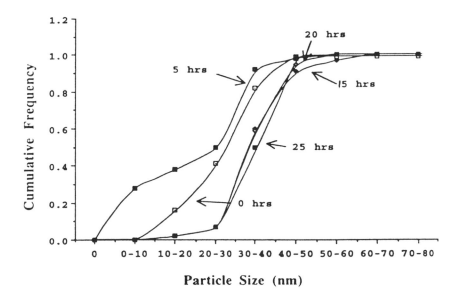

Figure 1. Cumulative frequency for gold particle size (TEM) for Au4 following sintering at 300 °C (times indicated in the figure).

Heating for 5 hours at 300 °C causes the appearance of a shoulder on the curve that corresponds to the formation of particles in the 0 to 30 nm size range. With further heating, the size approaches a limiting distribution; the curves for 15, 20 and 25 hours are essentially the same. Thus, it appears that the conversion of the gold to colloidal sized particles was not complete during the preparation; small, probably atomic, particles were present. Thus, it appears that atom migration is responsible for the formation of the small size particles in the 0 to 30 nm range. With further heating, it appears that the particles in the 0 to 30 nm range migrate to coalesce to form larger particles. However, a limiting distribution is attained after 15 hours of sintering and further growth of particle size is slower than can be ascertained on the 25 hour time scale.

Dispersion A4 was prepared at about 100 °C whereas dispersion A1 was prepared by reduction at room temperature. It is anticipated that the dispersion A4 would have smaller particles than A1, and this was indeed the case (Figure 2). The catalyst prepared from the A4 dispersion had a

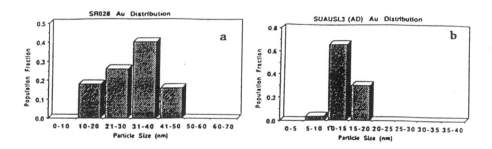

Figure 2. Histogram showing the distribution of gold particles in the indicated size ranges for the samples (a, A1; b, A4).

metal loading that was about twice that of the sample prepared from the A1 dispersion (0.84 x 10^{-3} for Au3 and 1.7 x 10^{-3} for Au4).

The particle sizes calculated from XRD line broadening and TEM should be the same provided the particles are a single crystal of the metal. This appears to be the case for sample Au3 (Figure 3). Initially this sample consisted of particles in the range of 5 to 20 nm, with an average size of 13.5 nm. The particle size returned from XRD and TEM measurements were essentially the same, and this remained the case during sintering at 700 °C to effect particle size growth for each of the five times. Thus, the data fit very well the parity line with a slope of one as anticipated provided the particles are indeed a single crystal of gold.

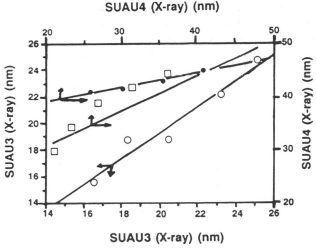

Figure 3. Comparison of the particle size of supported gold as measured by XRD and TEM following calcination for 5 to 25 hours at 700 °C for (1) Au3 (○), (2) at 400 °C for Au4 (□) and at 700 °C for Au2 (●).

A different pattern is observed for the sintering of the Au4 sample. At the lower sintering temperature (400 °C), the XRD and TEM average sizes increase rapidly but the particle size observed by TEM approaches asymptotically a limiting size whereas the particle size determined by XRD continues to increase with sintering time (Figure 3). This situation is more evident when sintering is accomplished at 700 °C. Here the particles, as measured by TEM, do not appear to sinter whereas the average crystallite size, as measured by XRD, continues to increase with increasing sintering time (Figure 3). It is noted that the "limiting" particle size of about 45 nm is reached irrespective of whether the sintering is conducted at 400 or 700 °C or whether the size of the sintered (after 25 hours) particle is determined by XRD or by TEM.

Furthermore, a similar limiting particle size is observed for both the Au3 and Au4 samples. It therefore appears that the larger gold particles that are present in the gold dispersion prepared at room temperature are polycrystalline. Thus, during the sintering process these polycrystalline particles do not undergo a significant growth in particle size but the particles do undergo crystallization to form particles consisting of a single crystal from the polycrystalline particles.

The dispersion depends on the calcination temperature for each calcination time. Likewise, for each calcination temperature, the dispersion depends upon the calcination time. To a first approximation, it appears that the sintering with time for Au4 is nearly linear with time; this is not unexpected since the data are for particle sizes determined by XRD and probably represents the formation of single crystals from polycrystalline particles.

In conclusion, it appears that the sample with small initial size does undergo particle growth during the sintering process. The sample with larger initial particle sizes appear to undergo a rapid, initial sintering process but at later sintering times the change in size is primarily limited to the formation of larger single crystal gold crystals from a similar size polycrystalline particle.

The dispersion decreases linearly with time and is only slightly dependent on temperature. This is consistent with the suggestion that for sample Au4 crystallization is the prime mechanism active.

References

1. E. Wanke and P. C. Flynn, <u>Catal. Rev.-Sci. Eng.</u>, <u>12</u>, 93-135 (1975).
2. P. Wynblatt and N. A. Gjostein, <u>Acta Metall.</u>, <u>24</u>, 1165-1174 (1976).
3. B. K. Chakraventry, <u>J. Phys. Chem. Solids</u>, <u>28</u>, 2401-2412 (1967).
4. P. C. Flynn and S. E. Wanke, <u>J. Catal.</u>, <u>34</u>, 390-399, 400-410 (1974).
5. E. Ruckenstein and B. Pulvemacher, <u>J. Catal.</u>, <u>29</u>, 224-245 (1973).
6. P. Wynblatt and N. A. Gjostein, in "Progress in Solid STate Chemistry," Vol. 9, 21-58 (1975).
7. E. Ruckenstein in "Metal-Support Interactions in Catalysis, Sintering, and Redispersion," (S. A. Stevenson, J. A. Dumesic, R. T. K. Baker and E. Ruckenstein, Eds.), Van Nostrand Reinhold, New York, 1987, pp. 141-308.
8. D. W. Thompson and I. R. Collins, <u>J. Colloid Interface Sci.</u>, <u>152</u>, 113 (1992).
9. S. Su, M. Eng. Thesis, University of Kentucky, 1995.

FE-Simulation of Compaction and Solid State Sintering of Cemented Carbides

Jan Brandt

Department of Solid Mechanics
University of Linköping, S-581 83 Linköping, Sweden

Aug 15 1995

Abstract

The process of hard metal products compaction and sintering (solid state) is simulated with a material model CAP88+SINT95 implemented in the Finite Element (FE) package LS-DYNA2D.

1 Introduction

We have further developed the FE material model for compaction simulation CAP88, from Sandvik Coromant AB, to also describe solid state sintering.

The model for sintering is visco-elastic, with a "build in shrinkage motor" describing the sintering activity. The model is primarily aimed at sintering without external load (that is not HIP-ing), but a specimen is generally not stress-free in its interior after the compaction operation, so considering the stress field in the creep response is quite important.

The material model has been implemented in the FE-code LS-DYNA2D (Hallquist, 1990), that uses explicit time integration, which proved quite robust for the compaction part.

2 Stress and Deformation update

The equation of motion is recast in a Galerkin form, which then is discretized by the finite element procedure:

$$M\ddot{U} = F^{ext} - F^{int} \tag{1}$$

$$\ddot{U} = M^{-1}(F^{ext} - F^{int}) \tag{2}$$

$$F^{int} = \sum_{elem\ e_i} F^{int,e_i} \tag{3}$$

$$F^{int,e_i} = \int_{elem\ e_i} B\sigma dV \tag{4}$$

$$\dot{\epsilon} = B\dot{U}^{e_i} \tag{5}$$

In the above formulas M is the diagonal mass matrix, U is the nodal displacement vector , F^{ext} is the external nodal (applied) force vector and F^{int} is the internal resisting nodal force vector. σ and ϵ are stress and strain tensors and B is the strain-displacement matrix. An overdot stands for time derivative. σ, ϵ and B all pertain to the one integration point of an element. U^{ei} is the nodal displacement vector of one element and $F^{int,ei}$ is the corresponding internal force vector. The internal state variable d is defined in (12).

The rational for the state, U, σ, d , update from time t to $t + \Delta t$ and $\dot{U}$ from $t - \Delta t/2$ to $t + \Delta t/2$ is:

1. update $\dot{U}, U$ according to the central difference scheme:

$$\dot{U}(t + \Delta t/2) = \dot{U}(t - \Delta t/2) + \ddot{U}(t)\Delta t \qquad (6)$$
$$U(t + \Delta t) = U(t) + \dot{U}(t + \Delta t/2)\Delta t \qquad (7)$$

2. calculate σ, d at $t + \Delta t$ according to section 3, $\dot{\epsilon}(t + \Delta t/2)$ according to (5) is used

3. calculate F^{int} at $t + \Delta t$ according to (4) and (3)

4. calculate $\ddot{U}$ at $t + \Delta t$ according to (2)

5. update the time $t \leftarrow t + \Delta t$ and go back to 1.

3 Material Model

The choice of mode for a certain element integration point e at an instant of time t is based on the temperature T:

1. $T < T^{sint}$, compaction mode: elasto-plastic, CAP-type; or tensile fracture, smeared crack bands

2. $T \geq T^{sint}$, sinter mode: visco-elastic, "sinter stress" and mechanical stress cause a Norton-type of creep

Except for the fracture submode we can state a unified description of the stress update:

$$\Delta\epsilon^{total} = \Delta\epsilon^{elastic} + \Delta\epsilon^{inelastic} + \Delta\epsilon^{thermal} \qquad (8)$$
$$T < T^{sint}, \Delta\epsilon^{inelastic} = \Delta\epsilon^{plastic}, \text{ fcn of } \sigma, \Delta\epsilon^{total}, d \qquad (9)$$
$$T \geq T^{sint}, \Delta\epsilon^{inelastic} = \Delta\epsilon^{creep}, \text{ fcn of } \sigma, \Delta t, d, L, T, \dot{T} \qquad (10)$$
$$\Delta\sigma = D\Delta\epsilon^{elastic} \qquad (11)$$
$$d \equiv \rho_{current}/\rho_{fully\ compacted} \qquad (12)$$
$$L \equiv \text{average (WC) particle size} \qquad (13)$$

D is the stiffness matrix. In the sequel we shorten superscripts according to t for total, e for elastic, i for inelastic, th for thermal, pl for plastic, cr for creep, s for sint, as in $\epsilon^t, \epsilon^e, \epsilon^i, \epsilon^{th}, \epsilon^{cr}$ and T^s .

In the case of tensile fracture during the powder compaction there is a degradation of the load carrying capacity after the tensile failure stress σ^u has been reached:

$$\Delta\sigma = \overline{D}\Delta\epsilon^t, \quad \Delta\sigma < 0, \quad \text{for a } \Delta\epsilon^t \text{ that further opens a crack} \qquad (14)$$

3.1 Compaction mode

The yield curve of the studied WC-Co powder at three different degrees of compaction is shown in Fig. 1 . J_1 is the first invariant of the stress tensor (15) that is proportional to the pressure. J_2' is the second invariant of the stress deviator tensor (16) that is proportional to the von Mises stress.

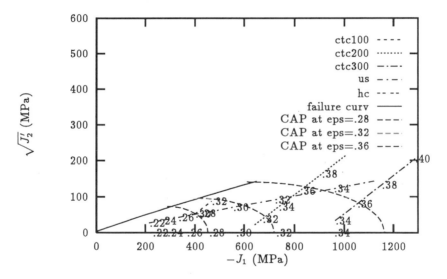

Figure 1: Volumetric plastic strain along stress paths; yield curves

3.2 Tensile fracture mode

$\overline{D}$ is based on: thin crack bands; experimentally measured crack opening energies; and the smearing out of crack bands over the (otherwise elastically responding) elements.

3.3 Sinter mode

A model for unloaded sintering (Ågren et al.,1995) has been extended to also model the influence of a mechanical stress field. We adopt the following notations: repeated indices implicate summation over the whole range of that index as in $\epsilon_{kk} = \sum_{k=1,3} \epsilon_{kk}$; $\delta_{ij} = 1$ for i=j, otherwise 0.

$$\sigma_m \equiv \frac{J_1}{3} \equiv \frac{1}{3}\sigma_{kk} \quad \text{(mean stress)} \tag{15}$$

$$\sigma'_{ij} \equiv \sigma_{ij} - \sigma_m\delta_{ij} \quad \text{(stress deviator)} \tag{16}$$

$$\epsilon^i_m \equiv \frac{1}{3}\epsilon^i_{kk} \quad \text{(mean inelastic strain)} \tag{17}$$

$$\epsilon^i_{ij}{}' \equiv \epsilon^i_{ij} - \epsilon^i_m\delta_{ij} \quad \text{(inelastic strain deviator)} \tag{18}$$

$$f_p = 1 - d \quad (f_p \text{ is volume fraction of pores}) \tag{19}$$

$$\dot{d} = -3\dot{\epsilon}^i_m d \tag{20}$$

The model tells how $\Delta\epsilon^i$ or d (12) shall be updated during a time increment Δt given $d, T, \dot{T}$. In a more elaborate version of the model also L is updated. We postpone the introduction of a mechanical stress field a little.

$$\dot{\epsilon}^i_m = -k_1 D_{eff}(T, \dot{T}^{aver})f_{bl}(T, t - t^{iso})\sigma^s \tag{21}$$

$$D_{eff}(T, \dot{T}^{aver}) = [D_v(T) + \omega D_s(T)][1 + f_{s2}(T)\dot{T}^{aver}]f_k(d) \tag{22}$$

$$\sigma^s = f_{s1}(d)\gamma/L \tag{23}$$

Where D_x are various diffusion coefficients of (WC) particles, ω is a blending factor and k_1 represents the self diffusivity of the binder (Co). f_{bl} is a blocking factor (due to friction between carbide particles) which is closely represented as an exponential decay by the time elapsed, $t - t^{iso}$, after isothermal (T^{iso}) conditions started. $f_k(d)$ is a factor on D_{eff} that represents the state of densification and the initial distribution of binder. $F_{s1}(d)$ is a factor on the sinter stress that represents the state of densification. $f_{s2}(T)$ describes the D_{eff} dependence on temperature and average (over time) temperature rate.

We introduce the notion of a viscous creep modulus K^*_{cr}:

$$\frac{1}{K^*_{cr}} \equiv 3k_1 D_{eff}(T, \dot{T}^{aver})f_{bl}(T, t - t^{iso}) \tag{24}$$

We can conclude the above roughly as having stated an equivalent (smeared) sinter stress σ^s and an equivalent creep modulus K^*_{cr} to work together in a Norton-type creep law with an exponent of one.

We have cared to dissect, based on physical reasoning and loaded and unloaded sinter experiments, the resulting mean sinter response rate $\dot{\epsilon}^i_m$ in a stress free state into a driving part σ^s and a resisting part K^*_{cr} and it is thus implicit how the addition of a stress field σ with mean stress σ_m shall affect the response:

$$\dot{\epsilon}_m^i = \frac{\sigma_m - \sigma^s}{3K_{cr}^*} \tag{25}$$

The remaining relation to be specified is the creep strain deviator response rate $\dot{\epsilon}_{ij}^{i'}$. We base this on the following assumption:

$$(\dot{\epsilon}_{22}^i - \dot{\epsilon}_m^i)/(\dot{\epsilon}_{11}^i - \dot{\epsilon}_i^s) = -\nu^*(d) \tag{26}$$

in uniaxial compression, where ν^* is the equivalent of the Poisson ratio ν in elastic response. ν^* is determined by a best fit to measured deformation at uniaxially loaded sintering of rod-like specimens. Based on (26) we get:

$$\dot{\epsilon}_{ij}^{i'} = \frac{\sigma_{ij}'}{3K_{cr}^*} \frac{1 + \nu^*}{1 - 2\nu^*} \tag{27}$$

4 Comparison of Simulations and Experiments

4.1 Compaction data for a Cylindrical Homogeneous Body

Compaction material parameters rely on assessments carried out under the supervision of Sandvik Coromant AB.

Figures 2 and 3 show how the compaction mode simulates the two important deformation modes hydrostatic compression and uniaxial strain.

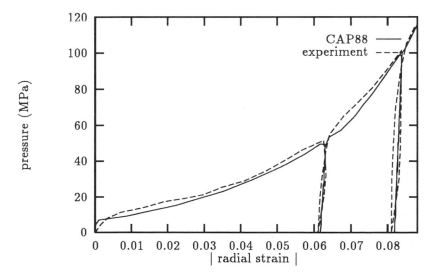

Figure 2: response of powder in hydrostatic compression

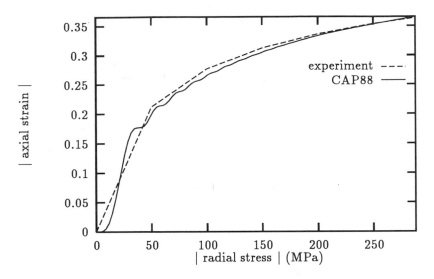

Figure 3: response of powder in uniaxial strain

4.2   Sinter data for a Cylindrical Homogeneous Body

A series of sinter experiments (Haglund et al.,1995) with circular cylindrical specimens are used to trim the material parameters of the sinter mode. We have $\gamma \approx 1N/m$ and L$\approx$ $.33\mu m$ which gives $\sigma^s \approx 2.6MPa$ at $d = .5$, decreasing linearly to 0 at $d = 1$.

Figure 4 shows experiments and simulations in terms of linear shrinkage for unloaded sintering with different isothermal temperature T^{iso} ($1100°$,$1175°$ and $1250°$).

Figure 5 shows experiments and simulations for sintering of specimens with various levels of axial load (1 N, 7 N and 27 N, constant in time). All with $T^{iso} = 1230°$. eps-r stands for radial strain and eps-ax for axial strain.

The specimens have a length of .022 m and a diameter of .006 m, such that 27 N gives an axial stress of -.95 MPa.

Both series of experiments have temperature cycles with the following characteristics: fairly constant temperature rate $\dot{T} \approx .3°/s$ until isotherm at different temperatures T^{iso}. The second series has a rapid transit to isotherm whereas the first has a slow transit (600 s at $1100°$, 1800 s at $1175°$ and 2400 s at $1250°$). Temperatures are given as degrees Celsius.

4.3   Compaction and Sinter of a Toolpiece, Simulation Compared to Experiment

We conclude with the compaction and sintering of a realistic WC-Co toolpiece. The geometry is axisymmetric. Figure 11 shows a cut through the piece and the axis of symmetry. Dimensions of the piece as compacted appear in Fig. 8. The simulation is carried out until 1 h of sintering, where the

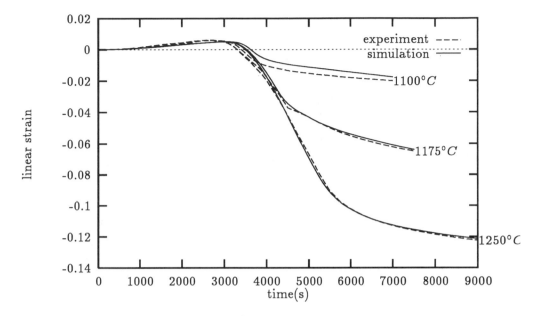

Figure 4: experimental and simulated sintering at various isotherms T^{iso}

relative density has reached approximately .75 . The compaction simulation starts out from a (homogeneous) relative density of .42 .

Figure 6 shows three instances of the deformation history.

The duration of the compaction phase has been largely prolonged such that, with the huge mass scaling ($\rho * 10^{10}$) that is necessary for the sinter phase, we do not introduce spurious inertia effects.

The temperature cycle of the sintering is shown in Fig. 7 .

There is performed a remap at t=1200 s (end of compaction phase) to improve the mesh quality.

The first results pertain to the specimen as compacted: Fig. 8 shows the relative density distribution and Fig. 9 shows the hoop stress distribution. The tensile strength is 3 MPa , so tensile fracturing is avoided. The peak stress magnitude 3 MPa (compression) is high compared to the maximum level of approximately 1 MPa at the loaded sintering of section 4.2. However the stresses relax to a considerably lower magnitude of .15 MPa during sintering, as can be seen in Fig. 10 .

Finally in Fig. 11 we see the shape after 1 h of sintering. The measured shape is marked with thick lines.

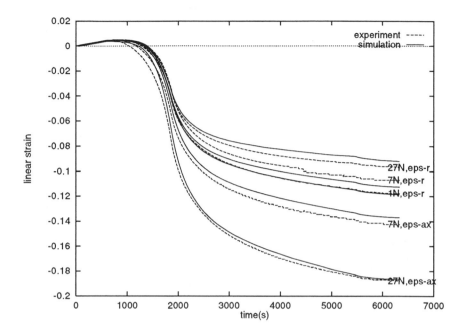

Figure 5: experimental and simulated sintering at various axial loads

5   References

1. J.S. Sandler and D. Rubin,"AN ALGORITHM AND A MODULAR
   SUBROUTINE FOR THE CAP MODEL", *Int. J. for Num. and
   Anal. Meth. in Geomechanics*, vol. 3 173- 186 (1979)

2. L. Nilsson and M.Oldenburg,"Nonlinear Wave Propagation in Plastic
   Fracturing Materials - A Constitutive Modelling and Finite Element
   Analysis" *Nonlinear Deformation Waves*,U.Nigul and J.Engelbrecht (Eds.),
   Springer, Berlin, 209-217, (1983)

3. J. Ågren, J. Brandt, S.A. Haglund, B. Uhrenius,"Modelling Solid State
   Sintering of Cemented Carbides" *Sintering 1995-1996*, R.G. Cornwall,
   R.M. German and G.L. Messing, Marcel Dekker Inc., (1996)

4. S.A. Haglund, J. Ågren, P. Lindskog, B. Uhrenius,"Solid State Sintering of Cemented Carbides" *Sintering 1995-1996*, R.G. Cornwall, R.M.
   German and G.L. Messing, Marcel Dekker Inc., (1996)

5. J.O. Hallquist,"LS-Dyna2d - An Explicit Two-Dimensional Hydrody-
   namic Finite Element Code",Livermore Software Technology Corp.,Livermore
   (1990)

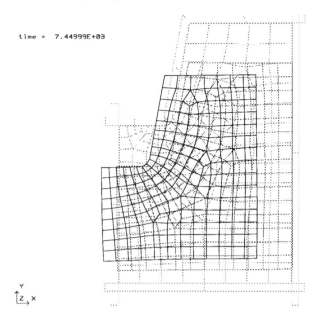

Figure 6: at outset (light grey); as compacted (medium grey); as sintered (dark grey)

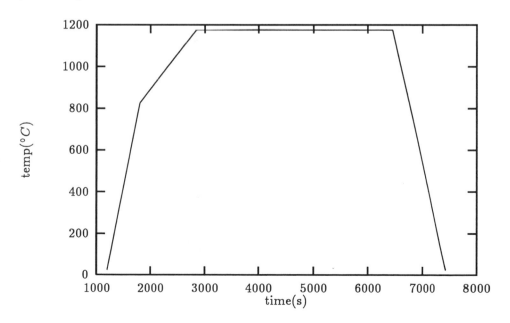

Figure 7: temperature cycle of the sintering

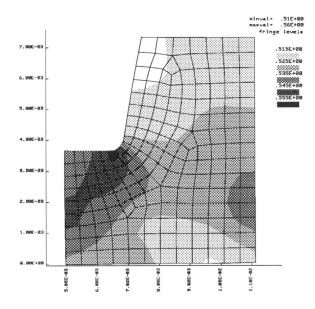

Figure 8: relative density distribution as compacted

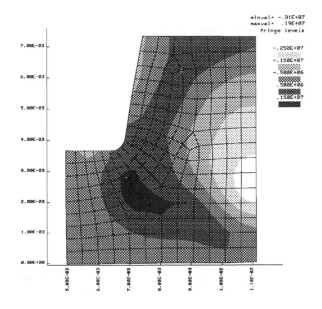

Figure 9: hoop stress distribution as compacted

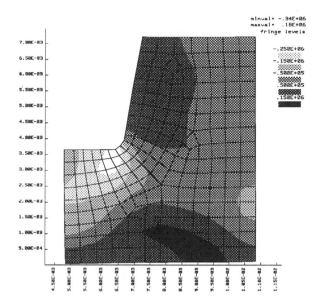

Figure 10: hoop stress distribution at start of sintering

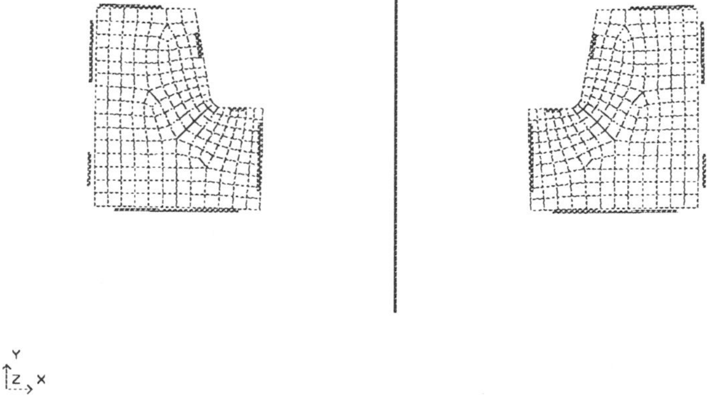

time =   7.44999E+03

Y
↑z ↗ x

Figure 11: simulated and measured (thick bars) shape after 1 h at 1175°

On the Sintering of Cemented Carbides

Björn Uhrenius[1], John Ågren[2] and Sven Haglund[2]

1) Sandvik Hard Materials, Stockholm, Sweden
2) Royal Institute of Technology, Department of Materials Science,
Stockholm, Sweden.

Abstract
A review of different subprocesses occuring during sintering of cemented
carbides is given. Information on the reduction of oxides during the
presintering stages is given as a result of mass spectrometric studies. The
transfer of carbon between the gas phase and the porous body during solid
state sintering is shown by sintering in gases of controlled carbon potentials.
The influence of minor amounts of reactive elements on the sintering as
measured by dilatometry is also indicated. The temperature for the formation
of the liquid phase on heating and the solidification of the binder on cooling is
indicated by thermal analysis and dilatometry.

1. Introduction
The sintering of cemented carbides is usually referred to as a liquid phase
sintering process. However, much of the reactions taking place during the
densification occur already in the solid state. Bengtsson, Johannesson and
Lindau (1973) showed that more than 50% of the densification was obtained
before the eutectic temperature was reached. Similar observations were made
by Exner et al. (1978) and Leitner, Jaenicke-Rössler and Wagner (1992). This
densification involves surface diffusion of the Co-binder metal and
agglomeration due to the surface tension between the metal and the carbide
particles. This process is to a high degree depending on the properties of the
surfaces of the carbide particles. Degassing of pressing aids and
contamination by minor amounts of reactive elements play an important role
in the early stages of sintering (Spriggs 1970, Uhrenius 1975). Reactions
between the gas phase in the sintering furnace and surface bound oxygen or
oxides in the compact and with the carbides influence solid state sintering. All
these processes are closely related to the industrial production routines used
and are in many cases not reported on. The technological properties of the
sintered products are also to some extent depending on the early stages of the

sintering process. Efforts made to collect information to be used for the simulation of the sintering of carbide compacts showed that more basic knowledge is needed on these early stages of the sintering process. In the present communication the results of some investigations by mass spectrometry, dilatometry and thermal analysis are presented. These were also made to study the temperature at which the liquid phase is formed on heating of the compact as well as when solidification after liquid phase sintering is taking place.

2. Influence of raw materials and milling

Milling of the carbide powders and the metal binder is made to obtain a good mixing of the carbide particles, the metal binder and the pressing aid.The latter usually consists of a paraffine wax or PEG (polyethylene glycol). The milling process also results in reduction of the size of the carbide particles, in deagglomeration and in creation of new active surfaces and in an increased defect structure of both carbides and metal binder. This makes sintering more efficient and facilitates obtaining a homogeneous and pore free body. The new surfaces created during milling are very reactive with the gaseous species in the environment. Although the surfaces of the particles are protected to some extent by the lubricant added an increased oxygen content is always obtained after milling. The mean carbide grain size after sintering is usually in a range from 0.5 μm to 10 μm depending on grade and application. Before sintering the mean grain size is much finer and the compact thus contains a large amount of submicron particles. Oxygen is then the major contaminant even if direct sintering in vacuum furnaces is common practice today. Both milling liquids and pressing aids contain oxygen and moisture which react with the carbides during dewaxing and presintering. Among the other impurities introduced during milling iron is often the major element. In addition the raw materials usually contain a number of trace elements (Ca, Al, Si, S,..) which might influence both sintering behavior and properties after sintering (Uhrenius et al. 1991).

3. Reactions during dewaxing and solid state sintering

Dewaxing is very often made in hydrogen and when PEG is used as pressing aid the hydrogen molecules facilitate the decomposition of the large organic molecules. PEG is then decomposed into easily evaporated gaseous species like CO, H_2O, CO_2 and CH_4 (Uhrenius 1975 and Leitner et al. 1992). These species might then also cause some oxygen contamination of the carbide particles. However, a large amount of this oxygen can be reduced by hydrogen if a flow of pure hydrogen is maintained to higher temperatures. In cases when more reactive elements like Ti, Ta and Nb are present oxygen is more strongly bound and is then only reduced by carbon. In Figure 1 the pressure of carbon monoxide is shown as measured by using a mass spectrometer during

Table 1. Composition of three cemented carbide grades presintered in vacuum. The evolution of carbon monoxide measured during heating is given in Figure 1.

Grade	#1	#2	#3
weight %			
WC	92	77	56
TaC	1	6	12
NbC	1	2	4
TiC	-	4	18
Co	6	11	10

presintering in vacuum of three cemented carbide grades. The vacuum in the furnace was obtained by using a simple rotary vacuum pump only and was because of that not better than 100 Pa. The composition of the carbide grades is given in Table 1.

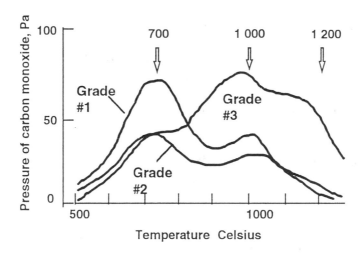

Figure 1. The pressure of carbon monoxide as measured by mass spectrometry during presintering of three cemented carbides of compositions given in Table 1.

It can easily be shown by thermodynamic calculations or by using an oxygen potential diagram that the oxides of the elements; W, Nb, Ta and Ti can be reduced by carbon at the pressure maintained if the temperature is above 400°C, 750°C, 900°C or 1150°C respectively. The pattern obtained from the mass spectrometer agrees fairly well with these temperatures and is also showing the very strong bond between oxygen and titanium. The open porosity still present at these temperatures facilitates the reduction of oxygen, but at the expense of the carbon content of the cemented carbide. It is thus necessary to make this reduction in the solid state as the formation of carbon monoxide at higher temperatures when the pores are closed would result in gas pores after sintering (Amberg, Nylander and Uhrenius 1976). The carbon content of the sintered carbide must though be kept within narrow limits and an uncontrolled contamination by oxygen of the compact will be detrimental because of the formation of brittle $(Co,W)_6C$-carbides (η-carbides) at higher temperatures if to much carbon is lost. The rapid interaction between the porous body and the furnace atmosphere was illustrated by Lundberg (1991) who investigated means for the control of the carbon content of cemented carbides by presintering in atmospheres of defined carbon potentials. Mixtures with different amounts of CH_4 and H_2 were chosen on basis of thermodynamic calculations. It is shown in Figure 2 that the gas can have a strong influence on the carbon content of the sintered carbide if the compact is subject to a short heat treatment in a temperature range of 800°C to 1000°C.

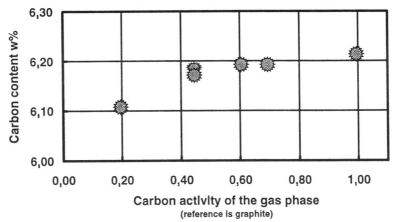

Figure 2. Carbon content of cemented carbides which were heat treated in CH_4-H_2 mixtures at presintering temperatures. Carbide composition in weight-%: 86WC, 3TaC, 2NbC, 3TiC and 6Co. (from Lundberg 1991).

The strong influence of the dewaxing process and of traces of reactive elements like titanium on the solid state sintering is illustrated in a recent investigation by Haglund (1994). In his investigation a comprehensive study of the solid state sintering of WC-Co and WC-TiC-Co grades was made by using dilatometry and interrupted sintering experiments. Powder mixtures with and without the addition of pressing aid (PEG) were used as well as different compaction techniques. Pure WC-Co grades were found to sinter readily already above 800°C and were densifying to about 90% of full density before the melting of the Co-binder. Grades containing titanium carbides showed much lower sintering rates and temperatures above the onset of liquid phase sintering had to be employed to reach high densities. A striking example of the influence of a contamination of the carbide surfaces is shown in Figure 3. In this figure the solid state sintering rates of two WC-Co grades are compared, one containing 200 ppm Ti and another without any detectable amount of titanium. Both grades start shrinking at 800°C but the shrinkage rate of the Ti-containing grade decreases after some time and at the end the shrinkage of this grade is less than half of that of the other grade.

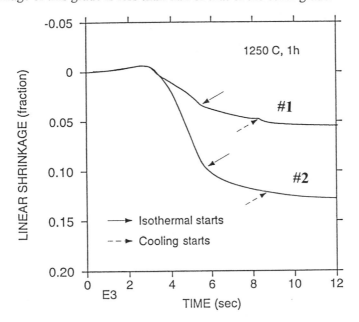

Figure 3. Linear shrinkage during sintering of two WC-10 w% Co grades. Grade #1 contains 200 ppm Ti, #2 contains < 40 ppm Ti (detect. limit). Heating rate 20°C/min, He, 0.1 MPa. Isothermal hold for 1h at 1250°C.

The only known difference between the two grades is their Ti-content and the resulting enrichment of oxygen and titanium to the surfaces of the carbide particles as analysed by ESCA and Auger. Microscopic investigations showed that cobalt diffuses out on to the carbide surfaces already at rather low temperatures (below 1000°C) and that agglomeration of the carbides starts at these temperatures. At 1100°C a pronounced change of the pore structure is obtained as shown in Figure 4. Surface diffusion of cobalt creates a thin film of the binder metal on the carbides and the capillary forces create an internal stress already in the solid state. It is believed that cobalt yields by creep deformation due to this stress at temperatures above 800°C. If oxidic layers are present on the carbide particles, surface diffusion and wetting of the cobalt metal is hindered and will not take place unless the oxides are reduced and oxygen is evaporated or dissolved in the carbides or the metal. This will not be the case until much higher temperatures are reached, which in the case of titanium would be above 1200°C as mentioned previously.

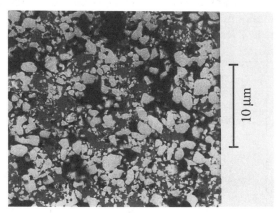

Figure 4. SEM-micrograph showing WC-particles (light grey), cobalt film (dark grey) and pores (black) after rapid heating to 1100°C and quenching. It is obvious that the cobalt metal has formed a film on the carbides resulting in agglomeration and densification of the body. 10000X. (from Haglund 1994).

Similar results were obtained by Bock (1992) on the sintering of fine grained WC-Co grades. He compared grades containing additions of small amounts of vanadium with undoped grades. He also found that sintering started at 800°C. The V-doped grades ceased shrinking at 1000°C. He also noted a strong influence of the origin of the carbide, WC-carbides produced from tungstic acid being more easily sintered than those produced from APT.

When an intimate contact between the binder and the carbides is established the finer carbides start to dissolve and the binder becomes alloyed with tungsten and carbon. Grain growth of coarser WC-particles takes place and a facetted WC structure of the coarser carbides is formed already at 1200°C (Haglund 1994). However, due to the much higher diffusion rate of carbon as compared to that of tungsten an equilibrium situation will not be reached at low temperatures and at normal heating rates unless the diffusion distances are very short. Carbon atoms will be drained off from the WC/Co interface and a tungsten rich zone is formed close to the WC-carbides. The time to reach equilibrium is then determined by the diffusion rate of tungsten. The dissolution rate might be even further reduced if in this zone thin layers of η-carbides are formed because of the high W/C-ratio. In such a case the dissolution will be even slower because of the much slower diffusion rate in the carbide layer.

4. Melting of the binder and microstructural changes in the liquid state
The melting temperature of the binder depends on its alloy content. A pure cobalt binder will melt at 1495°C, but the lowest melting point in the stable Co-W-C system is 1280°C (Uhrenius, Carlsson and Franzen 1976). The latter case represents the eutectic temperature corresponding to both graphite, cobalt and WC saturation of the liquid. If the carbon content of the metal is lower the melting point is increased considerably and in the presence of η-carbide instead of graphite the melting point will be close to 1375°C based on own experimental data from a DTA study of Co-W-C samples. If a true ortho-equilibrium with respect to both carbon and tungsten is not attained melting will start at a higher temperature, unless an excess of carbon is added to the powder. Hellsing (1988) made atom-probe and TEM analysis of the binder composition in WC-Co cemented carbides which had been liquid phase sintered and subject to different cooling rates. Based on a comparison with calculations of the diffusion of tungsten in the Co-binder he concluded that equilibrium will be maintained down to about 1000°C on cooling if the diffusion distances are less than 1 μm. His evaluations are based on the assumption that there are no hindering reactions at the WC/Co-interfaces and the results might thus have a limited value when considering reactions during heating of powder compacts. A few experiments were recently made by the present authors to find out when melting of WC/Co compacts started during

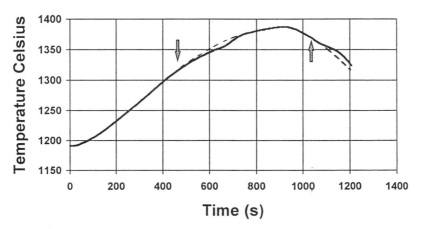

Figure 5. Result of thermal analysis of a WC-Co compact containing 9.86 w% Co and 5.52 w% C indicating melting in the interval 1320°C to 1360°C. Solidification begins at 1360°C on cooling.

normal heating rates. The results of dilatometric studies gave no distinct indications of the melting of the binder because of the sintering effects superimposed on any effect due to the melting of the binder. Because of difficulties of maintaining defined carbon potentials in the DTA equipment available, ordinary TA-experiments with rather large specimens were found to be more successful. The impact of the surrounding gas could be minimised during these experiments. A result is shown in Figure 5 indicating that melting is taking place in an interval between 1330°C and 1360°C and that after a short hold at 1390°C solidification on cooling of the same alloy begins at 1360°C. Chemical analysis of the sintered carbide showed that no changes of the carbon and cobalt contents had taken place during heat treatment.

An important property for the description of the microstructure of cemented carbides is their contiguity. Contiguity(C) is defined as the fraction of the total surface area of the carbide particles, that is carbide/carbide interfacial area. The contiguity of the present compacts sintered in the solid state were found to be approximately C=0.3 after sintering at 1250°C. The measurements also indicated a slow increase of contiguity with time and temperature during solid state sintering in the interval 1100°C to 1250°C. Measurements on the development of contiguity during liquid phase sintering made by Stjernberg (1970) and by Warren and Waldron (1972) show that contiguity is decreasing with time and temperature during liquid phase sintering in most cemented carbide systems. However the results of Stjernberg also show a slight increase at very short sintering times. After reaching a maximum, contiguity decreased monotonically down to a constant value after prolonged sintering times. German (1985) presented a model for the description of contiguity in liquid phase sintered systems. This model links together contiguity, fraction of liquid, dihedral angle and the size ratio of the particles of the solid phase. Although the model was developed for spherical particles it offered reasonable agreement with experimental data also for systems with particles of different shapes and might also be used to estimate the development of the contiguity of WC-Co compacts. Although this model does not directly involve any description of the kinetics of the microstructural changes it can be used to give a qualitative picture of the development at different stages of the process. An increase of contiguity with time during solid state sintering would thus be expected becuase of the loose structure after pressing and due to the contraction caused by the interfacial tension. When the liquid phase is formed, the solubility of the carbide is increased drastically. This results in a decrease of contiguity. At the first moment the liquid is not saturated with respect to the carbide forming elements and the surface energies do not have their equilibrium values. This will result in a high reactivity between carbides and binder metal. This is usually associated with a low surface energy beween liquid and solid, which would correspond to a low dihedral angle and low contiguity according to the model. Reaction rates are fast in the liquid and equilibrium values will very soon be reached. The increased surface energy between solid and liquid will then cause an increase of the dihedral angle and consequently in an increased contiguity. In addition wetting and spreading of the liquid binder will give rise to new and stronger internal stresses, resulting in contraction and in increased contiguity. When the equilibrium situation is reached this process will cease and a new somewhat higher contiguity has been reached. These are probably the initial reactions taking place on melting. On prolonged heating grain gowth due to coalescence will reduce contiguity, but at a much slower rate. On cooling the situation is changed again, due to the lower solubility of the carbides in the metal at lower temperatures. This results in an increased fraction of carbide and because of that in increased contiguity. This precipitation is difficult to suppress and is probably to be considered in any specimen at room temperature. It will thus be difficult to make direct experimental obsevations which describes the situation at sintering temperatures.

The onset of solidification is more easily estimated as it can be assumed that equilibrium is prevailing in the liquid phase. Figure 6 shows a discontinuity of the linear shrinkage of a WC-10w% Co specimen on cooling due to the freezing of the liquid phase. Similar observations were made by Gurland and Norton (1952). A larger contraction due to the solidification of the binder could be expected on basis of the densities of liquid and solid cobalt. Due to the rigid carbide skeleton the contraction of the sintered body is usually much less.

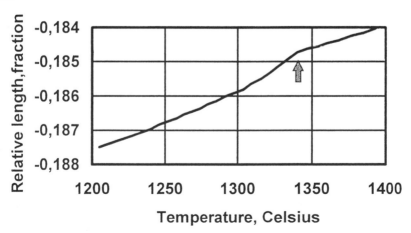

Figure 6. The relative change of length of a WC-10w% Co specimen on cooling from sintering temperature. The onset of solidification of the binder metal is indicated at 1340°C. Zero level of length is at RT before sintering.

The liquidus temperature indicated in this figure, 1340°C is lower than the 1360°C obtained by thermal analysis of another specimen from the same batch as shown in Figure 5 and it might be suspected that a slight carburization has influenced the results shown in Figure 6.

5. Conclusions

Results on the early stages of the sintering of cemented carbides show that as much as 90% of densification can be obtained in the solid state if carbide surfaces are clean and good wetting is obtained. However, small amounts of elements which are reactive with oxygen will drastically change sintering behaviour. In particular titanium will reduce shrinkage on sintering. The reduction of oxygen bound to the surfaces of the carbides, as investigated by mass spectrometry, indicates that the reduction takes place at temperatures close to what could be expected from equilibrium calculations. Thermal analysis of WC-Co compacts show that the binder metal both on heating and cooling is close to equilibrium at temperatures near or above solidus.

Acknowledgement

This work was supported financially by AB Sandvik Coromant, Seco Tools AB, Uniroc AB and NUTEK (the Swedish National Board for Industrial and Technical Development). The permission to publish data from a joint program between Sandvik Coromant and Jernkontoret is also gratefully acknowledged.

References

B. Bengtsson, T. Johannesson and L. Lindau, "Sintering of Milled WC Powder in the Presence of Cobalt", *Planseeberichte für Pulvermetallurgie* **21** 110-120 (1973).

H. E. Exner, J. Freytag, G. Petzow and P. Walter, "Sinterverlauf eines WC-12%Co Hartmetalls mit unterschiedlicher Bindephasenzusammensetzung", *Planseeberichte für Pulvermetallurgie* **26** 90-104 (1978).

G. Leitner, K. Jaenicke-Rössler and H. Wagner, "Processes During Dewaxing and Sintering of Hardmetals", *Advances in Hard Materials Production*, Bonn 1992, paper 13, ed. by MPR Publ Serv Ltd Shrewsbury England 1993.

G. E. Spriggs, "The Importance of Atmosphere Control in Hard-Metal Production", *Powder Metallurgy*, **13** 369-393 (1970).

B. Uhrenius, "Carbon Balance During Presintering and Sintering of Cemented Carbides" *Internal Report Sandvik Coromant*, Stockholm (1975).

B. Uhrenius, H. Brandrup-Wognsen, U. Gustavsson, A. Nordgren, B.Lehtinen and H. Manninen, "On the Formation of Impurity-Containing Phases in Cemented Carbides", *Refractory Metals and Hard Materials* **10** 45-55 (1991).

S. Amberg, E.-Å. Nylander and B. Uhrenius, "The Influence of Hot Isostatic Pressing on the Porosity of Cemented Carbides", *Powder Metallurgy International* **6** 178-180 (1974).

R. Lundberg, "Reaction Kinetics at Sintering of Powder Compacts", *Jernkontoret Report Nr 652 Serie D*, Stockholm (1991), (In Swedish).

S. Haglund, "Sintering of Cemented Carbides", *TRITA-MAC-0566*, The Royal Institute of Technology, Stockholm (1994).

A. Bock, "Sintering of Ultrafine WC/Co Hard Metals", Thesis, Technical University, Wien (1992).

B. Uhrenius, B. Carlsson and T. Franzen, "A Study of the Co-W-C System at Liquidus Temperatures", *Scand. J. Met.* **5** 49-56 (1976).

M. Hellsing, "High Resolution Microanalysis of Binder Phase in Cemented Carbides", *Materials Science Technology* **4** 824-829 (1988).

K. G. Stjernberg, "Some Relations Between the Structure and Mechanical Properties of WC-TiC-Co Alloys", *Powder Metallurgy* **13** [25] 1-12 (1970).

R. Warren and M. B. Waldron, "Microstructural Development During the Liquid-Phase Sintering of Cemented Carbides, I. Wettability and Grain Contact", *Powder Metallurgy* **15** [30] 166-180 (1972).

R. M. German, "The Contiguity of Liquid Phase Sintered Microstructures", *Met Trans A* **16A** 1247-1252 (1985).

J. Gurland and T. Norton, "Role of The Binder Phase in Cemented Tungsten Carbide Alloys", *J. of Metals* **14** 1051-1056 (1952).

SOLID STATE SINTERING OF CEMENTED CARBIDES

Sven A. Haglund*,John Ågren, Royal Institute of Technology
Per Lindskog, Björn Uhrenius, Sandvik AB, Sweden

Abstract

A large part of the densification of cemented carbides may take place in the solid state. In order to be able to formulate reasonable mathematical sintering models it is important to have a detailed knowledge also of this phenomenon. In this study we have examined solid state sintering of cemented carbides both in an ordinary dilatometer and in a dilatometer where an axial load is applied to the sample. The microstructure is analyzed and the most important sintering mechanisms are identified. The importance of the wetting of Co on WC surfaces during the densification is also analyzed.

1. Introduction

Cemented carbides are produced by liquid phase sintering and there are also numerous reports describing the sintering behavior when the binder is molten, see for example (Warren and Waldron 1985) or (Deshmukh and Gurland 1984). However, a lot of the total linear shrinkage may take place already in the solid state, which is often overlooked. This has been observed in some reports though, for example (Exner et. al. 1978), (Bengtsson et. al. 1973.) or (Schatt 1992) It is known from these references that a WC-Co powder may have between 50% and 90% of its total densification already in the solid state. Neglecting the solid state sintering part in the modeling of sintering, or model it without sufficient experimental knowledge would lead to incorrect models. The aim of this work is to clarify the sintering behavior of cemented carbides in the solid state, to serve as a background for further modeling, and since FEM calculations require knowledge of the deviatory behavior as well as the hydrostatic, sintering experiments were also carried out under under axial load.

2. Material

The experiments were conducted using a WC-Co powder with 9.87 wt% Co, 5.59 wt% C, 0.24 wt% O and <40 ppm Ti. The average WC grain size was 1.7 μm. Samples were pressed isostatically without pressing aid to bars approximately

6x8x23 mm. The density of the samples was 56% TD, which is a normal value in hard metal production. The bars used for dilatometry with an axial load were pre-sintered 10 minutes at 900°C in order to receive some mechanical strength and finally ground to rods with a length of 21.5 mm and a diameter of 5.9 mm. For the analysis of the influence of wetting two additional powders were used. One with the same chemical analysis as above but with 200 ppm Ti spread over the WC surfaces. The average grain size was 1 μm. The second additional powder contained 12.08 wt% Co, 6.80 wt% C and 10.63 wt% Ti.

3. Experimental

3.1 Dilatometry

Sintering experiments 'without load' were carried out in a dilatometer with helium atmosphere and an axial load of 40 g (-0.008MPa). The temperature readings and the length of each sample were recorded every 10 s by means of a computer logger. The dilatometer was calibrated with a platinum reference rod. All samples were covered with carbon paper in order to prevent decarburization. In three series of runs the temperature was increased to 1100, 1175 or 1250°C, where the samples were held isothermally. In the first two series the heating rate was 20°C/min up to the isothermal holding temperature. Due to limitations of the furnace the heating rate was lower at higher temperatures. Above 1200 °C the heating rate was only 5 °C/min. In addition, runs with heating rates of 10°C/min and 5 °C/min were performed. Later the furnace was changed in order to be able to maintain a heating rate of 20°C/min. In some samples the length changes obtained by dilatometry were checked against density measurements. A good agreement was observed.

3.2 Dilatometry with an axial load applied

Sintering experiments under compression were carried out in a MTS testing machine where a constant load could be maintained throughout the experiment. The sample was heated by induction heating with a graphite susceptor surrounding it. The atmosphere was pure argon at atmospheric pressure. The temperature was measured with a pyrometer and the pyrometer was also used to control the temperature. The heating rate used was 20°C/min up to 1230°C where the sample was held isothermally for 1 h. The axial shrinkage was measured using four LVDT's with pushrods acting on the sample. The radial shrinkage was measured with a laser scanner. The rig was calibrated with a platinum rod during an identical temperature cycle to the ones used during the sintering experiments. The loads used were 2, 6 and 26 N (-0.07, -0.22 and -0.95MPa).

3.3 Microstructural analysis

After sintering the samples from the 20 °C/min series from the ordinary dilatometer were cut in halves and infiltrated with epoxy plastic in vacuum. Thereafter
they were polished and examined in a SEM Jeol-840. Pictures were taken with a
video printer. The porosity was examined with an image analyzer on a polished
surface. However, it was not possible to make good determination of the distribution of WC particle size directly on the SEM pictures. Drawings were instead
made from these pictures which were subsequently analyzed. This procedure
might have caused an error when determining the size of the smallest WC
particles.

4. Results

4.1 Dilatometry

As an example, Fig. 1 shows a typical dilatometer curve of the linear shrinkage
during sintering of a body where the isothermal holding temperature was 1250°C.
The temperature curve is included in the figure. It was found that shrinkage starts
at low temperatures, at about 800°C. As can be seen, a large part of the densification takes place in the solid state since the first liquid appears in the temperature
range 1275 - 1325°C. In the case of 1 h holding at 1250 °C the total linear shrinkage is about 13% which should be compared with the 17.6% required to reach full
density. This means that the major part of the shrinkage to full density is reached
already at this temperature. Fig. 2 shows the linear shrinkage for different isothermal temperatures. Note that shrinkage rates decrease drastically when
isothermal conditions are established. Prolonged sintering does not result in full
density within reasonable time. One sample treated at 1175°C for 24 hours
reached the same density as a sample that had been heated to 1250°C and then
cooled.

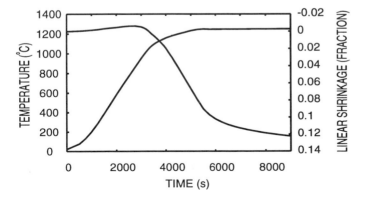

Fig 1. Dilatometer curve for a sample heated to 1250°C.

The microstructural evolution of the dilatometry samples is shown in figs 3 -5. Fig 3 shows the structure of a sample which was heated to 1100°C and subsequently cooled. As can be seen, Co has started to spread between WC grains. There is a large fraction of small WC grains and no grains are yet faceted.

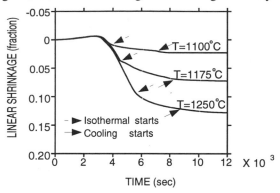

Fig 2. Dilatometer curve for samples heated to 1100°C, 1175°C and 1250°C.

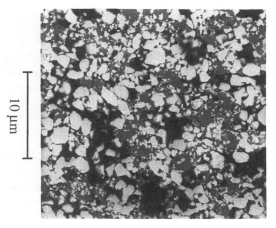

Fig 3. Microstructure for a sample heated to 1100°C.

Fig 4 shows the structure obtained after heating to 1250°C and subsequent cooling. There is a considerable agglomeration of the WC-grains. The smallest WC grains are dissolved in Co-phase. A strong tendency for faceting of the WC is observed. Fig 5 shows the structure obtained after heating to 1250°C, 1 hour isothermal holding and subsequent cooling. The fine WC-grains are almost totally dissolved. The structure is characterized by large agglomerates of faceted WC-grains. Image analysis was carried out on some SEM micrographs. The distributions of the grain size was measured and resulting average WC grain sizes and average pore sizes are given in tables 1 and 2.

Table 1. Average WC (µm) grain sizes after solid state sintering

temp	0'	60'	180'
1100°	0.22	0.30	0.30
1175°	0.26	0.32	0.37
1250°	0.35	0.40	0.44

Table 2. Average pore sizes (µm) after solid state sintering

temp	0'	60'	180'
1100°	0.51	1.06	0.88
1175°	0.52	0.84	0.69
1250°	0.59	0.59	0.64

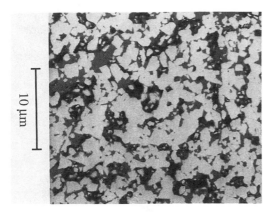

Fig 4. Microstructure for a sample heated to 1250°C.

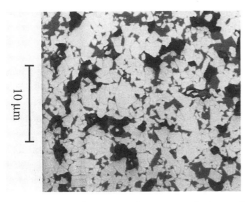

Fig 5. Microstructure for a sample heated to 1250°C and held isothermally for 1h.

The strong increase of the average WC grain size during heating does not seem to depend on normal grain growth but on the dissolution of the fine fraction of grains. Grain growth could be noticed at 1250°C.

4.2 Dilatometry with an axial load applied

Fig 6 shows a dilatometer curve for a sample which has been sintered with an axial load of 26N. The heating rate was 20°/min all the way up to isothermal temperature which was 1230°C. As a comparison an ordinary dilatometer curve with the same isothermal temperature is included in the figure. The temperature curves are also included. As can be seen, the axial shrinkage of the loaded sample is much larger than the radial shrinkage, which was anticipated, and the ordinary dilatometer curve is in between the two. There were slightly larger volume shrinkages on the loaded samples than the unloaded ones but this was noticed first at isothermal temperatures. These dilatometer curves can be used to later calibrate continuum mechanical models for sintering.

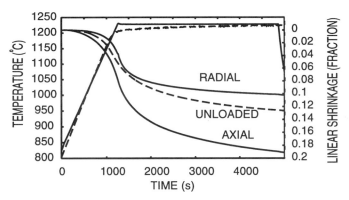

Fig 6. Dilatometer curves for a ordinary dilatometer sample and a sample with an axial load of 26N.

4.3 Powder with Ti dopant

The powders with Ti dopant were sintered in the ordinary dilatometer and the result can be seen in fig 7. It is notable that the powders containing Ti started to sinter at a much higher temperature than the pure WC-Co powder and that the powder with 200 ppm Ti started to sinter at the same temperatures as the pure WC-Co powder but that the sintering rate retards in a intermediate temperature range to later increase again at higher temperature. The microstructure of the 200 ppm powder after sintering (not shown) shows that Co have not spread by far to the same extent as in the pure WC-Co powder. After 1 h at 1250°C Co could still be seen as discrete particles. A closer analysis of the surfaces with Auger and ESCA showed that the Ti was enriched to the surface of the WC grains. From this information in addition to the fact that tungsten oxides are reduced at the temperatures where the WC-Co powder start to sinter and the fact that titanium oxides are reduced at the temperatures where the WC-TiC-Co powder starts to sinter (Uhrenius 1975) one may draw the conclusion that the higher sintering temperatures for Ti-containing powders are caused by oxides on the surfaces hindering Co from wetting and spreading across the carbide surfaces so that no capillary forces arise.

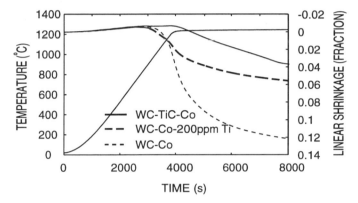

Fig 7. Dilatometer curves for samples with different Ti contents.

5. Discussion

The main microstructural changes during the solid state sintering of cemented carbides can be summarized as:

(1) Rapid wetting and spreading of Co, possibly under the action of capillary forces

(2) Rearrangement and formation of agglomerates of WC grains possibly under the action of capillary forces

(3) Dissolution of submicron size WC fragments

(4) WC particles become faceted

It can be noted that this is very much in line with what has been reported earlier (Åkesson 1978), in which WC was sintered with high amounts of a low melting additive. It seems like the wetting is a very important stage for the onset of sintering due to its strong influence on the formation of capillary forces. But it is not enough that the powder contains 200 ppm Ti to give the detrimental effect on sintering, it must be spread across the WC surfaces. When this effect was noticed powders with 0.1 wt% of TiC, TaC or NbC were prepared but these carbides did not spread on WC surfaces during milling resulting in no retarded sintering.

Since the rearrangement process takes place in the solid state it is probably not taking place by viscous flow. It is probably by some creep mechanism. The temperatures are quite high and the internal stresses are low so diffusional creep can probably be anticipated. An axial load on the sample will give rise to hydrostatic stress component which may in some way be added to the capillary forces giving rise to a slightly larger volume shrinkage. From the loaded experiments one may also draw the conclusion that the strong decrease in sintering rate at isothermal temperatures must at least partly be due to a change in the viscous modulus since the deviator strain rates decrease in the same manner as the volumetric sintering rates, maybe due to formation of a carbide skeleton. To quantify the effect of an axial load is however hard to without a model of the sintering process and therefore it will not be done in this work.

Acknowledgments

The financial support of the Swedish National Board for Industrial and Technical Development (NUTEK), Sandvik Coromant AB, Seco Tools AB and Uniroc AB is gratefully acknowledged. The authors also would like to thank Christer Fahlgren at Sandvik Coromant for all the help with image analyzing.

References

R. Warren and M. B. Waldron, Microstructural Development During the Liquid-Phase Sintering of Cemented Carbides, 1. Wetting and Grain Contact", Powder Metallurgy **15** [30] 1247-1252 (1985)

R. Deshmukh and J. Gurland, "A Study of Microstructural Development During the Liquid Phase Sintering of WC-Co alloys" Inst. Phys. Conf. Ser. No **75** Ch 4 347-358 (1984)

H. E. Exner, J. Freytag, G. Petzow and P. Walter, "Sinterverlauf eines WC-12%Co Hartmetalles mit Unterschiedlicher Bindephasenzusammensetzung", Planseeberichte für Pulvermetallurgie **26** 90-104 (1978)

B. Bengtsson, T. Johannesson and L. Lindau, "Sintering of Milled WC Powder in the Presence of Cobalt", Planseeberichte für Pulvermetallurgie **21** 110-120 (1973).

W. Schatt, "Sintervorgänge, Grundlagen", VDI-Verlag GmbH Düsseldorf (1992)

B. Uhrenius, "Carbon Balance During Presintering and Sintering of Cemented Carbides" Internal Report Sandvik Coromant, Stockholm (1975).

L. Åkesson, "A Sintering Study in the WC-Co System", Jernkontorets forskning 815/74 (1978)

Modeling of Solid State Sintering of Cemented Carbides

John Ågren *, Jan Brandt **, Sven Haglund * and
Björn Uhrenius ***,

* Royal Institute of Technology
** University of Linköping
*** Sandvik AB
SWEDEN

Abstract

Solid state sintering of cemented carbides is modeled by means of a sintering stress and a rearrangement of carbide particles. The rate determining process is diffusional creep of the cobalt binder. The influence of heating rate is modeled in terms of "free volume" created by the volume change caused by carbide dissolution upon heating and its enhancement of the rearrangement. A satisfactory agreement between calculations and recent experimental information is found.

1. Introduction

Cemented carbides are produced by liquid phase sintering, a process which has been thoroughly investigated, both experimentally and theoretically, by numerous authors, Gurland and Norton (1952), Kingery (1959), German (1985), Kaysser (1992). In the simplest analysis, Kingery (1959), the powder particles are approximated as monosized spheres and all structural changes prior to formation of liquid are neglected. As soon as the liquid has been formed, densification is assumed to take place rapidly under the influence of capillary forces. However, this picture is far too simple. For example, when sintering W heavy metals Li and German (1984) found that an appreciable densification occurs upon heating before melting. Although solid state sintering is generally very slow, it is well known that small additions of a lower melting substance may have a remarkable effect on the sintering rate, so called activated sintering, German (1983). Solid state sintering of alloys with larger additions of a lower melting substance has not been examined close to the extent of liquid phase sintering. It has been reported, Schatt (1992), that a considerable part of densification by sintering WC-Co cemented carbides takes part during the heating stage prior to Co melting.

This study is a part of a project where the aim is to model sintering of cemented carbides in order to predict, by finite element analysis, shape changes during sintering of complex bodies, Brandt (1995). Our modeling will be based on the microstructural changes during the sintering process. The aim of this study is to investigate theoretically the sintering behavior prior to temperatures where liquid phase forms and compare the behavior with recent experimental information, Haglund et al. (1995).

2. Summary of recent experimental observations

Solid state sintering of WC-Co mixtures was recently investigated experimentally by some of the present authors, Haglund et al. (1995), who applied dilatometry and metallographic examination of samples sintered under various conditions. In this section their observations will be summarized but the reader is referred to their report for details. Haglund et al. found that a large part of the densification takes place in the solid state. For example, in the case of holding at 1250 °C for 1h the total linear shrinkage is about 13%, which should be compared with the 17.6% required to reach full density in their particular case. This means that a major part of the shrinkage to full density is reached already at this temperature. They further noticed that the shrinkage starts at rather low temperature, around 800 °C.

Under isothermal conditions the sintering rate decreases drastically making it practically impossible to reach full density at low temperatures. It thus seems that sintering is enhanced by continuous heating.

In the green body, Co is present both as rather coarse particles and as fine highly deformed fragments. The WC particles are spheroidal with a diameter around 1 μm. In addition there are many small WC particles, < 0.1 μm. The porosity is very high and open, i.e. there is a connected network of pores and channels. The most striking difference between a sample cooled immediately after being heated to 1100 °C and the green body is the very rapid redistribution of solid Co by which Co has spread between the WC particles and pulled them together into aggregates. In samples heated to higher temperatures some of the smallest WC fragments have dissolved and the remaining ones appear faceted.

3. Solid-state sintering theory
3.1 The primary wetting and spreading of Co

As mentioned in the previous section, the densification is initiated by a very rapid rearrangement of Co whereby Co wets the WC particles and spreads between them. The spreading of Co causes an agglomeration of WC and is a necessary condition for subsequent WC rearrangement. When the wetting is poor there will be no spreading and consequently no capillary forces to cause rearrangement and densification. For the sake of simplicity, we shall assume that wetting and the subsequent Co spreading occurs instantly as the heating starts. Thus, the kinetics of wetting will not be further discussed here. Although, this may seem as a too crude assumption our experimental observations, see Haglund et al. (1995), indicate that it is satisfactory for Co-WC powders with sufficiently low impurity levels.

3.2 Rearrangement - concept of sintering stress

We shall only consider the rearrangement of WC particles and assume that the spreading of Co is instantaneous. The new model will be based on the ideas of

Cahn and Heady (1970), without attempt to develop a detailed geometrical model. Such a detailed model would necessarily be very complex and require too many simplifying assumptions. Our model will be based on the phenomenological concept of *sintering stress* applied previously by several authors, e.g. by Kingery (1959) and more recently by Riedel et al. (1990). The sintering stress is the driving force for densification of the material by eliminating the porosity. It is hydrostatic in its character and was recently defined by Svoboda et al. (1994) as "the mechanical hydrostatic stress which just balances the internal surface tension forces so that the porous solid does not shrink ." Kingery (1959) took for the sintering stress

$$\sigma_\otimes = -\frac{2\gamma}{r_p} \tag{1}$$

where γ is the surface energy between binder and pore and r_p is the pore radius.

A similar expression derived from a more specific geometrical model was applied by Riedel et al. (1990). An obvious problem with this choice of sintering stress is that it diverges as the pore vanishes which does not seem right. Regarding the sintering stress as the driving force for densification it seems more natural that it should vanish as the porosity vanishes. On the other hand, for the interparticle capillary forces between equisized spheres Heady and Cahn (1970) derived the following expression for the equivalent external pressure

$$P_x = 2\sqrt{2}\pi \frac{\gamma \cos\theta}{R} \tag{2}$$

where R is the radius of the spherical particles. It seems natural to identify this pressure with the quantity, $-\sigma_\otimes$, which would then remain finite as the porosity vanishes. It should be noticed that the sintering stress does not depend on the amount of the binder. Cahn and Heady (1970) made their analysis for a well defined geometry and two interacting particles. In general, we would like to consider a large number of particles in a body with porosity f_p defined as the volume fraction of pores. We may then consider the analysis of Cahn and Heady as a limiting case valid for very high porosity. As the porosity vanishes it is natural to expect that the sintering stress vanishes as well because in a dense body there should be no driving force for further densification. An extension of the result Cahn and Heady's result thus is

$$\sigma_\otimes = -const\, \frac{\gamma^{Co/p} \cos\theta}{L} f_p \tag{3}$$

where L is a characteristic distance representing the average size of the particles that undergo rearrangement. The constant should be on the order of unity. It should be noticed that the pore size does not enter into the expression for the sintering stress. However, as discussed by Park et al. (1984), a sufficiently small pore will be only bounded by the binder phase and the sintering stress would be given by eq. 1. The limiting pore size is proportional to the particle size and increases when the amount of binder increases for a given particle size. However, it should be noticed that filling of such small pores may occur without any rearrangement of WC.

3.3 Rearrangement - densification rate

Assuming that the sintering stress is known we shall now consider the rate of densification. Assuming isotropic shrinkage we have

$$3\dot{\varepsilon}_m = \frac{1}{V}\frac{dV}{dt} = \frac{1}{1-f_p}\frac{df_p}{dt} \tag{4}$$

where $\dot{\varepsilon}_m = \frac{1}{3}\dot{\varepsilon}_{kk}$ is the average volumetric strain rate and the usual convention of summation over repeated indices is applied. Eq. 4 is readily integrated and we obtain

$$f_p = 1 - (1-f_p^o)e^{-3\varepsilon_m} \tag{5}$$

where $\varepsilon_m = \int \dot{\varepsilon}_m dt$ is the total logarithmic strain. It is related to the technological strain, $e = \Delta \ell / \ell_o$ by the well known relation $\varepsilon = \ln(1+e)$. f_p^o is the volume fraction of pores at $t=0$.

Assuming that the rearrangement process is controlled by creep of the Co binder we can write

$$\dot{\varepsilon}_m = k_m \sigma_m \tag{6}$$

where σ_m is the hydrostatic component of the stress. In the absence of external load it is simply $-\sigma_\otimes$. As the temperature is high and the stress is quite low it would be expected diffusional creep rather than recovery creep as dominating deformation mechanism and k_m, the kinetic parameter, would contain an appropriate diffusion coefficient D. Moreover, since it is impossible to further compact an already fully dense body k_m must depend on the porosity and vanish in the limit when $f_p = 0$. Further, it is simply assumed a linear relation, i.e. $k_m = KDf_p$, where K is a constant.

We may expect that surface diffusion plays an important role at low temperatures and for the diffusivity we take

$$D = D_{volume} + \omega D_{surface} \qquad (7)$$

indicating that the two diffusion processes are coupled in parallel. ω will be treated as an adjustable parameter and will be evaluated to fit the experimental data.

As mentioned in section 2, the experimental information, Haglund et al. (1995), suggests that sintering is enhanced by continuos heating. In fact, the effect is very strong and if the shrinkage rate is plotted as a function of time there is a drastic drop under isothermal conditions, see Fig. 1. The heating data may be represented by the following relation

$$\dot{\varepsilon} = const \; D \frac{df^\alpha}{dT} \dot{T} \qquad (8)$$

where f^α is the equilibrium fraction of WC in the alloy and $\dot{T}$ the heating rate. The question then is whether the sintering stress or the kinetic parameter k_m is

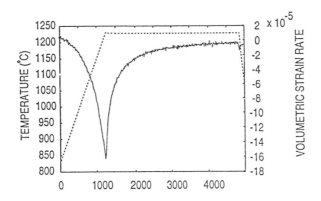

Fig.1: Volumetric strain rate as a function of time, Haglund et al. (1995).

affected by heating. Another important piece of experimental information comes from sintering under a small compressive axial load, Haglund et al. (1995). From their measurements it may be concluded that the deviatoric strain rate has a heating rate dependence very similar to that of the volumetric strain rate. The two rates ratio shows a smooth variation and no particular effect under isothermal conditions. These results strongly indicate that the heating rate dependence comes from k_m rather than from the sintering stress.

It may be noticed that as temperature increases there is a gradual dissolution of the WC particles. As the diffusion distances are very small, less than 1 μm, the equilibrium fraction of WC is approached almost instantly. The densities of the Co binder and WC are such that the dissolution is accompanied by a decrease in the volume of solid phases which must lead to an increase in porosity if the total volume of the powder compact is kept constant. This newly formed porosity should be distributed evenly around the WC particles. It seems reasonable to assume that this porosity increase would enhance rearrangement, i.e. increase k_m. Based on the free volume theory for relaxation of liquids, Cohen and Turnbull (1959), we shall call this type of porosity free volume and denote the fraction of free volume ξ. Under isothermal conditions the dissolution would rapidly came to a stop as equilibrium is approached and no more free volume is formed. By rearrangement the free volume would be redistributed and finally annihilated. Tentatively we may write the following balance equation for the free volume

$$\frac{1}{(1-f_p)}\frac{d\xi}{dt} = -\frac{\Delta V_m}{V_m}\frac{df^\alpha}{dT}\dot{T} - (1/\tau)\xi \tag{9}$$

where $\Delta V_m / V_m$ is the relative volume change accompanying the WC dissolution. The first term on the right-hand side thus denotes the creation of free volume by dissolution of WC upon heating and the second term free volume annihilation by means of rearrangement. For the annihilation we have rather arbitrarily introduced a relaxation time τ. We now adopt the following expression

$$k_m = KD\, f_p\, (1+a\xi) \tag{10}$$

As will be seen in section 4, the experimental information may be satisfactorily represented by a relaxation time

$$\tau = b(f_p)^{-4}\exp(1/f_p) \tag{11}$$

where b is a constant. A high porosity would thus cause a more rapid annihilation of free volume whereas in the limit of a dense material τ would be very large.

4. Comparison with experimental data

The above discussed model, as summarized by eqs. 3, 4, 6, 7, 9-11, was now implemented in the computer and the system of ordinary differential equations solved numerically. A thermal expansion term was introduced and eq. 6 was actually replaced by

$$\dot{\varepsilon}_m = k_m\, \sigma_m + (1-f_p)\,\alpha\,\dot{T} \tag{12}$$

Fig. 2 shows the result of a simulation for heating up to three different temperatures, 1100, 1175 and 1230 °C. The parameters used are listed in an appendix. The fraction f^{α} of WC was taken from the Thermo-Calc system, Sundman et al. (1985). It can be seen that the agreement between calculations and experimental results is not perfect but it should be considered as satisfactory. It seems likely that a more ambitious least square optimization, where some of the less well determined parameters are allowed to vary, would improve the fit. It is also interesting to notice that the predicted effect of heating rate on the sintering rate is quite strong and is in general agreement with the experiments, see Fig. 3.

5. Summary and conclusions

A phenomenological model for predicting the densification by sintering of Co-WC cemented carbides have been proposed. The model is based on a sintering stress, related to average carbide particle size and porosity, as well as rearrangement controlled by diffusional creep of the Co binder phase. The effect of heating rate is tentatively explained in terms of the volume changes accompanying WC dissolution upon heating. The model is represented by a set of differential equations. It has been possible to adjust the parameters in the model in order to obtain a satisfactory agreement between experimental data and simulations.

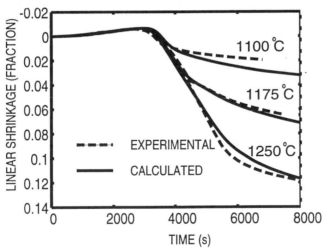

Fig. 2: Linear strain as a function of time for continuous heating to 1100, 1175 and 1230 °C. Solid line calculations, dashed line experimental data. Heating rate 20 Ks^{-1}

Acknowledgment

This work is a part of a project supported by Sandvik Coromant AB, Seco Tools AB, Uniroc AB and the Swedish National Board for Industrial and Technical Development (NUTEK).

References

J. Gurland and J.T. Norton, *Trans. AIME, Journ of Metals* 1051 (1952). W. D. Kingery, *Journ of Appl. Phys.* **30** 301 (1959).

R. M. German, Liquid Phase Sintering, Plenum, New York (1985)

W. A. Kaysser, Sintern mit Zusätzen, Gebrüder Borntraeger, Berlin-Stuttgart (1992)

C. J. Li and R.M. German, *Inter. J. Powder Metall. Powder Techn.* **20** 149 (1984)

R. M. German, *Sci. Sintering* **15** 27 (1983)

W. Schatt, Sintervorgänge, VDI Verlag, Düsseldorf (1992)

J. Brandt, FE-Simulation of Compaction and Solid State Sintering of Cemented Carbides - This Conference

S. Haglund, J. Ågren, P. Lindskog and B. Uhrenius, Solid State Sintering of Cemented Carbides - This conference

J.W. Cahn and R. B. Heady, *J. Am. Ceramic Soc.* **53** 406 (1970) and R. B. Heady an d J. W. Cahn, *Metall. Trans.* **1** 185 (1970)

H. Riedel, in Ceramic Powder Science III, Proceedings of the Third International Conference on Ceramic Powder Processing Science, Feb 4-7, San Diego 1990

Svoboda, H. Riedel and H. Zipse, *Acta Metall. Mater.* **42** 435 (1994)

H.H. Park, S.-J. Cho and D. N. Yoon: *Metall. Trans. A* **15A** 1075 (1984)

M. H. Cohen and D. Turnbull, *Journ. Chem. Phys.* **31** 1164 (1959)

B. Sundman , B. Jansson and J.-O. Andersson, *Calphad*, **9** 153 (1985)

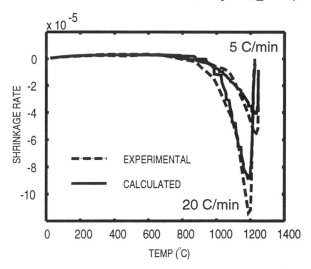

Fig. 3: Strain rate as a function of temperature for continuous heating to 1230 °C with the different heating rates indicated. Solid line calculations, dashed lines experimental data.

Appendix

$$\alpha = 1.4\ 10^{-5},\quad D = 5\ 10^{-4}\,(e^{-260\,000/RT} + 4.63\ 10^{-6}\ e^{-130\,000/RT})$$

$$K = 1.43\ 10^{3}\,/\,RT,\quad a = 10^{3},\quad b = 0.575,\quad f_{p}(t = 0) = 0.43$$

Grain Growth of Carbide-Based Cermets During Liquid Phase Sintering

Soon-Gi Shin and Hideaki Matsubara

Japan Fine Ceramics Center, 2-4-1 Mutsuno Atsuta-ku, Nagoya 456, Japan

Abstract

The grain growth of carbide phases during liquid phase sintering was investigated for TiC- and WC-base cermets sintered at 1673K for 3.6~144ks. The kinetics of grain growth behaviors of TiC-Ni, TiC-Mo$_2$C-Ni, WC-Co and WC-VC-Co cermets almost fitted to the cubic relation of $d^3 - d_o^3 = Kt$. The rate constant values of TiC-Mo$_2$-Co and WC-VC-Co cermets are approximately two or three orders magnitude lower than TiC-Ni and WC-Co cermets. The growth rate constant of carbide phases had a peak at a liquid phase content in TiC-Ni and WC-Co cermets, while it slightly decreased with decreasing liquid content in Mo$_2$C-and VC-added cermets. The growth rate constant tended to lower with higher contiguity of carbide particles. The grain growth behavior of these cermets could be explained by the mechanism that the existence of contiguous boundaries of carbides particles suppressed the movement of solid/liquid interfaces during liquid phase sintering.

I. Introduction

In liquid phase sintered TiC- and WC-base cermets, the grain size is of great importance since their mechanical properties are critically dependent on it. Thus, much attention has been paid to control grain growth during liquid phase sintering, especially in the case of fine or ultra-fine starting powders. Previous researches have shown that the addition of other carbides and nitrides restricts the grain growth in TiC- and WC-base cermets [1-7]. However, the detailed studies about grain growth of the cermets, especially concerning the growth mechanism, have not been reported yet.

The present study is aimed to examine the grain growth behaviors of carbide particles during liquid phase sintering of TiC-Ni, TiC-Mo$_2$C-Ni, WC-Co and WC-VC-Co cermets. The mechanism of grain growth in the cermets is discussed by taking account of the roles of solid/liquid interfaces and contiguous boundaries of solid particles.

II. Experimental Procedure

TiC- and WC-base cermets were prepared by usual powder metallurgy techniques. TiC, Mo$_2$C, VC and WC powders with the average particle size of about 1.5µm, Ni powder of 3.0µm and Co powder of 2.0µm were used as starting powders. In order to determine the effect of the metallic binder phase contents on the grain growth of the carbides, Ni and Co volume contents were changed from 2 to 60%. The atomic ratio of Mo/(Mo+Ti) or V/(V+Co) atomic ratio was fixed at 0.1 and 0.4, respectively. The carbon content of all cermets was higher side in the two phase region. The powders were ball-milled in ethanol for 172.8ks, dried in a vacuum and then granulated on a sieve of 250μm. Green compacts of the mixed powders were sintered at 1673K for 3.6ks in vacuum. The sintered bodies were re-sintered at 1673K for 32.4~144ks in Ar atmosphere in order to minimize the evaporation loss of binder phases. The cross sections were polished with diamond powders and then observed mainly by scanning electron microscopy to measure the size (intercept length) of about 1000 particles and the contiguity of carbides[8].

III. Results and Discussion

All sintered samples presented full densification. Typical microstructures of TiC-Ni, TiC-Mo$_2$C-Ni, WC-Co and WC-VC-Co cermets are shown in Figs.1~4, respectively. These photographs are commonly indicated for the cermets of 10, 30 and 60vol% Ni or Co and the two heating times of 3.6 and 72ks. The microstructures shown in the figures are normal features in grain size distribution regardless of binder contents and heating time. However, the carbide grains are partly in contact with each other and partly separated by binder phases, randomly. The grain size was measured by counting the contiguous boundaries of carbides. The shape of carbide particles well reflects the crystal structures of the carbides; cubic of TiC and hexagonal of WC. However, Mo$_2$C- or VC-added cermets show irregular grain shape comparing to those of TiC-Ni and WC-Co straight cermets. The grain size of the straight cermets have a peak at a binder content, 10% for TiC-Ni and 30% for WC-Co, but the grain size of Mo$_2$C- or VC-added cermets, much finer than the straight cermets, slightly increases with increasing Ni or Co content.

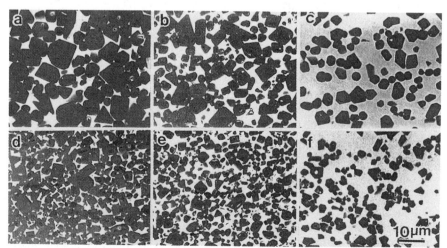

Fig. 1 Microstructures of TiC-Ni cermets sintered at 1673K for 72ks(a~c) and 3.6ks(d~f). Ni content; (a, d) 10%, (b, e) 30%, (c, f) 60%.

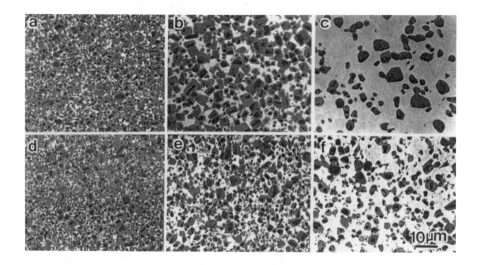

Fig. 2 Microstructures of TiC-Mo$_2$C-Ni cermets sintered at 1673K for 72ks(a~c) and 3.6ks(d~f). Ni content; (a, d) 10%, (b, e) 30%, (c, f) 60%.

Fig. 3 Microstructures of WC-Co cermets sintered at 1673K for 72ks(a~c) and 3.6ks(d~f). Co content; (a, d) 10%, (b, e) 30%, (c, f) 60%.

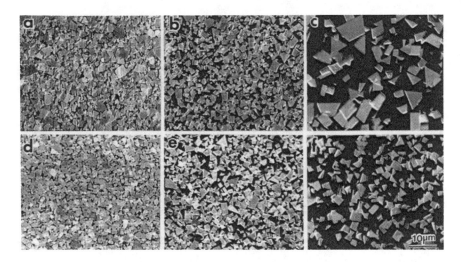

Fig. 4 Microstructures of WC-VC-Co cermets sintered at 1673K for 72ks(a~c) and 3.6ks(d~f). Co content; (a, d) 10%, (b, e) 30%, (c, f) 60%.

The kinetics for Ostwald ripening can be an equation of the form, $d^n - d_o^n = Kt$, where d_o is the initial mean grain size and d is that of at a time, t. The value of n is 2 for interfacial reaction control process or 3 for diffusion control process [9,10]. The cube of average particle size (intercept length) are plotted as a function of sintering time, as shown in Figs. 5~8 for TiC-Ni, TiC-Mo₂C-Ni, WC-Co and WC-VC-Co cermets. The time dependence of grain size in these cermets well fits to the equation of $n = 3$, having partly exceptions for short heating time

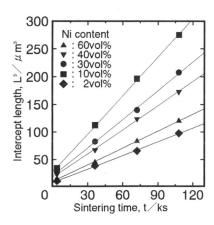

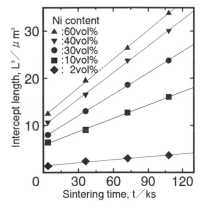

Fig. 5 Variation of mean particle size with time for TiC-Ni cermet.

Fig. 6 Variation of mean particle size with time for TiC-Mo₂C-Ni cermet.

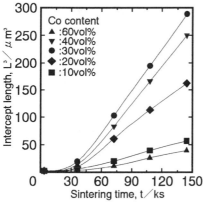

Fig. 7 Variation of mean particle size with time for WC-Co cermet.

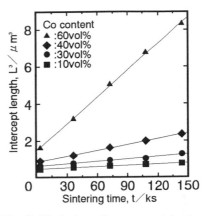

Fig. 8 Variation of mean particle size with time for WC-VC-Co cermet.

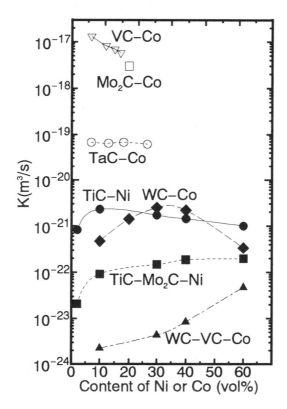

Fig. 9 Experimental growth rate constant of various carbide cermets.

in WC-Co cermet. The growth rate constants, therefore, can be obtained by measuring the slope of the plots in these figures.

In the diffusion control process of Ostwald ripening, the growth rate constant is related to inter-particle spacing in a system; the growth rate constant decreases with increasing content of liquid phase [11-13]. If the growth rate constant is independent of the liquid phase content, the grain growth is controlled by the reaction at solid/liquid interfaces.

The effects of liquid phase content in the range 2~60% on the growth rate constant of the various cermets are shown in Fig. 9. The data for VC-Co, Mo_2C-Co and TaC-Co cermets have been reported by Warren [14,15]. It is possible to consider three groups in growth behaviors of the cermets. The first group, VC-Co and Mo_2C-Co, has the characteristics of high growth rate and the decreasing effect with binder content. The values of the rate constants of VC-Co and Mo_2C-Co correspond with the theoretical values of the diffusion control process[14]. The second group consists of TiC-Ni, WC-Co and TaC-Co, having the feature that growth rates have a peak at a binder content; the results of TaC-Co are in too small range of binder content to judge the effect of binder content. The rate constant values of this group are approximately three orders of magnitude lower than the theoretical values of diffusion controlling process. In the third group, TiC-Mo_2C-Ni and WC-VC-Co cermets, the rate constants are much smaller than those of the other group and it has the tendency of an monotonous increase with binder content. In a higher binder content, the inhibitor effect of the addition carbides becomes fainter. These results mean that the grain growth of the cermets in this study can not explained by the pure theories of Ostwald ripening. For the mechanism of grain growth in the cermets, the present authors have proposed the process that the movement of solid/liquid interfaces is restricted by the existence of solid/solid boudaries [16,17].

Figure 10 shows the contiguity of carbide particle in the various cermets as a function of liquid phase content. The contiguity values commonly decrease with increasing liquid content. There is a rough tendency that the cermets with smaller growth rate have the larger values in contiguity, but it is not possible to understand the growth rate only from the viewpoint of the contiguity. It is considered that the grain growth rate and the movement of solid/liquid interface are affected both by the contiguity and by the mobility of contiguous boundaries in solid particles. Further investigations are need to clarify the mechanism, especially quantitative treatments in kinetics for the grain growth and characteristics of contiguous boundaries in carbide particles of the cermets.

IV. Conclusions

The grain growth behaviors of carbide phases during liquid phase sintering were investigated for TiC-Ni, TiC-Mo_2C-Ni, WC-Co and WC-VC-Co cermets sintered at 1673K for 3.6~144ks. The kinetics of growth behaviors of the cermets almost fitted to the cubic relation. The growth rate constant of carbide phases had a peak at a liquid phase content in TiC-Ni and WC-Co cermets, while it slightly decreased with decreasing liquid content in Mo_2C-and VC-added cermets. The growth rate constant tended to decrease with increasing contiguity

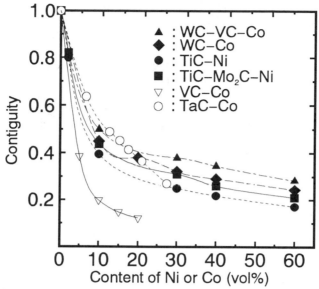

Fig. 10 Contiguity of carbide particles in various cermets as a function of binder phase content.

of carbide particles. The grain growth behavior of these cermets could be explained by the mechanism that the existence of contiguous boundaries of carbides particles restricted the movement of solid/liquid interfaces during liquid phase sintering.

References
1) D. Moskowitz and M. Humenik, Jr, "Cemented Titanium Carbide Tools", *Modern Development in Powder Metallurgy*, vol. 3, Plenum Press, NJ. pp 83-94, (1966).
2) H. Suzuki and K. Hayashi, "Effects of Addition-Carbide on the Properties of TiC-Ni Alloys", *J. Japan Soc. Powder and Powder Met.,* 17 [6] 28-32 (1971).
3) R. Kieffer, P. Ettmayer and Freudhofmeier, "About Nitrides and Carbonitrides and Nitride-Base Cemented Hard Alloys, *Modern Development in Powder Metallurgy* vol. 15, Plenum Press, NJ, pp 201-214, (1971).
4) K. Hayashi, Y. Fuke and H. Suzuki, "Effect of Addition Carbides on the Grain Size of WC-Co Alloy", *J. Japan Soc. Powder and Powder Met.,* 19 [2] 37-41 (1972).
5) H. Tulhoff, "On the Grain Growth of WC in Cemented Carbides", *Modern Development in Powder Metallurgy*, vol. 14, Metal Power Industries Federation Princeton, pp 269-268, (1981).
6) H. E. Exner, "Quanlitative and Quantitative Interpretation of Microstructures in Cemented Carbides", *Science of Hard Materials*, Plenum Press, 233-262, (1983).

7) H. Doi, "Advanced TiC and TiC-TiN base Cermets", *Inst. Phys. Conf. Ser. No 75, Chap. 6, Science Hard Mater.*, Adam Hilger Ltd, pp 489-523, (1986).

8) J. Gurland, "An Estimate of Contact and Continuty of Dispersions in Opaque Samples", *Trans.AIME* 236 [3] 642-646 (1966).

9) C. Wagner, "Theorie der Alterung von Niederschlagen durch Umlosen". *Z. Elektrochem.*, 65 [Nr7/8] 581-591 (1961).

10) I.M. Lifshitz and V.V. Slyozov, "The Kinetics of Precipitation from Supersaturated Solid Solutions", *J.Phys.Chem.Solids* 19 [Nos.1/2] 35-50 (1961).

11) S. Sarian and H.W. Weart, "Kinetics of Coarsening of Spherical Particals in a Liquid Matrix", *J.Appl.Phys,* 37 [4] 1675-1681 (1966).

12) A.J. Ardell, "The Effect of Volume Fraction on Particle Coarsening: Theoretical Considerations", *Acta Met.*, 20 [1] 61-71 (1972).

13) P.W. Voorhees and M.E. Glicksman, "Ostwald Ripening during Liquid phase sintering-Effect of Volume Fraction on Coarsening Kinetics", *Metall. Trans.*, 15A [6] 1081-1088 (1984).

14) R. Warren and M.B. Waldron, "Carbide Grain Growth", *Power Met.,* 15 [30] 180-201 (1972).

15) R. Warren, "Microstructural Development during the Liquid-Phase Sintering of VC-Co alloys, *J. Mater. Sci.*, 7 1434-1442 (1972).

16) H. Matsubara, S-G Shin and T. Sakuma, "Growth of Carbide Particles in TiC-Ni and TiC-Mo$_2$C-Ni Cermets during Liquid Phase Sintering, *Trans.Japan Inst. Met.*, 32 [10] 951-956 (1991).

17) H. Matsubara, S-G Shin and T. Sakuma, "Grain Growth of TiC and Ti(C,N) Base Cermets during Liquid Phase Sintering", *Solid State Phenomena*, Vol.25&26, pp 551-558, (1992).

Gas Formation in Fe-Mo-C Below the α-γ Transition

Herbert Danninger and Gert Leitner*

Institut für Chemische Technologie anorganischer Stoffe,
Technische Universität Wien, Vienna, Austria
*Fraunhofer-Institut für Keramische Technologien und
Sinterwerkstoffe, Dresden, Germany

Abstract

The sintering behaviour of PM carbon steels was studied by thermal analysis. Dilatometry and thermogravimetry combined with mass spectrometry and DTA revealed weight loss due to gas formation, indicated in the dilatometric graph by slight shrinkage, clearly below the ferrite-austenite transition temperature. Mass spectrometry showed the gas formed to be carbon monoxide. Deoxidation of the Fe surfaces through reaction with carbon thus apparently does not require dissolution of all the carbon in the Fe matrix - which is possible only with γ-Fe - but may start already with α-Fe.

1. Introduction

Sintered steels containing carbon are frequently used for precision parts, in particular for those loaded mechanically or under wear. As in cast and wrought steels carbon increases strength and hardness and enables heat treatment. Most common are the steels of Fe-C, Fe-Cu-C and Fe-Cu-Ni-C type, but also Mo and Cr alloyed steels have increasingly found applications.

For precision parts, carbon is added as elemental powder, i.e. as fine graphite. Natural graphite is mostly used, but also artificial graphite types have been found to be suitable if the SiO_2 content is low and the material is well crystallized. Carbon is dissolved in the iron/steel matrix during heating to sintering temperature. This process has so far been studied rarely despite its technical importance. Houdremont (1956) states that carbon is dissolved in iron via gas phase transport, i.e. as carbon monoxide. Dautzenberg and Hewing (1975) described the dissolution of carbon occurring rather through solid state diffusion.

It is well known that the actual carbon content of sintered steels is slightly lower than that calculated from the amount of graphite added. This is due to reaction between carbon and the oxide layers covering the iron powder particles. Also here, however, the mechanism has been hardly ever studied. It is mostly assumed that carbon is first dissolved in the austenite and then this dissolved carbon reacts with the oxide layers to form carbon monoxide. With cemented carbides the process of deoxidation has been thoroughly studied

(Leitner, Jaenicke-Rößler, Wagner, 1993), and it was found that only after the oxide layers have been removed the solid state sintering process, discernible through shrinkage, proceeds noticeably.

Within this work, gas formation during sintering of carbon containing PM steels has been studied and the consequences on the dimensional behaviour have been investigated. By using mass spectrometry combined with thermo-balance, the gases set free have been identified.

2. Gas Formation and Dimensional Change

In order to identify the temperature ranges within which gas is formed during sintering, compacts were dilatometrically investigated in vacuum and the pressure in the vacuum system was recorded. The specimens were mixed from iron powder - water atomized powder ASC100.29 supplied by Höganäs - and fine natural graphite Kropfmühl UF4. The alloy metals were added as elemental powders, in particular Mo <25 µm (Plansee) and Cr (aluminothermic quality, <10 µm). The powder mixtures were compacted in a pressing tool with floating die to form tensile test specimens. In part the specimens were pressed without admixed lubricant, only with die lubrication, in part 0.5% Microwax C (Hoechst) was added as lubricant, and the compacts were then de-waxed for 30 min at 550°C in flowing hydrogen.

From the head sections of the compacts pieces of about 10 mm length were cut and tested in a dilatometer Bähr 801 with alumina measuring system, the dimensional change being measured perpendicular to the pressing direction. The test runs comprised heating at 10 K.min^{-1} up to a given level, holding at this temperature for 60 min and then cooling at 10 K.min^{-1} to 300°C. All dilatometric graphs are shown here without compensation for thermal expansion of the specimen or the alumina sample holder system.

The tests were run in rotary pump vacuum, initially at about 6.10^{-3} mbar, and the pressure in the vacuum system was recorded during each entire test run. This was to give at least a qualitiative indication of gas formation processes during the sintering cycle.

A typical dilatometric graph is plotted in Fig.1 together with the respective pressure curve. In order to avoid dimensional effects caused by alloy metals, plain Fe-0.8%C was selected for the first runs, and the compacts were prepared without adding any pressing lubricant. It is clearly evident that after some gas formation in the range of 200°C, which is probably due to desorption of gases and moisture, a very pronounced peak is discernible at about 720°C, which is clearly below the α-γ transition. In that respect it must be noted that the eutectoid point for the stable system Fe-graphite is slightly higher than in the

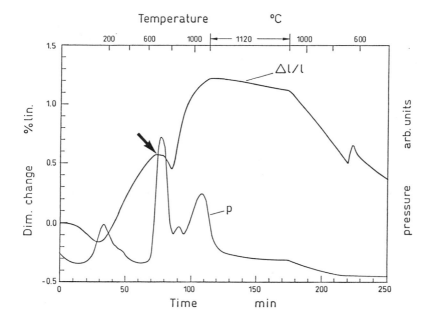

Fig.1: Dilatometric and pressure graphs for Fe-0.8%C.

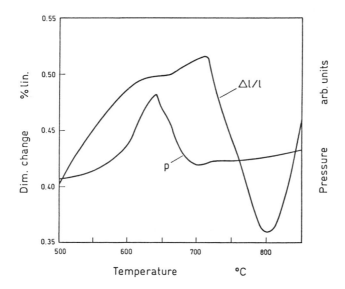

Fig.2: Section of dilatometric graph and pressure graph for Fe-2%Cr-0.7%C.

case of the metastable system Fe-Fe₃C, 738°C compared to 723°C after Klemm (1987). Some de-gassing occurs during further heating, in the temperature range between 1000 and 1100°C, but the peak is much less pronounced than that at 720°C.

When observing the latter temperature range more closely a slight break in the dilatometric graph is found just below the α-γ transition. (indicated by an arrow in Fig.1).This break becomes very pronounced if dilatometric graphs are plotted at higher magnification, as shown in Fig.2. Here it is evident that the gas formation is linked to the onset of a slight shrinkage, and the maximum of the pressure peak coincides with the shrinkage maximum. The exact position of the peak maximum apparently depends on the material: while in the case of Fe-0.8%C with all specimens tested the gas peak maximum was in the range 720-730°C, in the case of alloy steels - as with Cr steels shown in Fig.2 - gas formation occurs at slightly lower temperatures.

However, the shrinkage closely below the α-γ transition has been found with all types of carbon steels investigated so far - Fe-C, Fe-Mo-C, Fe-Cr-C, and Fe-Cr-Mo-C. Typically the break in the curve is more pronounced with high density specimens since with parts compacted at low pressure the „natural" shrinkage occurring in the α range near 700°C partly masks the „abnormal"shrinkage linked to the gas formation. The reason for this shrinkage can be found in the activation of sintering through presence of freshly reduced iron surfaces.

It can thus be concluded that with all PM carbon containing steels, gas formation is most pronounced in the α range. Thus, deoxidation through reaction of the oxides with carbon seems to take place already at temperatures at which no austenite has yet been formed and at which the solubility of carbon thus is still very small. However, it was still to be checked if the gas formation is actually due to reaction of oxide and carbon.

3. Analysis of the Gases Formed

For chemical analysis of the gases generated during the sintering process mass spectrometry was employed which has been used to advantage with cemented carbides (Leitner, Heinrich, Görting, 1995). It was combined with DTA / thermobalance using a high temperature skimmer coupling system (Netzsch STA 409/QMG 420) in order to check the mass loss through gas formation and to define the α-γ transition clearly.

The material tested was Mo alloy steel Fe-1.5%Mo-0.7%C prepared from elemental powder mixture. The compacts were pressed with addition of lubricant and de-waxed afterwards. The test runs were performed by heating the specimens at 10 K.min⁻¹ up to 850°C in helium (purity 99.999%).

The thermobalance graphs show that the most pronounced mass loss also here occurs at about 710°C (Fig.3a) which is well below the α-γ transition which in the DTA graph is found to occur above 760°C (Fig.3b). This agrees very well with the results obtained on Fe-0.8%C.

When the MS graphs (ionization currents for selected mass numbers) are evaluated it can be stated also here that the most pronounced peaks are found at about 710°C (m28 and m44 in Fig.4). Other measured peaks - such as that found at 360°C in the m44 curve - correspond mainly to dewaxing effects at lower temperatures.

The intensity curves allow a qualitative discussion of the existence of different volatile components; a quantitative comparison of the contents of gases is more complicated (because of the temperature dependence of the viscosity for the different gases). The highest intensities are found with mass numbers m28 (CO or N_2) and, though markedly less pronounced, m44 (CO_2). Intensity curves for other mass numbers also reveal clear peaks at 710°C: m12 (C) - ever found when CO is present - and small peaks for m14 (N) and m16 (O). In addition it must be considered that, caused by methodical reasons, the base lines of the intensity curves are quite different for the various mass numbers. E.g. the decreasing intensity for the m28 curve up to temperatures of 600°C is found in each run and does not represent a gas evolution. The pronounced peak of the m28 curve at 710°C corresponds to the formation of CO because also the m12 curve (C) shows a pronounced peak at that temperature while the intensity of m14 (which is proof for the existence of nitrogen) shows only a very small effect.

Furthermore it must be taken into consideration that C, CO and CO_2 are linked through the Boudouard equilibrium. Under standard conditions, i.e. at 1 bar pressure, the equilibrium ratio of the partial pressures of CO and CO_2 is about 60:40 at 700°C but at low pressures the equilibrium is shifted towards CO. (Frohberg, 1981).

The CO_2 graph (mass 44) exhibits a pronounced peak also in the temperature range 350-500°C (Fig.4b) without any counterpart at mass 28. This indicates that in this temperature range Boudouard's equilibrium is actually in favour of CO_2.

If the specimens are run twice without any contact of the sample with air, in the second run the mass loss and the MS peaks at mass 28 are markedly smaller, only the peak at mass 44 is comparable to that in the first run. The temperature at which these peaks are found is markedly lower, about 670°C,

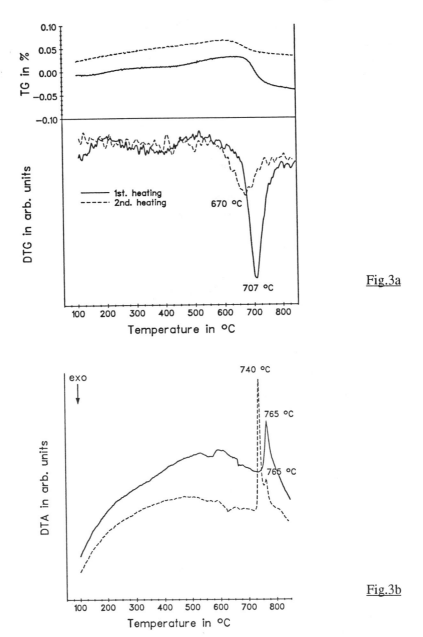

Fig.3a

Fig.3b

Fig.3: Thermal analysis (thermobalance / DTA) on Fe-1.5%Mo-0.7%C prepared from elemental powders.

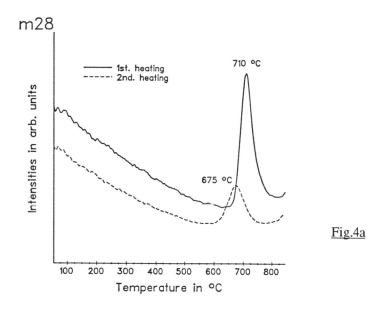

Fig.4a

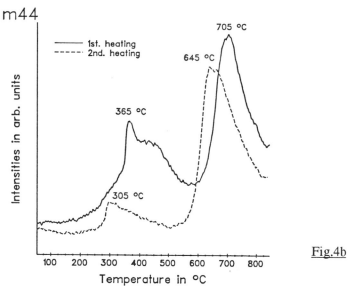

Fig.4b

<u>Fig.4:</u> MS graphs of Fe-1.5%Mo-0.7%C as a function of the temperature.

which is understandable since also the α-γ transition here occurs at lower temperatures (737°C), due to carbon here being present already as fine and evenly distributed cementite. Also here however gas formation occurs definitely in the α range.

4. Conclusions

During sintering of carbon containing steels, the most pronounced gas formation is observed closely below the α-γ transition, in the temperature range 700-730°C, but definitely still in the α range. This gas formation coincides with a slight shrinkage, indicating activation of sintering through freshly reduced Fe surfaces. TG/DTA analysis combined with mass spectrometry showed that the gas formed in this temperature range is mostly CO with some CO_2, fulfilling Boudouard's equilibrium.

This indicates that the reduction of oxide layers covering the iron particles by formation of CO/CO_2 not necessarily requires dissolution of carbon in austenite but that apparently already the very low solubility in ferrite - enhanced probably by surface diffusion - is sufficient to cause reaction with the surface oxides.

5. Acknowledgment

The authors thank Mr. K. Jaenicke-Rößler for carrying out the mass spectrometry measurements.

6. References

Ed. Houdremont: Handbuch der Sonderstahlkunde (Handbook of Special Steels), 3rd Ed., Springer-Verlag Berlin (1956)

N.Dautzenberg, J.Hewing: Proc. 4th Europ. PM Symp. Grenoble (1975) 3-9-1

G.Leitner, K.Jaenicke-Rößler, H.Wagner: Proc. Int. Conf. on Advances in Hard Metals Production, Bonn 1992, MPR Publ. Services ed., Shrewsbury (1993) Paper No.13

H.Klemm: Die Gefüge des Eisen-Kohlenstoff-Diagramms (The Microstructures of the Iron-Carbon System). 7th Ed., VEB DVG, Leipzig (1987)

G.Leitner, W.Heinrich, K.Görting: Proc. 1995 Int. Conf. Powder Metall. and Partic. Mat.,. Seattle, MPIF ed, Princeton (1995) in press

M.G.Frohberg; Thermodynamik für Metallurgen und Werkstofftechniker (Thermodynamics for Metallurgists and Materials Scientists). VEB DVG, Leipzig (1981)

Effects of WB Addition on Sinterability of Cr_3C_2 Cermets

Kazunori Nakano, Masaru Inoue, and Ken-ichi Takagi [*]

Technical Research Laboratory, Toyo Kohan Co. , Ltd.
Kudamatsu, Yamaguchi, JAPAN,
[*] Technical Department, Toyo Kohan Co. , Ltd. , Tokyo, JAPAN

Abstract

Coating of Cr_3C_2 cermets on steels has been attempted by using a sinter-bonding technique. This coating process requires that the sintering of the cermet and the bonding to the steel occur simultaneously. Lowering the sintering temperature is hence essential to minimize the degradation of the steel properties. Additions of WB to the Cr_3C_2 cermets result in the densification temperatures lower than 1450K, while the pseudo-binary Cr_3C_2 -Ni is fully densified at as high as 1550K. The liquid phase sintering behavior of Cr_3C_2-Ni-WB cermets was investigated by means of thermal dilatometry. The lower densification temperatures were attributed to liquid phase formation as the result of the WB-Ni eutectic-like reaction prior to the Cr_3C_2-Ni reaction.

I. Introduction

In spite of outstanding corrosion resistance, particularly oxidation resistance at high temperatures[1, 2], Cr_3C_2 cermets have been found in limited applications due to the poor mechanical properties. Cr_3C_2 powder has been mainly utilized in a thermal-spraying process for surface protection, but the problems of porosity and chipping from base material have not been sufficiently solved[3]. An exploitation of forming a Cr_3C_2-Ni cermet layer on various steels by a powder metallurgical technique, so-called sinter-bonding has been attempted to utilize the attractive properties of Cr_3C_2. Sinter-bonding requires that the sintering of the cermet and the bonding to the steel take place in a similar temperature range. The degradation of the steel during the processing is a great concern, and reducing the sintering temperature is hence indispensable for this purpose. It has been known that boron, in general, is effective to reduce the sintering temperature of cemented carbides[4]. Moreover, the WB-Ni pseudo binary system has been reported to generate a liquid phase at a relatively low temperature[5]. The authors had attempted to produce a Cr_3C_2 cermet with a Ni metal matrix. The full densification of the cermet required heating up to at least 1550K, which is too high for low carbon steels to be sinter-bonded with the cermet. It is anticipated that additions of WB to the Cr$_3$C$_2$-Ni cermet can lead to decrease in the sintering temperature.

In this paper, the sintering behavior of pseudo-ternary Cr$_3$C$_2$-Ni-WB cermets was investigated by means of thermal dilatometry, and thus the sinter-bondability to low carbon steels was assessed.

II. Experimental Procedures

Commercially available Cr$_3$C$_2$ (mean particle size:2. 8 μ m), WB(3. 9 μ m), Ni(3~7 μ m) and Cr(80 μ m) powders were mixed as shown in Table 1. Additions of five levels of WB were made to examine the effects of WB on the sintering behavior of the Cr$_3$C$_2$-Ni system. A small amount of Cr was added to compensate for free carbon. The premixed powders were ball milled in acetone to an average particle size of about 1 μ m. The ball milled powders were compacted in a hydraulic press at 98 MPa. The green compacts were sintered in a vacuum ($(2\sim3) \times 10^{-1}$ Pa) for 1. 2ks at temperatures between 1375K and 1575K with an interval of 25K. The sintered bodies were subjected to transverse rupture testing, hardness and density measurements and metallographic observations. The thermal dilatometry were performed to investigate the sintering mechanisms[6]. Sinter-bonding to a low carbon steel was assessed by observasions of the interface between the cermet and the steel.

Table 1 Compositions of the cermets(mass%).

Cr$_3$C$_2$	Ni	WB	Cr
69. 0	30. 0	0. 0	1. 0
67. 5	30. 0	1. 5	1. 0
66. 0	30. 0	3. 0	1. 0
64. 0	30. 0	5. 0	1. 0
61. 0	30. 0	8. 0	1. 0

III. Results and Discussion

Strength and Microstructure

Figure 1 shows the transverse rupture strength(TRS), hardness and the sinterable temperature range of the cermets, as a function of WB content. The TRS and hardness are plotted for the highest values of each alloy obtained at sintering temperatures between 1375K and 1575K. The sinterable temperature range is defined as a temperature range where more than 90% of the highest TRS value for a given alloy was obtained. TRS slightly increases with a small addition of WB up to 3%, while hardness increases monotonically with WB content. The increase in hardness may also be attributed to solution hardning of the matrix phase by tungusten.

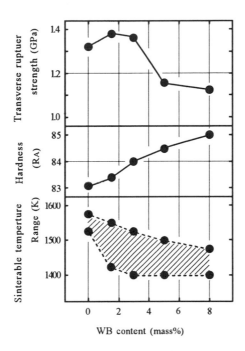

Fig. 1 Transverse rupture strength,
hardness and sinterable temperature
range as a function of WB content.

The sinterable temperature drastically decreases with WB additions and the range extends more than 100K. Based on the results of the mechanical properties and the sinterable temperature ranges, the optimum amount of the WB addition to the Cr_3C_2-Ni system is 3mass%.

Figure 2 shows the microstructures of the cermets with five different WB contents sintered at the optimum sintering temperatures which gives the highest TRS value. For the alloy without WB addition, the black particles represent Cr_3C_2, while the white area is the Ni metal matrix. The Cr_3C_2 particles in the alloys containing WB are surrounded by a gray area which increases with WB content.

Figure 3 shows the result of EDS analysis at three points for the Cr_3C_2-30mass%Ni-1mass%Cr-3mass%WB alloy. Tungsten was detected in the gray area(Point 2)and the Ni matrix phase(Point 3), while the core of the Cr_3C_2 particles(Point 1) does not contain tungsten. XRD results revealed that even the sintered 8mass%WB alloy consisted of the Cr_3C_2 phase and the Ni metal phase, and none of WB. It can be thus assumed that decomposition of WB and substitution of W with Cr in the Cr_3C_2 phase occurred during sintering.

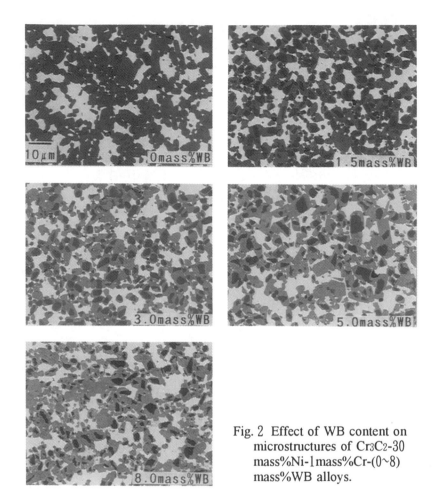

Fig. 2 Effect of WB content on microstructures of Cr3C2-30 mass%Ni-1mass%Cr-(0~8) mass%WB alloys.

Thermal Analyses

Figure 4 shows the results of the dilatometric measurements during a sintering cycle from the green state. The 3mass%WB alloy which exhibited a wider sinterable temperature range is shown as well as the 0mass%WB alloy for a reference. Both alloys commence gradual shrinkage at 1000K. The full density was obtained after roughly 20% shrinkage of the initial green compact dimension. The sharp contractions for the 0mass%WB and 3mass%WB alloys at 1300K and 1500K respectively indicate
the formation of a liquid phase. The liquid phase of the 0mass%WB alloy was produced by the eutectic-like reaction between Cr3C2 and Ni. On the otherhand, the 3mass%WB alloy can be sinterd to near the full density by the

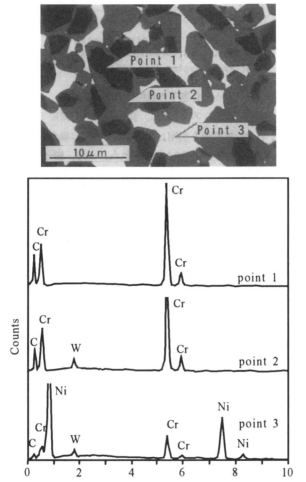

Fig. 3 SEM image and EDS spectra at three
analysis points from Cr₃C₂-30mass%Ni
-1mass%Cr-3mass%WB alloy.

liquid phase produced by the reaction between WB and Ni prior to the Cr₃C₂
-Ni reaction.

Sinter-bondability
        Figure 5 is a micrograph showing the sinter-bonded interface
between the Cr₃C₂-30mass%Ni-1mass%Cr-3mass%WB alloy and a low carbon
steel, JIS S25C(AISI 1025 equivalent). The carbide grains are etched dark by

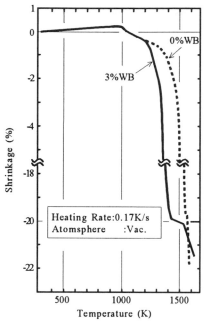

Fig. 4 Dimensional changes durring a sintering
cycle of Cr₃C₂-30mass%Ni-WB alloys from
the green state.

an etching reagent which colors carbides preferentially. Perfect metallurgical
bonding without any defects was achieved. On the other hand, in case of the
alloy without WB, a satisfactory sintered layer can not be obtained, because
an excessive reaction of the liquid with the steel occurred due to the high
sintering temperature (above 1550K).

IV. Conclusions

An addition of WB to Cr₃C₂-Ni cermets leads to densification at
temperatures lower than 1450K, providing the feasibility of a sinter-bonding
with a low carbon steel. A liquid phase formed by the WB-Ni eutectic-like
reaction prior to the Cr₃C₂-Ni reaction considerably reduces the sintering
temperature, and WB additions results in more than twice as wide sinterable
temperature ranges as the alloy without WB.

In the alloys containing WB, a structure with Cr₃C₂ particles
surrounded by a diffusional area containing W was formed, and W exists in
the Ni matrix phase as well. Transverse rupture strength increased with
additions of WB up to 3mass%. Based on the TRS values and the sinterable
temperatures, the optimum amount of the WB addition is 3mass%. Hardness
increases monotonically with WB content.

100 μ m

Fig. 5 Photomicrograph of sinter-bonded interface between
Cr_3C_2-1mass%Cr-30mass%Ni-3mass%WB alloy and JIS S25C.
(Etchant :$K_3[Fe(CN)_6]$, 10g+KOH, 30g+H_2O, 100ml)

References

1. Edmund K. Storms, " The Refractory Carbides ",
Academic Press, 102(1967)

2. Paul Schwarzkopf, Richard Kieffer, " CEMENTED CARBIDES", THE
MACMILLAN COMPANY, 177-181(1960)

3. I. Kvernes, E. Lugscheider, " Thermal Spraying ", P. M. I. **24**[1]7-13(1992)

4. P. Goeuriot, F. Thevenot, " Boron as Sintering Additive in Cemented
WC-Co(or Ni)alloys ", Ceramic Int. [13]99-103(1987)

5. Yu. B. Kuz'ma, M. V. Chepiga, " AN X-RAY DIFFRACTION
INVESTIGATION OF THE SYSTEMS Ti-Ni-B, Mo-Ni-B, AND W-Ni-B",
Poroshkovay Metalurgia, **82**[10]832-835(1969)

6. T. Ide, K. Nakano, K. Takagi, " Sintering Mechanisms of Iron Containing
Multiple Boride Base Hard Alloys ", J. Japan Soc. Powder and Powder Met.
39[4]247-253(1992)

SINTERING OF BINDER MODIFIED $Mo_2Fe(Ni)B_2$-Fe(Ni) CERMETS AND THEIR PROPERTIES

G.S. Upadhyaya[*] and P.K. Bagdi[**]

* Department of Materials & Metallurgical Engineering
** Materials Science Program
Indian Institute of Technology, Kanpur 208 016, INDIA

ABSTRACT

$Mo_2Fe(Ni)B_2$-Fe(Ni) (up to 37 vol.% binder) cermets were prepared by conventional powder metallurgy route after milling MoB and elemental powder premixes, compacting and sintering at $1350^{\circ}C$ in vacuum (10^{-5} MPa) for 1 hour. In addition to elemental binder a mixed binder of iron and nickel with equiatomic ratio was also selected. The densification process relates to typical reactive liquid phase sintering. Densification, and mechanical behaviours are described. It is found that nickel bonded cermets exhibit best room temperature transverse rupture strength (TRS).

1. INTRODUCTION

Molybdenum-Boron-Metal (Fe,Co,Ni) boride based hard materials possess, high hardness and excellent wear resistance properties. Their intrinsic physical, mechanical and tribological properties have attracted the attention of many researchers who utilized them in developing the next generation of hard materials [Takagi et al. 1986a,b]. Takagi et al. [1986 a,b] and Ide et al. [1988] investigated in detail the preparation and characterization of Mo_2FeB_2-Fe cermets. Their starting powders were FeB, MoB and iron. They sintered various compositions so as to obtain resultant phase Mo_2FeB_2, iron and Fe_2B.

Ide et al. [1988] found cermets containing 26-28 vol. % iron binder had a maximum TRS as high as 1900 MPa and hardness of 88 Ra. Komai et al. [1992] investigated Mo_2NiB_2-Ni cermets and showed that MoB, Ni and Mo below $955^\circ C$ formed the ternary boride by solid state sintering. However at temperature greater than $955^\circ C$, a quasi-eutectic reaction between Mo_2NiB_2 and nickel binder took place giving rise to considerable densification.

In the present investigation an attempt is made to prepare $Mo_2Fe(Ni)B_2$-Fe, Mo_2NiB_2-Ni cermets using MoB and elemental powder milled premixes and over a range of binder contents (0-37 vol.%). The densification, and mechanical properties were investigated.

2. EXPERIMENTAL PROCEDURE

$Mo_2Fe(Ni)B_2$ boride based cermets were prepared from the carbonyl iron, nickel and molybdenum boride (MoB) powders. The powder characteristics of the above powders used in the present research are given in Table 1.

Table 1: Various Powder Characteristics of the Carbonyl Iron, Nickel and Molybdenum Boride Powders

Property	Fe	Ni	MoB
Supplier	GAF	INCO	CERAC
Grade	SF	123	–
Av Particle size, μm	2.17	1.59	2.16
App. Density, Mg/m^3	2.26	2.03	1.98

The powder mixture compositions were adjusted in such a manner that the Mo_2FeB_2-α Fe, Mo_2NiB_2- Ni and $Mo_2(Fe_{0.5}Ni_{0.5})B_2$-$\gamma$(FeNi) cermets contained approximately equivalent volume fraction of the binder

phases.   The binder content of these cermets varied
from 0 to 37 vol. %.   Powder premixes along with 2 mass
% micronized wax were prepared by wet ball milling in
acetone in 'Fritsch Pulverisette-5' centrifugal type
ball mill using 19.95 mm ϕ WC balls for 38 hrs.   The
ratio of powder-to-ball by mass was kept at 1:5.   The
powder slurries were dried in a vacuum desiccator at
room temperature.   Rectangular green compacts of the
size 25.1 x 8.2 x 5.9 mm^3 were prepared from the dried
powder mixtures in a 50 Tons electric driven hydraulic
press using 330-340 MPa to achieve a 70% theoretical
density.   In the initial stage vacuum de-waxing
(pressure 1.01 x 10^{-5} MPa) was carried out at 380°C for
45 mins.   slow heating rate of 6°C/min was used for
this stage.   The second stage was vacuum sintering at
1350°C for 1 hr in the same furnace.   The heating rate
was 7°C/min.
        The Transverse Rupture Strengths (TRS) of the
sintered specimens, after final polishing to 2.5 μm
diamond finish, were evaluated in the three point
bending test using a fixture containing 3 mm ϕ WC
rollers having 15 mm length with a 15 mm span.
Indentation fracture toughness (K_c) of the sintered
cermets was measured according to the expression given
by [Palmquivst, 1957].

3. RESULTS

        As shown in Fig. 1 it is evident that the %
total sintered porosity decreases with the increasing
binder contant.   This is maximum in case of α Fe
binder.   The porosity values of cermets containing γ
(FeNi) binder is similar to that of Ni binder,
particularly in low volume fraction binders.
        $Mo_2Fe(Ni)B_2$-Fe(Ni) cermets exhibit a
significant increase in TRS with the increase in the
binder content as shown in Fig. 2.   The maximum value
of TRS in various cermet systems decreases in the
following order Ni $\longrightarrow$ γ(FeNi) $\longrightarrow$ α Fe binders.   The
scatter in the values is more in case of γ(FeNi) bonded
cermets as compared to straight elemental binders.   The
variation in the indentation fracture toughness of the

$Mo_2Fe(Ni)B_2$-Fe(Ni) cermets with the binder content is shown in Fig. 3. In all the cermet systems indentation fracture toughness increases with increase in the volume fraction of binders. Like TRS variation, the γ(FeNi) bonded cermets exhibit maximum scatter. It is interesting that after 24 vol. % γ(FeNi) binder there is a sudden drop in the value.

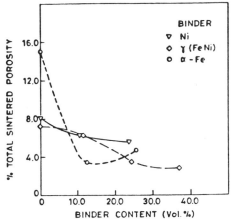

Fig. 1 : % total sintered porosity variation of $Mo_2Fe(Ni)B_2$-Fe(Ni) cermets with respect to binder content.

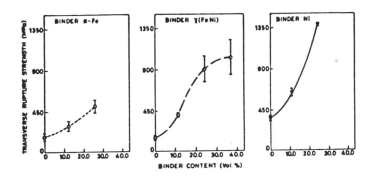

Fig. 2 : Transverse rupture strength variation of $Mo_2Fe(Ni)B_2$- Fe(Ni) cermets with respect to binder content.

4. DISCUSSION

Densification Behaviour : During reaction sintering
starting with powder mixture of MoB binary boride and
metal binder Fe(Ni), a ternary boride is formed at the
contacting points.   A quasi-binary liquid phase is
formed between the ternary boride (Mo_2MB_2) particles
and the metal binder phase.   This liquid phase so
formed promotes densification through the liquid phase
sintering mechanism.

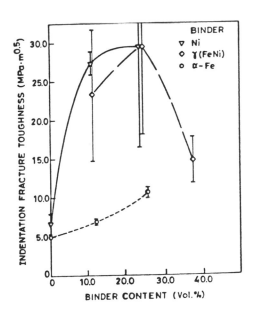

Fig. 3 : Indentation fracture toughness variation of
$Mo_2Fe(Ni)B_2$- Fe(Ni) cermets with respect to binder
content.

During synthesis of boride based cermets following
steps can be considered [Ide and Ando, 1989]:

* Formation of $Mo_2Fe(Ni)B_2$ from MoB and Fe(Ni) by
   the solid state reaction.
* Formation of liquid phase, which helps in particle
   rearrangement during densification by the
   reaction.

$$\text{Austenite} + L_1 + Mo_2MB_2 \longrightarrow L_2 + Mo_2MB_2$$

The temperature at which this reaction takes place is dependent on the type of binder. In case of iron binder the temperature is $1202°C$ [Ide and Ando, 1989], while in case of nickel it is above $955°C$ [Komai et. al. 1992]. This indicates that after sintering at $1350°C$ densification in nickel- bonded cermets should be better than that in iron bonded ones. As a matter of fact this is evidenced by better Ostwald ripening in case of nickel bonded cermets. However this is not reflected by the sintered porosity variation, (Fig. 1). One of the probable reason appears the fact that the calculation of theoretical densities assuming the rule of mixture in such reactive systems is on erroneous presumption. Another supporting feature for the better sinterability in case of nickel bonded cermets can be guessed from the fact that the contact melting temperature of MoB_2-Ni ($1000°C$) is lower than in case of MoB_2-Fe ($1100°C$) [Yurchenko, 1971].

Transverse Rupture Strength (TRS) and Fracture Toughness : The improvement in the TRS and fracture toughness of cermets with increase in binder content irrespective of the type of binder can be justified by two reasons. A better densification with an increase in the binder volume fraction contributes in enhancing the TRS of the cermets. It is interesting that TRS of the cermets decreases in the sequence Ni $\longrightarrow$ Fe/Ni $\longrightarrow$ Fe binder, which is also generally similar in case of fracture toughness. It is worth noting that the ultimate tensile strength of nickel (310 MPa) is greater to that of iron (193-276 MPa) [Everhart, 1971]. In addition, the % elongation of the former is somewhat higher than the latter. This suffices to conclude that the transverse rupture strength and fracture toughness of the Mo_2NiB_2-Ni cermets, in principle, should be greater than the MO_2FeB_2-Fe cermets. As a matter of fact, this has been experimentally confirmed. FCC nickel is known to be responsive to a large number of slip systems during plastic deformation as compared to solid solution strengthened Fe-Ni binary alloy [Everhart, 1971].

Betteridge [1977] although shows better strength of Fe-Ni alloy as compared to straight nickel. However such a feature is not supported from the present TRS variation (Fig. 3), where the fall in TRS is in the sequence Ni $\longrightarrow$ γ(Ni/Fe) $\longrightarrow$ Fe. This can be related with the probable segregation of intermetallics around the stacking faults in Fe-Ni alloy, thus lowering the energy. Such a feature of solute segregation was noticed by Howie and Swann [1961].

5. CONCLUSIONS

Sintered porosity of ternary borides decreases in order Mo_2FeB_2 - Mo_2NiB_2 $\rightarrow$ $Mo_2(FeNi)B_2$, while in their based cermets the order is reverse. The transverse rupture strength and indentation fracture toughness of cermets increases in the binder sequence as Fe $\longrightarrow$ (FeNi) $\longrightarrow$ Ni.

REFERENCES

W. Betteridge, Nickel and its Alloys, Macdonald and Evan, Plymouth, p. 65, (1977).

J. L. Everhart, Properties of Ni- and Ni- alloys, Plenum Press, New York, 196 (1971).

A. Howie and P.R. Swann, Phil. Mag., Series 8, Vol. 6, 1215, (1961).

T. Ide, K. Nakano and T. Ando, 'Effect of Mo Contact and sintering temperature on the strength, hardness and density of Fe- 6-mass% B-Xmass % Mo Alloys', Powder Metallurgy Int., 20[3] 21-24 (1988).

T. Ide and T. Ando, 'Reaction study of an Fe-6 wt % B-48 wt% Mo alloy in presence of liquid phases', Metal Trans., 20A, 17-24 (1989).

M. Komai, Y. Yamasaki and K. Takagi, 'Sintering behaviour of a reactive sintered ternary boride base cermet', Sintering '91, ACD Chakladar and J. A. Lind (Eds.), Trans Tech Publication, Aedermannsdorf, 531 (1992).

S. Palmquivst, Jernkont Ann., Vol. 141, 300 (1957).

K. Takagi, M. Komai, T. Ide, T. Watenabe and Y. Kondo, 'Characterization of the Mo_2FeB_2 type hard phase in an iron containing multiple boride base hard alloy', Harizons in Powder Metallurgy', Part II, W.A. Kaysser and W.J. Huppmann (Eds.), Verlag Schmid, Freiburg, pp 1077 - 1082 (1986a).

K. Takagi, T. Watanabe, T. Ando and Y. Kondo, 'Effect of molybedenum and carbon on the properties of iron molybdenum boride hard alloys', Int. J. of Powder Metall, 22[2] 91-96 (1986b).

O. C. Yurchenko, "Investigation of iron and nickel stability during heating in contact with refractory compounds", Poroshkovafya Metallurgia, [1]45-49(1971) [in Russian].

Sintering of SiCN -Nanocomposite Materials

M. Herrmann, E. Rupp and Chr. Schubert

Fraunhofer Institute of Ceramic Technologies and Sintered Materials,
Winterbergstr. 28 D-01277 Dresden , Germany

Abstract

Silicon nitride materials reinforced by nanozised silicon carbide are one of the promising candidates for high temperature applications. The densification of these materials is complicated due to the chemical interaction of the SiC with the silicon nitride and the oxynitride liquid phase. The influence of the amount and grain size of SiC on the densification was analysed based on thermodynamic considerations, experimental observed sintering curves and the grain growth during sintering. It will be shown that it is possible to produce dense nano-composite materials with excellent creep resistance by gas pressure sintering.

I. Introduction

Silicon nitride materials have been intensively studied for many years for their potential application as structural ceramics at both room and elevated temperatures. This is due to the combination of excellent mechanical properties with a good corrosion resistance. However the application at temperatures above 1000°C is still limited by the relative high creep rate and subcritical crack growth compared to other ceramic materials such as SiC [Niihara, 1991, 1992]. This is the consequence of the softening of the glassy phase at the grain boundaries. A possibility to overcome this problem is the development of Si_3N_4/SiC nano-and microcomposites. These materials containing finely dispersed SiC particles in the Si_3N_4-matrix exhibit high mechanical properties at room and elevated temperature. The densification of these materials is retarded due to the interaction between SiC and silicon nitride and the oxynitride liquid phase during sintering. That is why hot pressing and HIP-ing was used for the densification up to now [Niihara 1992; Watari 1989].

The purpose of the paper is to investigate sintering behaviour during gas pressure sintering and hot pressing of Si_3N_4-SiC nanocomposites from different powders taking in account the interaction with the sintering atmosphere.

2. Experimental methods

The Si_3N_4/SiC-composites were produced by mixing a commercial Si_3N_4-powder (SN E10 UBE, Japan) with different SiC-powders (B20: HC Starck, BET 20.5 m^2/g, O-content 1.6 %; UF25: HC Starck, BET 25.3 m^2/g, O-content 2.3 %; UF45: HC Starck, BET 43.7 m2/g, O-content 4.7 %). Additionally a plasmachemical synthesized SiC-powder [Zalite 1992] (SiC 80: 84 % SiC, 16 % Si_3N_4; BET 10.5 m^2/g; O- content 0.8 %) from the Institute of Inorganic Chemistry in Riga, Latvia was used. Sintering additives are Y_2O_3 (grade fine, HC Starck, Germany) and Yb_2O_3 (Johnson Matthey Company, United Kingdom). A homogeneous mixture was carried out by milling in an isopropanolic solution in a planetary mill. The powder mixtures were calcinated and hot pressed at 1800 °C for 1h or cold isostatic pressed (25x25x60 mm) and gas pressure sintered at 1900 °C for 1.5h. The sintering was realised using a gas pressure sintering furnace (up to 100 bar). In order to minimise the evaporation the specimens were placed in RBSN-crucibles.

The grain size of the Si_3N_4- and SiC- grains was examined at plasmachemically etched polished sections. The etching time for the determination of SiC and Si_3N_4-grains were different. The evaluation of the grain size were carried out using SEM micrographs (magnification 5000x and 10000x; number of grains: 400-800). The phase content was determined by X-ray diffraction analysis of the grinded samples.

3. Results and Discussion
3.1 Gas pressure sintering and interaction with the sintering atmosphere

A condition necessary for a reproducible production of Si_3N_4/SiC nano- and microcomposites is the thermodynamic stability of the composites and a controled interaction of the materials with the atmosphere during sintering. A detailed analysis of the thermodynamic calculations of the interaction is given in [Herrmann 1995]. SiC and Si_3N_4 are stable in the temperature and nitrogen pressure region between the curves I and II in Fig 1. These lines are limits of the thermodynamic stability of SiC and Si_3N_4 and are given by the equilibrium:

$$Si_3N_4 \text{ (s)} \Leftrightarrow 3 \text{ Si (l)} + 2 N_2 \text{ (g)} \quad (1) \text{ (lower limit, s- solid,}$$
$$\text{l - liquid, g,- gas)}$$

and

$$3 \text{ SiC (s)} + 2N_2\text{(g)} \Leftrightarrow Si_3N_4 \text{ (s)} + 3 \text{ C (s)} \quad (2) \text{ (upper limit)}$$

The limits may differ slightly due to the different thermodynamic data, the state of the carbon or the grain size of the component [5]. Additionally the interaction between SiC and Si_3N_4 with the sintering additives and SiO_2, existing on the surface of silicon nitride and taking part at the formation of the liquid, must be taken into account. The interactions of the SiC with the liquid during sintering can be described in a good approximation by the following reactions [Herrmann 1995]:

$$2 \text{ SiC (s)} + SiO_2\text{(l)} + 2 N_2\text{(g)} \Leftrightarrow Si_3N_4 \text{ (s)} + 2 \text{ CO (g)} \quad (4)$$
$$Si_3N_4 \text{ (s)} + 3SiO_2\text{(l)} \Leftrightarrow 6 \text{ SiO (g)} + 2 N_2 \text{ (g)} \quad (5)$$
$$SiC \text{ (s)} + 2SiO_2\text{(l)} \Leftrightarrow 3 \text{ SiO(g)} + \text{ CO (g)} \quad (6)$$

The reaction (6) follows from the reaction (4) and (5) by subtraction. All of the reactions (4) - (6) reduce the amount of liquid phase necessary for sintering and result in the formation of a gas phase stabilising the pores during sintering.

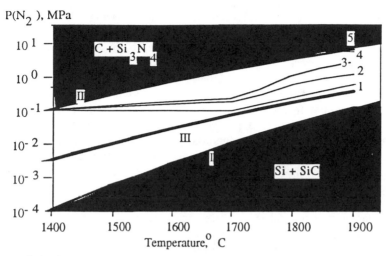

Fig 1: Calculated stability range of Si_3N_4/SiC-composites (between line I and II) and the calculated nitrogen pressure, at which the decomposition of the oxynitride liquid is minimised (curve III). (1-5 - temperature depending nitrogen gas pressure during sintering cycles 1-5)

The partial pressures of CO and SiO increase with rising activity of SiO_2 in the oxynitride liquid, however the ratio of the partial pressures is independent of the activity of SiO_2. According to equation (4) the CO-partial pressure increases with increasing nitrogen pressure. The decomposition of the liquid phase and the formation of SiO according reaction (5) decreases with increasing nitrogen pressure. This means that there must be a temperature depending nitrogen pressure, at which the amount of CO and SiO in the sintering atmosphere, e.g. the amount of decomposed liquid phase, is minimal. This nitrogen pressure corresponds to the curve III in the Fig 1. A detailed deducing of the pressure is given elsewhere [Herrmann 1995]. At decreasing temperatures the amount of decomposed liquid phase rapidly decreases. At temperatures below 1500 °C the partial pressure of CO and SiO are very low and on the other hand the reactions are hindered by kinetic reasons, e.g. the decomposition at temperatures below 1500 °C can be under normal sintering conditions neglected.

Different sintering cycles with variation of the time-temperature-pressure-regime were tested to confirm the thermodynamic calculations (Table 1; Fig. 1). It is visible, that with increasing SiC content the densification is retarded. On the other hand the weight loss grows with increasing SiC content. The sintering cycle 1 which is near the ideal regime shows the lowest weight loss during sintering (Fig. 2). The weight change and the decomposition is higher for all other sintering cycles. The highest degree of decomposition takes place at sintering cycle 4 and 5, when the nitrogen pressure is higher than the equilibrium pressure for the SiC decomposition (curve II fig 1). In this case we found a nearly complete decomposition of the SiC into carbon (Table 1). The samples sintered at this cycles (cycle 4,5) shows an increase in weight. This is in good agreement with the theoretical increase of weight according to reaction (2) (15,5wt.%). Additionally nearly all SiO_2 is reduced by the carbon and after sintering a oxynitride grain boundary phase is crystallized (Table 1).The sintering cycles 2a..c differ only in the second sintering step. Increasing nitrogen pressure during the second sintering step decrease the weight loss (Fig. 2) . This seems not to be in agreement with predictions based on thermodynamic calculations. The reason of this discrepancy is the overlapping of the sinterhip-effect with the decomposition effect. At the time when the gas pressure is built up, the samples have mostly closed porosity and only a small surface area for decomposition exist. During the period of a high porosity where the decomposition more intensive takes place all three sintering cycles 2a-3c have the same temperature-pressure-time regime. Dense materials tempered at high temperatures and nitrogen pressures decompose only at the surface region (cycle 2a+5 Fig. 2) because the nitrogen necessary for decomposition must deffuse through the body. Probably the diffusion of nitrogen limit the rate of reaction. This is also in agreement with the results of post-HIPing of Si_3N_4/SiC-composites [Ukyo 1993].

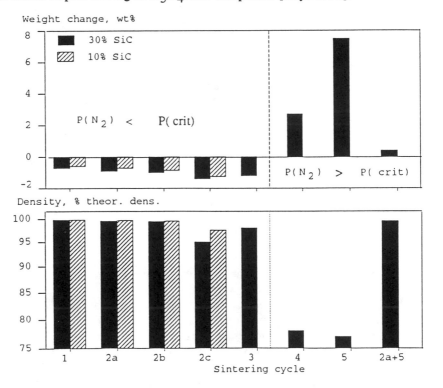

Fig. 2: Weight change and relative density of the sintered samples with different SiC -content and sintering cycles (additive Yb_2O_3; P(crit).- decomposition pressure of SiC)

Table. 1: Composition of the gas pressure sintered materials for different sintering cycles (Cycle 2a- c have only a different pressure in the second sintering step.
The maximum pressure was built up during the isothermal sintering period at
1900°C. The gas pressure during the heating up period is the same (curve 2 Fig.
1); cycle No. 1-5 - curve No 1-5 in the Fig 1.)

Material	Additive wt %	SiC wt-%	Sintering cycle No.	P_{max}, MPa	Phase composition of grain boundary
Yb1	15 Yb_2O_3	10 (B20)	2a	10	amorphous
			2b	40	amorphous
			2c	60	amorphous
Yb2	15 Yb_2O_3	30 (B20)	1	60	amorphous
			2a	10	amorphous
			2b	40	amorphous
			2c	60	amorphous
			3	60	amorphous
			4	60	SiC (reduced amount), $Yb_4Si_2O_7N_2$
			5	80	SiC(traces), $Yb_4Si_2O_7N_2$
Y1	10 Y_2O_3	30 (B20)	2a	10	$Y_2Si_2O_7$
Y2	10 Y_2O_3	30 (B20)	2a	10	$Y_2Si_2O_7$(tr), amorphous

In furnaces with carbon heater additionally CO can be produced by oxygen and water vapour impurities of the sintering atmosphere. If the CO- partial
pressure in the furnace was produced by other sources then by reaction (5) and at
least near the equilibrium pressure of reaction (5) the decomposition of the
samples will be deminished. In our investigation these conditions could not be
found due to the low impurity content in the nitrogen and the protection effect of
the RBSN crucibles. The necessary sintering atmosphere was built up only in the
crucible which allows a low exchange. A indication for such an interpretation of
the results is also, that in all sintering cycles the samples have a surface layer
with a reduced SiC-content.

3.2 Hot pressing

Due to the fast densification during the hot pressing the interaction with
the atmosphere is reduced. The densification behaviour is very similar to that of
the materials without SiC (Fig. 3, 4). After formation of the liquid and the application of the load, the densification starts. The densification in this region (1500-
1600 °C) depends on the amount and viscosity of the formed liquid [Herrmann
1995].

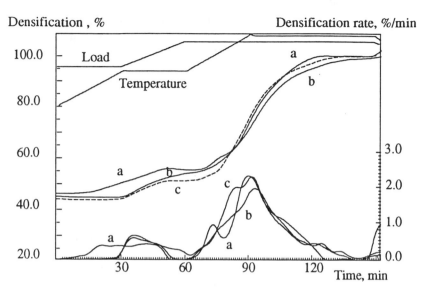

Fig. 3: Densification and densification rate during hot pressing of the materials containing 30 wt-% of different SiC powders a) without SiC b) UF25, c) UF45 with 8 wt.-% Y_2O_3 as sintering additive

Table 2: Composition and microstructure of the hot pressed materials ($d_{50}(Si_3N_4)$ thickness of the Si_3N_4-grains, $d_{50}(SiC)$ -equivalent circle diameter of SiC grains; [1] additionally heat treated at 1800°C for 90 min,[2] intensity ratio of the x-ray peaks of $YSiO_2N$ / Si_3N_4

Material	Additive wt.- %	SiC wt.-%	$d_{50}(SiC)$, µm	$d_{50}(Si_3N_4)$, µm	Phase composition of grain boundary
7Y	7 Y_2O_3	-		0.12	$YSiO_2N$ $(0,63)^{[2]}$
7Y5S	7 Y_2O_3	5 (SiC80)	0.19	0.12	$YSiO_2N$ $(0,45)^{[2]}$; $Y_5(SiO_4)_3N($ tr)
7Y10S	7 Y_2O_3	10 (SiC80)			$YSiO_2N$ $(0,27)^{[2]}$, $Y_5(SiO_4)_3N$
7Y15S	7 Y_2O_3	15 (SiC80)	0.16	0.12	$YSiO_2N$ $(0,46)^{[2]}$ $Y_5(SiO_4)_3N$
7Y20S	7 Y_2O_3	20 (SiC80)			$YSiO_2N$ $(0,6)^{[2]}$
8Y25S	8 Y_2O_3	25 (SiC80)	0.16	0.12	$YSiO_2N$ $(0,5)^{[2]}$
8Y25St[1]	8 Y_2O_3	25 (SiC80)	0.26	0.38	$YSiO_2N$ $(0,5)^{[2]}$
8Y	8 Y_2O_3	-			$YSiO_2N$
8Y30S1[1]	8 Y_2O_3	30 (UF25)			$Y_5(SiO_4)_3N$, $Y_2Si_2O_7$
8Y30S2	8 Y_2O_3	30 (UF45)			$Y_5(SiO_4)_3N$, $Y_2Si_2O_7$

The materials exhibit an additional maximum of the densification rate at 1600 to 1640 °C. This maximum is connected with a secondary rearrangement of the powder after starting of the solution-precipitation process of the Si_3N_4 or with the decreasing densification rate behind the maximum is the result of the formation of oxynitrides reducing the amount of the liquid phase. The main maximum in the temperature range of 1750 -1800 °C is connected with the solution-precipitation process of the Si_3N_4.

The materials with a SiC content of 5 to 15 % made from the powder SiC80 densify faster then the materials with a higher SiC-content or the SiC free materials. This is connected with the changing oxygen content on the one hand and the retarded densification due to the SiC content on the other hand.

The highest oxygen content was found in materials with 10 wt.% SiC. An indication for this is also the changing amount of crystallized $YSiO_2N$ in the materials (Tab. 2). The material with 10% SiC has the lowest amount of crystallized $YSiO_2N$. Additionally in this material the more oxygen rich phase with apatit structure crystallize.

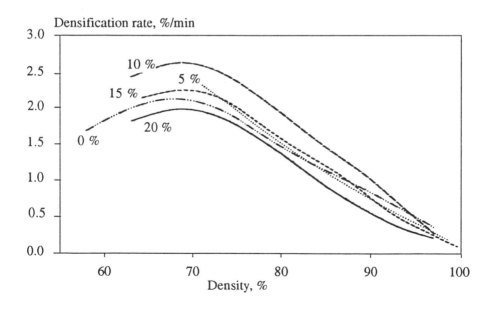

Fig. 4: Densification rate during the isothermal period of hot pressing of the materials containing different amounts of SiC powders SiC80 (7 wt. % Y_2O_3 as sintering additive).

By increasing SiC content (Fig. 4) and decreasing SiC particle size the densification rate decreases (under constant conditions, e. g. additive and oxygen content). However a smaller grain size of the SiC usually corresponds with a higher oxygen content (UF25, UF45). This increase is the reason for the better densification of the material containing UF45 in comparison to the material with UF 25 and the SiC free material (Fig. 3). The investigations of the microstructure of the as hot pressed and the heat treated materials show that the SiC grows only with a smaller rate then the silicon nitride grains, but takes part at the solution-diffusion-precipitation process. With increasing grain size of silicon nitride the amount of SiC inside the grains increases too. However the amount of the SiC inside the grains is lower then 5% of the total amount of SiC e.g. most of the SiC -particles are located in the grain boundary of the materials, where they are most effective for creep resistance. The investigated materials show creep rates at 1450 °C and 100 MPa load lower than 3×10^{-9} s^{-1} (under bending conditions).

4. Conclusions

The densification of silicon nitride materials reinforced by silicon carbide is influenced by the chemical interaction of the SiC with the silicon nitride and the oxynitride liquid phase. The preparation of dense nanocomposite materials with excellent creep resistance by gas pressure sintering is possible.

Increasing amount and reducing grain size of SiC retard the densification. The ratio of the sintering additive to the total silica content in the mixture has an important influence on the densification.

5. Acknowledgment

The work was supported by the DFG (contract number He 1958 3/3).

6. Literature

1.   K.Niihara, „New Design Concept of Structural Ceramics- Ceramic Nanocomposites" , J. Cer. Soc. Jap., Int.End., **99**, 945-953 (1991)

2.   K. Niihara, T. Hirano, A. Nakahara, K. Ojima, K. Izaki, T. Ka-wakami:"High-Temperature Performance of Si_3N_4-SiC Composites from fine, amorphous Si-C-N Powder", Proc. Int. MRS Meeting on Advanced Materials Tokio 107-122 (1988)

3.   K.Watari, K. Ishizaki, „Influence of gas pressure on HIP sintered Silicon nitride and stability of carbon impurity", J. Ceram. Soc. Jpn. Inter. Ed. **96** ,535-540 (1988)

4.   I.Zalite, G.Boden, Chr.Schubert, A.Lodzina, J.Plitmanis, T.Millers, „Sinterimg of Fine Silicon Nitride Powders" ,Latvian Chemical Journal, **2**, 152-59 (1992)

5.   M. Herrmann, Chr, Schubert, A. Rendtel, H. Hübner, „Si_3N_4/SiC Nanocomposite Materials", Submitted to J. Am. Ceram Soc.

6.   Y.Ukyo, T. Kandori, S. Wada, „Si_3N_4 -SiC Composite Consolidated by Post-HIPing Process", J.Cer.Soc. Jap. , Int. Edition, **101**, 1398-1400 (1993)

Strength and Microstructure Evaluation of Si_3N_4/SiC Composites

Masakazu Watanabe and Katsura Matsubara

R&D Center, NTK Technical Ceramics, NGK Spark Plug Co., Ltd.
2808 Iwasaki, Komaki, Aichi, 485, Japan

Abstract

Dense Si_3N_4/SiC (10, 15, 20 and 25 weight%) composites were prepared by hot–isostatic–pressing after gas pressure sintering, using commercially available silicon carbide powder with the average particle size of 0.3 μm. The grain growth of β silicon nitride during the densification was suppressed by the addition of SiC. As the result, the microstructure became finer and the flexure strength was improved. However, the strengths of the Si_3N_4/SiC composites and the Si_3N_4 monolith began to deteriorate at a same temperature and the creep strain of the Si_3N_4/SiC composite was larger than that of the Si_3N_4 monolith. In this study, it was concluded that the addition of SiC is effective for the flexure strength improvement but not for the heat resistance.

I. Introduction

Since silicon nitride ceramics have an excellent heat resistance, they have been already applied to high temperature structural parts, such as turbocharger rotors and glow plugs. From a viewpoint of environmental pollution, it is to be desired that engines should be run at higher temperatures, so that the strength of the silicon nitride ceramics at elevated temperatures must be further improved. The improvement has been attempted by adding SiC to Si_3N_4 as an approach.

Lange (1973) has shown that the strength of the hot–pressed Si_3N_4/SiC composite with 5 wt% MgO is twice as high as that of the Si_3N_4 monolith at 1400°C. Niihara et al. (1989a, 1989b, 1994) have reported that the hot–pressed Si_3N_4/30vol%SiC composite obtained from amorphous Si–C–N with 8 wt% Y_2O_3 maintains over 1000 MPa strength up to 1200°C. However, hot–pressing was, in many cases, necessary to obtain the dense Si_3N_4/SiC composites owing to the poor sinterability of these systems.

Akimune (1990) has succeeded in densifying the Si_3N_4/SiC composite with a relatively large amount of sintering additives, 10 wt% Y_2O_3 and 5 wt% Al_2O_3, by hot–isostatic–pressing process after pressureless sintering (post–HIPing process), but its high temperature properties have not been reported. Recently,

Ukyo et al. (1994) have prepared dense Si_3N_4/SiC composites containing up to 30 wt% SiC, from very fine SiC powder with the average particle size of about 0.03 μm, by post–HIPing process. The Si_3N_4/30wt%SiC composite with 10 wt% Y_2O_3 was very stable to temperature and maintained the r.t. strength up to 1400°C.

Some effects of an addition of SiC particles to silicon nitride have been suggested by many workers. One of them is to strengthen the Si_3N_4 monolith. This is because the dispersed SiC particles suppress the grain growth of β silicon nitride during the densification, and hence the microstructure is fined. Another effect is an improvement of the creep resistance. Several mechanisms for this creep behavior improvement have been proposed: devitrification of the glassy phase (K. Ramoul–Badache and M. Lancin, 1992), direct bonding between Si_3N_4 and SiC (Niihara et al., 1994), and forming of a three–dimensional network with increasing SiC contents (K. Yamada and N. Kamiya, 1994).

It is known that the Si_3N_4/SiC composites show the excellent high temperature properties compared with the Si_3N_4 monolith. However, the effects of the addition of SiC particles are still unclear. In the present report, we discuss the relationship between flexure strength and microstructure of the Si_3N_4/SiC composites densified in a post–HIPing process.

II. Experimental Procedure

Commercially available silicon nitride powder (specific surface area (SSA) $10 m^2/g$, SN–E10 made by UBE Kosan Co., Ltd., Tokyo, Japan) and silicon carbide powder (SSA $15 m^2/g$, Betarundum UF–grade made by IBIDEN Co., Ltd., Gifu, Japan) were used as starting powders. The 0, 10, 15, 20 and 25 wt% SiC powder was added to the Si_3N_4 powder with the same amount of sintering additives. As sintering additives, the total 10 wt% of rare earth oxide Er_2O_3 and transition metal oxide V_2O_5 and WO_3 of Va and VIa groups were used. The powders were wet–mixed, dried and granulated by sieving. The resultant powder mixtures were shaped into plates 60 x 60 x 30 mm by cold isostatic pressing. The powder compacts were pre–sintered at 1900°C for 4 h under 1 MPa nitrogen atmosphere and then hot–isostatic–pressed at 1800°C for 2 h under 100 MPa nitrogen atmosphere.

Densities were measured by Archimedes' immersion method. Specimens for flexure strength and creep behavior measurements were cut out from the plates and ground to rectangular bars of 3 x 4 x 35 mm with a 140–grid diamond wheel. The flexure strength was measured in 4–point bending according to Japanese Industrial Standard (JIS) R–1601 (The inner and outer spans were 10 mm and 30 mm, respectively.) at a cross head speed of 0.5 mm/min. The creep behavior was measured in 4–point bending under 250 MPa at 1300°C.

Specimens for SEM observation were ground and polished with a 3 μm diamond paste used for the final finish, and etched for 1 min by r.f. plasma of 500 W at a frequency of 13.56 MHz under the etching atmosphere of CF_4 + 10% O_2 mixture. The microstructure of the cross section was characterized with SEM and was digitized with an image analyzer (LUZEX III. produced by NIRECO, Tokyo, Japan). The maximum grain length and the area of the cross section of each grains were quantitated. Then, the grains were classified by the maximum grain length, and the value of (the sum of the grain area of the class)/(the sum of the total grain area) was adopted as a frequency of the grain size distribution. The values of m and σ were determined, as this grain size distribution is best–fitted to a log–normal distribution function in the range of $\ln m - \ln \sigma \leq \ln a \leq \ln m + \ln \sigma$. The parameter a is the particle size, the parameter m is the geometric mean size, and the parameter σ is the geometric standard deviation indicating the range of sizes.

III. Results and Discussion

Figure 1 shows the relative densities after pre–sintering and HIPing. When the SiC content was increased, the density after pre–sintering became lower. This indicates that the addition of SiC retards sintering. After HIPing, the Si₃N₄/SiC composites containing up to 25 wt% SiC could be fully densified. During the densification, the transition metal oxides of Va and VIa groups were found to react with silicon nitride and form silicides which have relatively high melting points (Y. Tajima et al., 1994). The XRD analysis for these composites

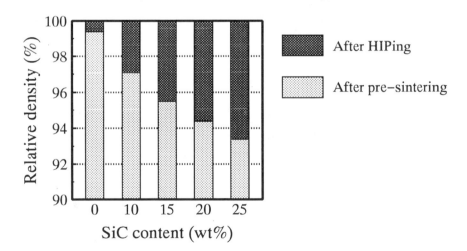

Fig. 1. Relative densities after pre–sintering and HIPing.

revealed the presence of β–SiC. A TEM micrograph of the Si₃N₄/20wt%SiC composite is shown in Fig. 2. It was found that SiC particles exist both in the Si₃N₄ grains and in the grain boundaries from the carbon analysis with EDS. The sizes of the SiC particles dispersed in the Si₃N₄ grains were below 300 nm. The Si₃N₄/SiC composites in this work had "nano–composite" microstructure.

Figure 3 shows SEM micrographs of the Si_3N_4 monolith and the $Si_3N_4/20wt\%SiC$ composite after 1 min of plasma–etching. The microstructure of the Si_3N_4/SiC composite was finer than that of the Si_3N_4 monolith. It suggests that the addition of SiC suppressed the grain growth of $\beta-Si_3N_4$ during the densification. In the composite microstructure, several particles were observed in the Si_3N_4 grains. These particles would be SiC on the basis of the microstructure observation with TEM and the report that the etching rate for SiC is considerably lower than that for Si_3N_4 (O. O. Ajayi et al., 1993). The grain size distributions

——— 500 nm

Fig. 2. TEM micrograph of $Si_3N_4/20wt\%SiC$.

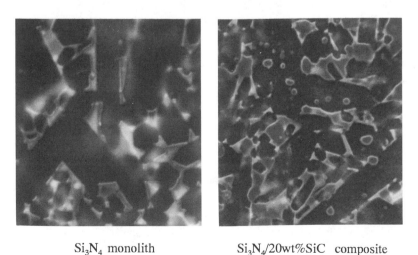

Si_3N_4 monolith $Si_3N_4/20wt\%SiC$ composite

——— 1 μm

Fig. 3. SEM micrographs of Si_3N_4 and $Si_3N_4/20wt\%SiC$.

of these samples are shown in Fig. 4. In determining the microstructure, SiC particles dispersed in the Si_3N_4 grains were neglected. In order to reflect the large Si_3N_4 grains, which were a few in the analyzed area, on the grain size distribution, area% of the cross section was adopted as a frequency of the grain size distribution. Table 1 summarizes the mean length m and the standard deviation σ determined from fitting the log–normal distribution function. The grain size distribution of the composite was narrow compared with that of the Si_3N_4 monolith, and the σ of the composite was somewhat smaller than that of the Si_3N_4 monolith. These results clarified that the addition of the SiC particles contributed to fining the microstructure.

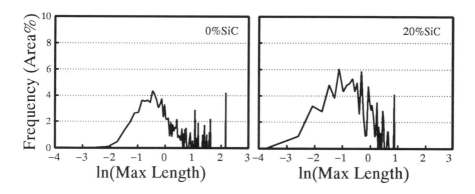

Fig. 4. Grain size distributions for the Si_3N_4 and the $Si_3N_4/20wt\%SiC$.

Table 1. Log–normal parameters for the Si_3N_4 and the $Si_3N_4/20wt\%SiC$.

LN Parameters	Si_3N_4	$Si_3N_4/20wt\%SiC$
m (μm)	1.05	0.61
σ (μm)	2.25	2.12

Figure 5 shows the relationship between the SiC content and flexure strength. As the SiC content was increased up to 20 wt%, the flexure strengths both at room temperature and at 1250°C were improved. The addition of 25 wt% SiC resulted in the deterioration of the strength at the both temperatures. This may be because the SiC particles are apt to agglomerate in the grain boundaries when the SiC content is large. The flexure strengths at various temperatures up to

1400°C for the Si₃N₄ monolith and the Si₃N₄/15wt%SiC composite are shown in Fig. 6. The flexure strength of the Si₃N₄/15wt%SiC composite was higher by 200 MPa than that of the Si₃N₄ monolith up to 1200°C, however, it deteriorated above 1200°C and became closer above 1300°C. The creep behavior of these samples at 1300°C under 250 MPa is shown in Fig. 7. The Si₃N₄ monolith fractured in a relatively short time, but the strain of the Si₃N₄ monolith was smaller than that of the Si₃N₄/SiC composite. It indicates that the addition of SiC lowers the creep resistance. The XRD analysis showed that the intergranular phases in the both samples were glassy. The results could be explained from the difference of microstructures rather than grain boundary chemistry; the addition of SiC leads to an increase of the number of fine equiaxed grains and a decrease of the number of elongated grains. As the result, the flexure strength is improved whereas the creep resistance is lowered.

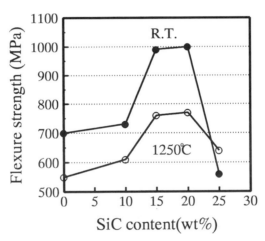

Fig. 5. Flexure strength as a function of SiC content.

For the Si₃N₄/SiC composite, it has been reported that the addition of SiC was effective in preventing the deterioration of flexure strength at high temperatures (Niihara et al., 1994, Ukyo et al., 1994) and/or improving the creep resistance (Niihara et al., 1994, K. Ramoul–Badache and M. Lancin, 1992, K. Yamada and N. Kamiya, 1994). However, in this system, although the Si₃N₄/SiC has the nano–composite microstructure, the flexure strength of the Si₃N₄/SiC composite decreased rapidly above 1200°C as that of the Si₃N₄ monolith with the same amount of the sintering additives did. This decrease can be attributed to softening behavior of the intergranular phases above 1200°C for the both samples. It can be concluded that the nano–composite microstructure is not a sufficient condition for improving heat resistance. Therefore, the heat resistivity of the intergranular phases is the key for improving the high temperature properties of the Si₃N₄/SiC composites.

Evaluation of Si_3N_4/SiC Composites

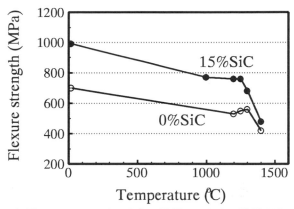

Fig. 6. Flexure strength of the Si_3N_4 and the Si_3N_4/15wt%SiC.

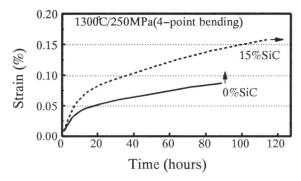

Fig. 7. Creep behavior of the Si_3N_4 and the Si_3N_4/15wt%SiC.

IV. Summary

Dense Si_3N_4/SiC composites were prepared by hot–isostatic–pressing after gas pressure sintering, using commercially available silicon carbide powder with the average particle size of 0.3 μm. They had so–called "nano–composite" microstructure; the nano–sized silicon carbide particles were dispersed both in the silicon nitride grains and in the grain boundaries. The grain growth of β silicon nitride during sintering was suppressed by the addition of silicon carbide. As the result, the microstructure became finer, and the flexure strength was improved. However, the strength of the Si_3N_4/SiC composite and the Si_3N_4 monolith deteriorated at the same temperature and the creep strain of the Si_3N_4/SiC composite was larger than that of the Si_3N_4 monolith. In this study, it can be concluded that the nano–composite microstructure is not a sufficient condition for improving heat resistance. Moreover, the heat resistivity of the intergranular phases is the key for improving the high temperature properties of the Si_3N_4/SiC composites.

References

1. F. F. Lange, J. Amer. Ceram. Soc., **56**,445–50(1973).
2. K. Niihara, T. Hirano, A. Nakahira, K. Ojima, K. Izaki and T. Kawakami, "High–Temperature Performance of Si$_3$N$_4$/SiC Composites from Fine, Amorphous Si–C–N Powder"; pp.107–12 in Proceedings of MRS International Meeting on Advanced Materials, vol.5, Materials Research Society, Pittsburgh, 1989.
3. K. Niihara, T. Hirano, A. Nakahira, K. Suganuma, K. Izaki and T. Kawakami, J. Jpn. Soc. Pow. & Pow. Mett., **36**,243–46(1989).
4. K. Niihara, T. Hirano, K. Izaki and F. Wakai, "High Temperature Creep/Deformation of Si$_3$N$_4$/SiC Nanocomposites"; pp.207–19 in Silicon–Based Structural Ceramics. Edited by Brian W. Sheldon and Stephen C. Danforth. American Ceramic Society, Westerville, OH, 1994.
5. Y. Akimune, Nippon Seramikkusu Kyoukai Gakujyutsu Ronbunshi, **98**[5]424 –28(1990).
6. Y. Ukyo, T. Kandori and S. Wada, "Si$_3$N$_4$–SiC Composite Consolidated by Post–HIPing Process"; pp.581–84 in Advanced Materials '93,I/A, vol.14A. Edited by N. Mizutani, M. Yoshimura, H. Kawamura, K. Kijima and M. Mitomo. Trans. Mat. Res. Soc. Jpn., 1994.
7. K. Ramoul–Badache and M. Lancin, J. Euro. Ceram. Soc., **10**,369–79(1992).
8. K. Yamada and N. Kamiya, "Dispersion Morphology of SiC particles in Si$_3$N$_4$/SiC Composites"[in Japanese]; pp.33–36 in 13th Kouon zairyou kiso touronkai youshisyu. Edited by Ceramic Society of Japan, 1994.
9. Y. Tajima, K. Mizuno, K. Matsubara and M. Watanabe, "Fabrication and High Temperature Properties of Silicon Nitride Ceramics with Transition Metal Silicides Dispersion"; pp.443–50 in Tailoring of Mechanical Properties of Si$_3$N$_4$ Ceramics. Edited by M. J. Hoffmann and G. Petzow. Kluwer Academic Publishers, 1994.
10. O. O. Ajayi, R. H. Lee and R. E. Cook, Mat. Sci. and Eng., **A169**,L5–L7(1993).

Sintering of Aluminum Nitride with Y_2O_3 by Secondary Phase Composition Control

Youngmin Baik[*] and Robin A.L. Drew

Department of Mining and Metallurgical Engineering
McGill University, Montreal, Quebec, CANADA

Abstract

Aluminum nitride is emerging as an important electronic substrate material for thermal management in advanced integrated circuits. Liquid-phase sintering is commonly used to fabricate AlN ceramics. The liquid-phase sintering of AlN with Y_2O_3 has been studied by controlling the secondary phase composition using the Y_2O_3-Al_2O_3 phase equilibria. Sintering of AlN was carried out over a temperature range of 1750-1950°C for 1 hour under N_2 atmosphere in an AlN/BN powder bed. Full densification of the different compositions was achieved at 1900°C for one hour. The effects of temperature and composition on the sintering behavior and the secondary phase formation as well as thermal conductivity are discussed.

1. Introduction

AlN has excellent potential for application as a substrate material due to its superior thermal properties that are suitable for substrate materials in advanced integrated circuits. The thermal conductivity of an AlN single crystal (320 W/m·K)is ~15 times higher than that of Al_2O_3 at room temperature, and the CTE of AlN ($4.4x10^{-6}$/°C) is close to that of Si ($3.2x10^{-6}$/°C).[1]

Since AlN is a highly covalently bonded material, it has a low diffusivity and requires very high temperatures for sintering.[2] Moreover, in the presence of oxygen impurities on the powder surface and in the lattice, it is necessary to use sintering additives to enhance the sinterability and thermal conductivity of AlN. The role of the additives is to form a liquid phase by reacting with the oxide (Al_2O_3) layer on the surface of AlN particles and to act as an oxygen getter to remove oxygen from the AlN lattice.[3] The sintering of AlN with an additive is usually performed at 1450-2000°C for 1-6 hours under N_2 atmosphere. Many oxide materials can be used as additives, but among them, Y_2O_3, CaO, SiO_2 and a combination of Y_2O_3 and CaO or SiO_2 are the most commonly used.

The objectives of this paper were to investigate sintering behavior of AlN by the secondary phase compositional control as well as the effect of different compositions on the thermal conductivity.

2. Experimental Procedure

All the experiments were performed using AlN powder (Grade F, Tokuyama Soda, Japan) containing <0.9 wt% of oxygen, <400 ppm of carbon, and trace amounts of other impurities (<60 ppm of Ca, <10 ppm of Fe, <15 ppm of Si). The powder had a mean particle size of ~0.3 μm and a specific surface area of 3.3 m^2/g. A Y_2O_3 powder (Grade 5630X, Union Molycorp, U.S.A) used as a sintering additive had a mean particle size of 1.8 μm and specific surface area of 33 m^2/g. The AlN powder was mixed with various amounts of Y_2O_3 powder. The pre-mixed compositions were ball milled for 24 hours using 3 mm diameter ZrO_2 (Tosho U.S.A Inc.) media in reagent grade isopropanol. The amount of Y_2O_3 required as a sintering additive depends on the quantity of oxygen present in AlN powder. Analysis of the oxygen content after milling without Y_2O_3 addition was used to determine the amount of Al_2O_3 associated with the AlN powder, assuming that all the oxygen is present in the form of Al_2O_3. Using the Y_2O_3-Al_2O_3 equilibrium phase diagram (Figure 1),[4] four compositions (first, second and third eutectic compositions and one composition equivalent to $3Y_2O_3 \cdot 5Al_2O_3$(YAG) compound) were determined by varying the amount of Y_2O_3.

After milling, the mixtures of AlN and Y_2O_3 were dried in a microwave oven, and then uniaxially pressed into disc shaped compacts 32 mm in diameter, and 3 mm in height. These green compacts were heated to 500°C in air for 4 hours to remove residual moisture and isopropanol. After heating, the green compacts were then isostatically pressed at 200 MPa.

All sintering experiments were performed in a horizontal, graphite element resistance furnace equipped for atmospheric control. The green compacts were placed in a BN crucible and surrounded by a powder bed of 50% BN/50% AlN. The green compacts were then sintered in a N_2 (101 kPa) atmosphere at the temperatures in the range of 1700-1950°C for 1 hour. The specimens were heated at a rate of 17°C/min to the desired sintering temperatures, maintained at temperature for 1 hour and then furnace cooled to room temperature at an initial rate of approximately 200°C/min.

The total oxygen level in the AlN powder was analyzed using a TC136 LECO EF100 inert gas fusion furnace and associated analyzer (by ELKEM Metals, Niagara Falls, New York). Densities of the sintered AlN compacts were measured by the Archimedes' principle according to a version of the

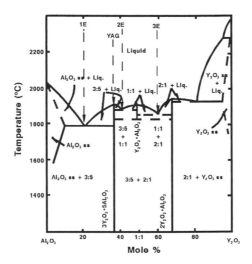

Figure 1: The Phase Diagram of the Y₂O₃-Al₂O₃ System.[4]

ASTM standard C373-72. The secondary phase formed after sintering was detected using a X-Ray Diffractometry (APD 1700 using Cu-Kα radiation with the tube operating at 40 kV and 20 mA, Philips, Netherlands).

3. Results and Discussion

3.1 Densification

Four different temperatures were chosen for each composition according to the liquidus/solidus temperature obtained from the Y_2O_3-Al_2O_3 phase diagram. The plot of relative density versus temperature for each composition is shown in Figure 2.

In general, the relative densities increased steadily with the sintering temperature until approximately 1850°C, at which point the relative density exceeded 97.5%. The 1E (20.9% Y_2O_3) and YAG (37.5% Y_2O_3) samples reached maximum densities at 1850°C, of 97.8% and 97.6%, respectively, while the densities of 2E (40.7% Y_2O_3) and 3E (61.1% Y_2O_3) approached full density at 1900°C. Interestingly, the relative density exceeded 97% for all the compositions sintered at 1850°C, with the exception of 1E, this temperature being much lower than that necessary for liquid formation according to the Y_2O_3-Al_2O_3 phase diagram. This suggests that liquid formation occurs below the melting temperatures of the yttrium aluminate secondary phases.

The shrinkage behavior of the four compositions is shown in Figure 3. The volume shrinkage at 1800°C of 1E was the highest (40.5%) followed by 2E (38.3%), and those of YAG and 3E were 32.2% and 31.4%, respectively. The

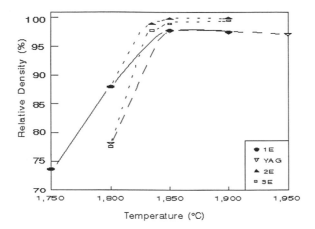

Figure 2: The Relative Density Change versus Sintering Temperature.

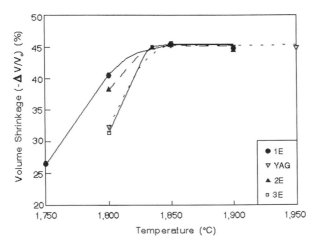

Figure 3: The Volume Shrinkage versus Sintering Temperature.

volume shrinkage for all the compositions at 1850°C was a similar value of approximately 45%, with no significant change with increasing temperature.

These density and shrinkage results permit the investigation of the effect of liquid phase formation on the densification. Other investigations[5-7] on the sintering of AlN without additives revealed that a liquid can form and assist sintering at temperatures as low as 1200°C in the AlN-Al$_2$O$_3$[7] (on the surface of

AlN particles) system. Therefore, it is suggested that liquid yttrium aluminum oxynitride could form and promote densification at the temperatures lower than the melting temperatures of the secondary phases obtained from the Y_2O_3-Al_2O_3 phase diagram, and then transform to yttrium aluminates and AlN phases upon further cooling. However, it is only possible to estimate the liquidus temperatures of yttrium aluminum oxynitrides by comparing the Y_2O_3-Al_2O_3 and AlN-Al_2O_3 phase diagrams because no equilibrium diagram for the AlN-Al_2O_3-Y_2O_3 ternary system has been proposed to date. Furthermore, it is difficult to draw a direct comparison of the shrinkage behavior between the four compositions, since the melting temperatures of the secondary phases (aluminum yttrium oxynitride) and the amount of the liquid phase of each composition all differ. Moreover, due to the different chemical compositions of each secondary phase, the viscosity of the liquid phases in the four compositions might differ. Hence, it can be concluded that the extent of liquid phase contribution to the densification of each composition might be controlled by the combination of its melting point, amount, and viscosity at the sintering temperature.

Figure 4 is a typical microstructure of fully densified sample (YAG) showing that most grains are interconnected and the grain edges have dissolved in the liquid phase. It also shows that the secondary phase is only present at the grain triple junctions.

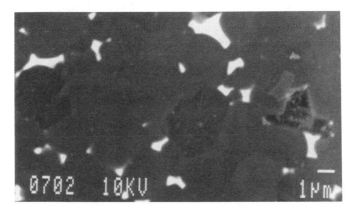

Figure 4: Backscattered Image of YAG Sintered at 1900°C for 1 hour.

3.2 Formation of the Secondary Phases

The secondary phase formation during sintering were detected by XRD using crushed sintered compacts. However, it was difficult to identify and quantify the peaks clearly due to peak overlapping over most of the scanning range and absence of standards for secondary phases. Instead, a comparison of the highest intensity peaks of AlN and the major secondary phases is presented as R_q to semi-quantify the change of the secondary phase contents:

$$R_q = \frac{I_{MS}}{I_{MS}+I_{AlN}} \times 100 \tag{1}$$

where I_{MS} is the relative intensity of the highest peak of the major secondary phases and I_{AlN} is the relative intensity of the highest peak of AlN. The values of R_q for 1E, YAG, 2E, and 3E are shown in Table 1.

Table 1: Relative Contents (R_q) of the Four Compositions Sintered at Different Temperature for 1 Hour.

Temp. (°C)	R_q			
	1E	YAG	2E	3E[†]
1750	6.0			
1800	6.2	7.7	8.3	13.9
1835			8.8	11.2
1850	6.6	9.4	10.7	9.8
1900	6.7	10.9	11.2	7.2
1950		16.6*		

*: R_q of the high temperature polymorph of $3Y_2O_3 \cdot 5Al_2O_3$.
†: R_q of $2Y_2O_3 \cdot Al_2O_3$, the remainder is that of $3Y_2O_3 \cdot 5Al_2O_3$.

The secondary phase formed in the 1E was $3Y_2O_3 \cdot 5Al_2O_3$ only. This phase was found in all the compacts sintered in the temperature range of 1750-1900°C, and increased in quantity as the temperature increased. For the YAG, $3Y_2O_3 \cdot 5Al_2O_3$ was detected as the major secondary phase with a trace amount of $Y_2O_3 \cdot Al_2O_3$. The amount of $3Y_2O_3 \cdot 5Al_2O_3$ increased as the temperature increased, whilst the amount of $Y_2O_3 \cdot Al_2O_3$ decreased. For the samples sintered at 1950°C, the peaks of $3Y_2O_3 \cdot 5Al_2O_3$ were not present, but those corresponding to the high temperature polymorph, $3Y_2O_3 \cdot 5Al_2O_3$,[8] appeared. The three secondary phases found in the 2E were the major secondary phase $3Y_2O_3 \cdot 5Al_2O_3$, and traces of $Y_2O_3 \cdot Al_2O_3$ and $2Y_2O_3 \cdot Al_2O_3$. As the temperature increased from 1800°C to 1900°C, the amount of $3Y_2O_3 \cdot 5Al_2O_3$ increased, and the peak intensities of both $Y_2O_3 \cdot Al_2O_3$ and $2Y_2O_3 \cdot Al_2O_3$ decreased. The secondary phases found in 3E were $2Y_2O_3 \cdot Al_2O_3$ and $Y_2O_3 \cdot Al_2O_3$. The major secondary phase in the compacts sintered at 1800°C was $2Y_2O_3 \cdot Al_2O_3$. As the temperature increased to 1900°C, the amount of $2Y_2O_3 \cdot Al_2O_3$ decreased while the peak intensity of $Y_2O_3 \cdot Al_2O_3$ increased.

As the XRD data reveal, all the secondary phases detected do not contain any phase with nitrogen even though the liquid phase at the sintering temperature is believed to consist of a Al-Y-N-O phase. It is assumed that the stable phases at room temperature upon cooling are AlN and yttrium aluminate phases. Moreover, the results of the secondary phases detected by XRD agree with the phase equilibria of Y_2O_3-Al_2O_3 system quite well, with the only discrepancy being $Y_2O_3 \cdot Al_2O_3$ phase which probably results from rapid, non-equilibrium cooling of the samples.

3.3 Thermal Conductivity

Thermal conductivity measurements were performed on the fully densified materials which were sintered at 1900°C for 1 hour. Among the four compositions, YAG resulted in the highest thermal conductivity (148.5 W/m·K), followed by 3E (140.4 W/m·K). The thermal conductivity values of the 1E and 2E were lower than the other two compositions, being 107.9 and 127.6 W/m·K, respectively. These thermal conductivity values are somewhat lower than those achieved by Watari *et al.*[9] for the AlN sintered for 1 hour at 1900°C with 1 mole% Y_2O_3 (180 W/m·K), but comparable to those obtained by Buhr *et al.*[10] (150 W/m·K).

Since the samples were all fully densified ($>97.5\%$), it was not appropriate to relate their thermal conductivity values to their density. Hence, it is hypothesized that the main factor controlling the thermal conductivity of AlN containing a secondary phase is the oxygen absorption capability of the secondary phase. Additionally, oxygen absorption may also be dependent upon the amount of liquid phase.

Conclusions

The use of the Y_2O_3-Al_2O_3 phase diagram to control the secondary phase compositions of the sintered AlN ceramics was successfully performed and resulted in full densification at ≥ 1850°C for 1 hour. It is observed that the formation of the liquid phases takes place at temperatures below the melting point of the secondary phases shown in the phase diagram. This is believed to be due to the formation of aluminum yttrium oxynitride liquid instead of yttrium aluminates, and the melting point of the former is assumed to be <1850°C. The XRD results reveal only yttrium aluminates ($3Y_2O_3 \cdot 5Al_2O_3$, $Y_2O_3 \cdot Al_2O_3$ and $2Y_2O_3 \cdot Al_2O_3$) as the secondary phases on cooling to room temperature, and none of the compounds from the Al_2O_3-AlN system were found. The thermal conductivity values of YAG and 3E (>140 W/m·K) were superior to those of 1E and 2E (<130 W/m·K). The difference is thought to be caused mainly by the oxygen absorption capability of the secondary phases.

References

1. K. Kurokawa, K. Utsumi, H. Takamizawa, T. Kamata and S. Noguchi, "AlN Substrates with High Thermal Conductivity," *IEEE Trans. Components, Hybrids, Manuf. Technol.*, CHMT-8[2], 247-252 (1985)

2. J.C. Schuster, "Phase Diagrams Relevant for Sintering Aluminum Nitride Based Ceramics," *Revue de Chimie Minerale,* 24, 676-686 (1987).

3. A. V. Virkar, T.B. Jackson and R.A. Cutler, "Thermodynamic and Kinetic Effects of Oxygen Removal on the Thermal Conductivity of Aluminum Nitride," *J. Am. Ceram. Soc.,* 72[11], 2031-2042 (1989).

4. E.M. Levin and H.F. McMurdie, "Phase Diagrams for Ceramists," *Am. Cearm. Soc.,* 1975 Supplement, Published in Columbus, Ohio, 132 (1964).

5. N. Hashimoto and H. Yoden, "Sintering Behavior of Fine Aluminum Nitride Powder Synthesized from Aluminum Polynuclear Complexes," *J. Am. Ceram. Soc.,* 75[8], 2098-2106 (1992).

6. T. Sakai, M. Kuriyama, T. Inuka and T. Kijima, "Effect of Oxygen Impurity on the Sintering and the Thermal Conductivity of AlN Polycrystal." *J. Ceram. Soc. Japan,* 86[4], 30-35 (1978).

7. J.W. McCauley and N.D. Corbin, "Phase Relations and Reaction Sintering of Transparent Cubic Aluminum Oxynitride Spinel (ALON)," *J. Am. Ceram. Soc.,* 62, 476-479 (1979).

8. H.S. Yodder and M.L. Keith, "Complete Substitution of Aluminum for Silicon: the System $3MnO \cdot AlO_3 \cdot 3SiO_2$-$3Y_2O_3 5AlO_3$," *J. Mine. Soc. Am.,* 36[7-8], 519-533 (1951).

9. K. Watari, M. Kawamoto and K. Ishizaki, "Sintering Chemical Reactions to Increase Thermal Conductivity of Aluminum Nitride," *J. Mater. Sci.,* 26, 4727-4732 (1991).

10. H. Buhr, G. Müller, H. Wiggers, F. Aldinger, P. Foley and A. Roosen, "Phase Composition, Oxygen Content, and Thermal Conductivity of $AlN(Y_2O_3)$ Ceramics," *J. Am. Ceram. Soc.,* 74[4], 718-723 (1991).

The Microstructures of Liquid Phase Sintered Materials

Randall M. German

P/M Lab, Penn State University, University Park, PA 16802-6809

Abstract

Microstructure rules have been identified for liquid phase sintered materials. These rules prove accurate in predicting grain growth, grain size, contiguity, and connectivity from basic thermochemical, composition, and processing factors. New grain growth and grain size distribution functions have been identified for liquid phase sintered materials. A new relation emerging from microgravity studies gives the grain coordination number in terms of the effective pressure on the structure.

1. Introduction

Liquid phase sintering (LPS) is the most widely applied technique for the densification of inorganic particulate materials (German, 1985). The sintered microstructure consists of solid grains dispersed in a matrix that was liquid at the sintering temperature. In the classic systems the solid has solubility in the liquid, with minimal liquid solubility in the solid. Densification is observed in most useful LPS systems, including W-Ni-Fe, WC-Co, Si_3N_4-Y_2O_3, Al_2O_3-SiO_2, TiC-Ni, and Fe-Cu. This compositional variety results in a spectrum of sintered microstructures, but there are some common rules applicable to LPS materials as summarized here.

2. Microstructure Parameters

The dihedral angle ϕ is formed where a grain boundary intersects the liquid and represents a balance of surface energies,

$$2 \, \gamma_{SL} \cos(\frac{\phi}{2}) = \gamma_{SS} \quad (1)$$

It is characteristic of the energy ratio of the grain boundary γ_{SS} and solid-liquid γ_{SL} interfaces. If this energy ratio is greater than 2, then the dihedral angle is $0°$ and the liquid penetrates the grain boundaries. Such an event is observed in the instant after liquid formation. There is a lower energy with two solid-liquid interfaces as compared to a grain boundary. Thus, initially no equilibrium exists involving grain boundaries in the presence of the newly formed liquid and the system lacks rigidity with high liquid contents. The other extreme occurs when the interfacial energy ratio is small and densification is inhibited by lack of grain boundary penetration by the liquid.

Contacts between grains grow during LPS to a stable size dictated by the dihedral angle. The diameter of the grain contact divided by the grain diameter *(X/G)* relates to the dihedral angle as follows:

$$\frac{X}{G} = \sin(\frac{\phi}{2}) \quad (2)$$

During LPS the intergranular bond grows to satisfy this ratio and further bond growth depends on the rate of grain growth. This differs from solid-state sintering, where the neck size ratio continuously enlarges with sintering time.

The grain coordination is the number of contacting grains in three dimensions. Usually only the two-dimensional connectivity is apparent in micrographs. The coordination influences many properties. If the mean number of contacts per grain exceeds 2, then a continuous chain-like structure is expected. However, a grain coordination of 4 is needed to hold system rigidity. The coordination number N_c is related to the volume fraction of solid V_S and the dihedral angle ϕ by an empirical relation,

$$V_S = -0.8 + 0.8\, N_c - 0.06\, N_c^2 + 0.002\, N_c^3 - 0.4A + 0.008\, A^2 \quad (3)$$

where the parameter $A = N_c \, cos(\phi/2)$. An important result from experiments on gravitationally settled microstructures shows the grain coordination number is linearly dependent on the effective pressure on the grains (Liu *et al.*, 1995),

$$N_c = N_{c_o} + K_C P \quad (4)$$

The three-dimensional grain packing coordination starts at an initial value near 4 and increases linearly with pressure P, with K_C being a factor similar to that observed in compaction. For a freely settled microstructure, the pressure is generated by the gravitational acceleration acting on the solid-liquid density difference times the height of the solid structure: Packing coordination changes with height due to self-compression.

Typically a LPS microstructure is analyzed using cross sections. The connectivity per grain C_g from a section relates to the dihedral angle as,

$$C_g = 0.68 \ N_c \ \sin(\frac{\phi}{2}) \quad (5)$$

assuming a steady-state grain size distribution. Since a two-dimensional microscopic section has a random encounter with the underlying three-dimensional structure, the connectivity depends on the dihedral angle and solid content, which in turn relates to the three-dimensional coordination.

The grain coordination increases with solid content, but depends on the dihedral angle. An upper limit is 14 contacts per grain at 100% solid, and 4 to 6 contacts are expected at near 55 to 60 vol.% solid. For low dihedral angles, there will be approximately 8 to 12 contacts per grain at 75% solid. For a high dihedral angle, there will be 4 to 6 contacts per grain at 75% solid.

The surface area of solid-solid contacts S_{SS} as a fraction of the total interfacial area is termed the contiguity C_{SS},

$$C_{SS} = \frac{S_{SS}}{S_{SS} + S_{SL}} \quad (6)$$

where the solid-liquid (matrix) surface area per grain is S_{SL}. The contiguity is measured in a microscope on a two-dimensional cross section using the number of intercepts per unit length of test line. It increases with volume fraction of solid and dihedral angle. The behavior can be approximated as,

$$C_{SS} = V_S^i \ (0.43 \ \sin(\phi) + 0.35 \ \sin^2(\phi)) \quad (7)$$

assuming no shape accommodation, making it invalid at high solid levels.

The shape of a solid grain depends on several factors, but is most affected by the dihedral angle, liquid content, and surface energy anisotropy. Nearly flat contacts form between neighboring grains. These contacts allow the grains to change shape to attain better packing. For dihedral angles over 60° and small volume fractions of liquid, the liquid structure is dispersed along grain edges and is not continuous. Beere (1975) and Wray (1976) calculated the equilibrium liquid shape under various conditions of 0% porosity and a coordination of 14. With sufficient liquid the structure is connected along the grain edges, and is connected for all dihedral angles below 60°, independent of the amount of liquid. Alternatively, for low liquid contents and large dihedral angles, a disconnected liquid is expected. At a low solid contents the grains are often rounded, approaching a spherical shape. However, a high solids content causes the grains to flatten along neighbor faces because of grain shape accommodation. Usually, the particle shape before LPS has no significant influence on the final grain shape.

As LPS progresses, the large grains grow at the expense of the smaller grains. Dissolution makes the smaller grains spherical. However, growing grains tend toward shapes dictated by either grain shape accommodation or anisotropic surface energies. Low energy crystallographic orientations are favored, leading to faceting of the grains. Further, in systems where the surface energy varies with crystallographic orientation by more than approximately 15%, angular grain shapes are expected (Warren, 1968).

Usually the quantity of liquid is insufficient to fill all pore space on melt formation. Solution-reprecipitation allows growing solid grains to deviate from a minimum energy shape to better fill space. In turn, this releases liquid into the remaining pores. For a given grain volume, the adjusted grain shape has a higher solid-liquid surface area (as compared to a sphere), but the elimination of pores and the associated surface energy provides for a net energy decrease. The grains attain better packing by selective dissolution of the solid with reprecipitation at points in the microstructure removed from the grain contacts. Transport is through the liquid surrounding the solid grains. Coalescence also contributes to grain coarsening and shape accommodation.

The net energy decreases during solution-reprecipitation; thus, the reduction in surface energy due to pore filling exceeds the surface energy increase due to the formation of nonspherical grains. A compact with grain shape accommodation is not at the lowest energy condition as shown by Kaysser *et al.* (1982). If a liquid reservoir is available, then a full density compact with grain shape accommodation will siphon liquid. The added liquid allows the solid-liquid interface to relax to a spherical grain shape thereby eliminating excess solid-liquid surface energy associated with nonspherical grains (Lisovsky, 1987) or gravity effects (Kipphut *et al.,* 1988).

3. Microstructural Coarsening

During LPS there is grain growth, even as densification slows in the final stage. The peak density depends on the characteristics of the pores and any internal gases trapped in the pores. Furthermore, the solid skeletal microstructure provides rigidity to the compact, and inhibits pore elimination. With continued sintering, the grain size enlarges by solution-reprecipitation. Coupled with transport events through the liquid phase, there are simultaneous solid-state sintering events, but typically liquid diffusivities dominate.

For most systems, coarsening occurs in parallel with densification and dominates the final stage. The driving force for coarsening is a decrease in the interfacial energy at the solid-liquid interface. Grains of small dimensions are more soluble in the liquid than grains of large dimensions. A reduction in chemical potential occurs by enlarging the microstructure scale;

thus, a measure of microstructural coarsening is the grain size. The mean grain size increases since the large grains grow and the small grains disappear over time, even though the volume fraction of solid remains constant. This leads to a decrease in the solid-liquid interfacial area.

The kinetics of grain growth give the mean grain size versus time as,

$$\bar{G}^n \sim t \quad (8)$$

where t is the isothermal time and n varies from 2 to 4 for most materials. Analytic models are available for the case where the volume fraction of solid approaches infinite dilution (Voorhees, 1992). However, the situations typical to LPS are poorly treated by such models, because they assume spherical grains without contacts. For solution-reprecipitation controlled growth $n = 3$.

One approach is to assume a continuum with no details on diffusion. For a dilute solid concentration, the mass loss and gains give the rate of change in size for an individual spherical grain of size G as follows:

$$\frac{dG}{dt} = \frac{K}{G}(\frac{1}{\bar{G}} - \frac{1}{G}) \quad (9)$$

where K is a kinetic rate constant estimated as follows:

$$K = \frac{2 D_L C \Omega \gamma_{SL}}{kT} \quad (10)$$

where D_L is the diffusivity through the liquid, C is the solubility in the liquid, Ω is the solid atomic volume, γ_{SL} is the solid-liquid surface energy, k is Boltzmann's constant, and T is the absolute temperature. The units of K are volume per unit time. Because both diffusivity and solubility have exponential temperature dependencies, K is very sensitive to temperature. The rate of grain growth for a particular grain depends on its relative difference from the size of its neighbors. A wide grain size distribution shows more rapid initial grain growth. Over time the grain size distribution converges to a steady-state form and the growth kinetics converge to steady-state behavior.

Nominally, grains with a size smaller than average shrink, while grains larger than average grow. The maximum growth rate declines as coarsening continues since the overall solid-liquid surface area declines. For diffusion controlled growth with no neighbor-neighbor interactions, several assumption are made concerning the coarsening system. These include an isotropic surface energy, no contact between grains, spherical grains, quasi-stationary diffusion field, and a mean concentration of solid in the liquid.

Such models indicate key parameters, but are inapplicable to LPS. The assumed conditions of dispersed grains at large separations are far from the reality of a highly connected solid-liquid mixture. Critical tests using concentration profiles around growing and shrinking grains fail to detect the assumed gradients. Finally, coalescence is an important contributor to coarsening that is ignored by most models. However, diffusion controlled growth is active in most cases; although, there are a few reports of reaction controlled growth. Systems exhibiting diffusion control exhibit rounded grains versus angular grains for reaction controlled systems with complex compositions. Simple systems exhibit diffusion control and rounded grains.

One of the fundamental assumptions of coarsening theory is that small grains shrink while the large grains grow. However, this is not totally accurate (Voorhees and Schaefer, 1987). The local environment plays a role, so some large grains shrink and some of the small grains grow. Shrinking grains are spherical. Although there is a neighbor effect on the growth or shrinkage of an individual grain, the system converges to an average behavior independent of the initial size distribution (Fang *et al.,* 1992).

The effect of a high solid volume fraction is to accelerate the rate of grain growth because of the shorter diffusion distance and greater number of coalescing contacts. An analysis of grain growth rate dependence on the volume fraction of liquid has been the subject of several theories (Voorhees, 1992). The basic result is a kinetic law with the grain size cubed varying with time. The rate constant K is modified to account for the effect of the shorter diffusion distance as the liquid content decreases. Several forms of this modification have been suggested, but the most successful for explaining grain growth during LPS is as follows (German, 1995):

$$K = K_o + K_L / V_L^{2/3} \quad (11)$$

where $K_o + K_L$ is the infinite dilution rate constant applicable to $V_L = 1$, K_L is a parameter sensitive to the microstructure, and V_L is the liquid fraction.

In multiple component solid systems, grain growth is controlled by diffusion of the slower moving species. To preserve stoichiometry in the reprecipitated material, all species must have coupled fluxes. The effective diffusion coefficient depends on the abundance of each component and the relative diffusivity. Situations arise where an impurity can control grain growth; especially if the impurity is soluble in the solid and has a slow diffusion rate. In W-Mo-Ni-Fe alloys, Mo reduces the solubility of W in the liquid, resulting in slower grain growth. Similar inhibition effects have been observed in two solid phase systems; each phase interferes with growth.

Grain growth by coalescence occurs in conjunction with other coarsening processes. Coalescence broadens the grain size distribution and increases the rate constant (Takajo *et al.*, 1984). The coalescence contribution increases with the dihedral angle. The exponent n is 3 even with a substantial coalescence contribution. Typically coalescence occurs simultaneously with solution-reprecipitation. An intercept size distribution function has been isolated for several LPS systems. The cumulative normalized distribution is,

$$F(L) = 1 - \exp(-0.7 \, L^2) \quad (12)$$

where $F(L)$ is the cumulative fraction of grains having a size L, where $L = G/G_{50}$ is the normalized intercept size (Yang and German, 1992, Fang *et al.*, 1993). Here G is the grain size and G_{50} is the median grain size (50% point on the cumulative distribution). This distribution fits data from several studies.

Parallel microstructural changes occur as the grain size enlarges, . The number of grains per unit volume decreases. Assuming zero porosity and volume conservation, the cubic grain growth law leads to the conclusion that N_V, the grain density, will vary with time as

$$N_V \sim t^{-1} \quad (13)$$

Likewise, the grain-matrix surface area per unit volume depends on the grain size. Accordingly, surface area per unit volume S_V will vary approximately as,

$$S_V = \pi \, N_V \, G^2 \quad (14)$$

with G being the grain size. Since the grain population decreases with inverse time and the mean grain size increases with time to the 1/3 power, the interfacial surface area decreases with time to the -1/3 power.

4. Summary

A majority of all sintering is performed in the presence of a liquid phase. In many materials the liquid phase is present by design, but there are several cases where it forms as a byproduct of incidental impurities. Solubility dictates the chemical interactions that determine mass flow and consequential densification behavior. Classic LPS systems have a low level of chemical interaction between constituents such that surface energy dominates the microstructural evolution during sintering. For these cases there are some important laws as captured in this paper.

This report summarizes basic studies on liquid phase sintering as supported by the U. S. Army Research Office, National Aeronautics and

Space Administration, and National Science Foundation.

5. References

W. Beere, "A Unifying Theory of the Stability of Penetrating Liquid Phases and Sintering Pores," *Acta Metall.*, 28 131-138 (1975).

Z. Fang, B. R. Patterson and M. E. Turner, "Influence of Particle Size Distribution on Coarsening," *Acta Metall. Mater.*, 40 713-722 (1992).

Z. Fang, B. R. Patterson and M. E. Turner, "Modeling Particle Size Distributions by the Weibull Distribution Function," *Mater. Char.*, 31 177-182 (1993).

R. M. German, *Liquid Phase Sintering*, Plenum Press, New York, NY, (1985).

R. M. German, "Microstructure of the Gravitationally Settled Region in a Liquid Phase Sintered Dilute Tungsten Heavy Alloy," *Metall. Mater. Trans.*, 26A 279-288 (1995).

W. A. Kaysser, O. J. Kwon and G. Petzow, "Pore Formation and Pore Elimination During Liquid Phase Sintering," *Proceedings P/M-82*, Associazione Italiana di Metallurgia, Milano, Italy, 23-30 (1982).

C. M. Kipphut, A. Bose, S. Farooq and R. M. German, "Gravity and Configurational Energy Induced Microstructural Changes in Liquid Phase Sintering," *Metall. Trans.*, 19A 1905-1913 (1988).

A. F. Lisovsky, "On the Imbibition of Metal Metals by Sintered Carbides," *Powder Met. Inter.*, 19 18-21 (1987).

Y. Liu, D. F. Heaney and R. M. German, "Gravity Induced Solid Grain Packing During Liquid Phase Sintering," *Acta Metall. Mater.*, 43 1587-1592 (1995).

S. Takajo, W. A. Kaysser and G. Petzow, "Analysis of Particle Growth by Coalescence During Liquid Phase Sintering," *Acta Metall.*, 32 107-113 (1984).

R. Warren, "Microstructural Development During the Liquid-Phase Sintering of Two-Phase Alloys with Special Reference to the NbC/Co System," *J. Mater. Sci.*, 3 471-485 (1968).

P. J. Wray, "The Geometry of Two-Phase Aggregates in which the Shape of the Second Phase is Determined by its Dihedral Angle," *Acta Metall.*, 24 125-135 (1976).

P. W. Voorhees, "Ostwald Ripening of Two-Phase Mixtures," *Annual Reviews in Materials Science*, 22 197-215 (1992).

P. W. Voorhees and R. J. Schaefer, "*In Situ* Observation of Particle Motion and Diffusion Interactions During Coarsening," *Acta Metall.*, 35 327-339 (1987).

S. C. Yang and R. M. German, "Generic Grain Size Distribution for Liquid Phase Sintering," *Scripta Met.*, 26 95-98 (1992).

Theoretical Analysis of Final-Stage Liquid-Phase Sintering

Suk-Joong L. Kang,   Kwan-Hyeong Kim[*]   and   Sung-Min Lee

Department of Materials Science and Engineering
Korea Advanced Institute of Science and Technology, Taejon 305-701, KOREA
[*]Now, with Samsung Electromechanics Co., Suwon   441-743, Korea

Abstract

Based on the pore filling mechanism, analysis of the densification and shrinkage during liquid phase sintering has been made for compacts containing isolated pores with size distribution.   The volume of available liquid for pore filling has been calculated by taking into account grain-shape accommodation and microstructure homogenization.   The grain growth determines the liquid filling of pores and homogenization of microstructure.   The calculated densification curves for compacts containing pores with various sizes are similar to those typically observed in real powder compacts.   With increasing pore volume, the densification time increases considerably.   The exponent of scale in the scaling law is equal to that of grain growth equation.   The present theory is an alternative to Kingery's theory in reflecting real phenomena.

1. Introduction

The process of final densification during liquid phase sintering involves the elimination of isolated pores which are different in size and randomly distributed in a grain-liquid mixture(Kang et al. 1991).   For pore elimination in conventional liquid phase sintering, two mechanisms may be operative, namely, contact flattening(Kingery 1959) and pore filling(Kwon and Yoon 1980, Park et al. 1986).   The contact flattening mechanism proposed by Kingery assumes no grain growth during sintering and a mono-size distribution of pores. These two assumptions, however, can never be applicable to real powder compacts.   In fact, microstructural observation and theoretical calculation show that grain growth is much faster than contact flattening(Kang et al. 1985).

221

Some experiments on model systems and fine powder compacts demonstrated that the liquid filling of pores is the essential process for densification during liquid phase sintering(Kwon and Yoon 1981, Kang *et al.* 1984, Park *et al.* 1989).   Theoretical analysis also showed that the liquid can flow into the pore by breaking a pressure balance of liquid between specimen surface and pore surface, when the liquid completely wets the surface of the grains surrounding the pore(Park *et al.* 1986).   The complete wetting of the grains can occur by the growth of grains.   The grain growth thus appears to be essential for the pore filling(Kang *et al.* 1991, Park *et al.* 1986).

In the present investigation, a theory of liquid-phase sintering has been developed, based on the previously proposed liquid-phase sintering model(Kang *et al.* 1991).   The kinetics of densification and shrinkage has been analyzed for powder compacts containing pores of various sizes.

2. Theoretical Model and Calculation

At a certain moment of final-stage sintering, the pore size distribution is believed to show an upper part of a continuous distribution curve with a minimum size, because the pores are eliminated sequentially in size by liquid filling.   In the present analysis , as an example, the pore size distribution is assumed to be a part of a log-normal distribution curve, namely, the upper 88% of the distribution, as shown by the shaded area in Fig. 1.   The log-normal distribution curve in Fig. 1 has the interval of 0.01 in initial pore size/grain size ratio, and starts at the ratio of 10 and ends at the ratio of 40.   Any size distribution of pores may be taken for the calculation; however, the calculation procedure is the same and similar result is obtained.

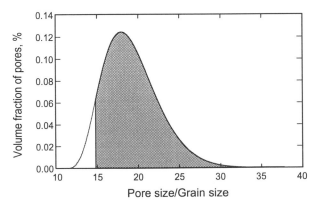

Fig. 1. Pore size distribution used for the calculation of densification and shrinkage.

For a given pore size distribution, the evaluation of the critical condition for liquid filling of individual pores is the basic step for the quantitative evaluation of densification kinetics.   The estimation of liquid meniscus radius during grain growth and pore filling is therefore prerequisite for the analysis.

The radius of liquid meniscus is determined by grain size and shape, packing geometry of grains, liquid volume fraction, wetting angle and dihedral angle(Park *et al.* 1986).   For the calculation, mono-size grains and close-packing(face-centered cubic packing) are assumed, as in the previous investigation(Park *et al.* 1986).   The wetting and dihedral angles are assumed to be zero, for simplicity.   Then, the liquid meniscus radius as a function of liquid volume fraction in a dense body can easily be calculated.   In the present analysis, the previously obtained result for $\gamma_{lv}/\gamma_{sl} = 7$ is used.   In the case of non-zero wetting and non-zero dihedral angles, the estimation of the liquid meniscus radius is much more complicated(Park *et al.* 1986).   The basic scheme of the calculation, however, is unchanged and the physical meaning of the results being obtained must be the same.

The effective volume fraction of liquid, f_l^{eff}, for pore filling is determined by the initial liquid volume fraction, the amount of liquid filling of pores and the homogenization of liquid pockets formed by pore filling.   The estimation of f_l^{eff} is also critical for the present calculation, because the liquid meniscus radius depends on f_l^{eff}.   Figure 2 shows typical microstructures of a liquid pocket formed right after pore filling (a)   and a liquid pocket being homogenized during sintering (b).   The etch boundaries formed in grain A during cyclic sintering (Fig. 2(b)) show that the grain grows towards the liquid pocket, which was initially similar to that in Fig. 2(a) after the liquid filling, resulting in microstructure homogenization.   With the growth of grains towards liquid

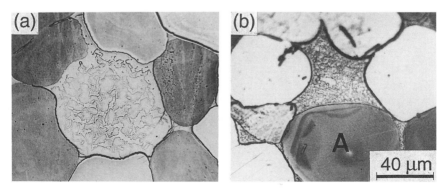

Fig. 2. Microstructures showing   (a) a liquid pocket formed right after pore filling and (b) liquid pocket being homogenized.   96Mo-4Ni specimen sintered at 1460°C in cycle of 60-30-30 min(Kang *et al.* 1984).

pocket, a fraction of the liquid in the pocket is squeezed out to neighboring grain-liquid bulk and contributes to f_l^{eff}.   The volume of liquid squeezed-out from the liquid pocket is thus determined by grain growth.   The contribution of microstructure homogenization of liquid pocket to f_l^{eff} would proceed until the complete elimination of liquid pocket to reach a microstructure similar to neighboring grain-liquid bulk structure, *i.e.* a structure with uniform grain-matrx distribution.

The shrinkage of specimen may occur   with the microstructure homogenization of liquid pockets because the liquid meniscus changes with the homogenization.   In the calculation, it is thus assumed that the homogenized volume of liquid pockets is equivalent to the volume shrinkage.

As explained so far and also in the previous investigation(Kang *et al.* 1991), the grain growth is believed to determine the densification and shrinkage at the final stage sintering.   The grain growth is assumed to follow a cubic law, as in many real systems, *i.e.*

$$G^3 - G_o^3 = Kt \qquad (1)$$

where G is the grain size at the time of observation at $t = t$, G_o the grain size at the time of observation at $t = 0$,   and K the kinetic constant.   Then, the homogenized volume of liquid pocket under microstructural homogenization, V_{homo}, can be expressed as :

$$V_{homo} = \int_o^t 2\pi r_\tau^2 \cdot dG \qquad (2)$$

where r_τ is the radius of remaining liquid pocket at time τ.   The effective liquid volume fraction f_l^{eff}, relative density ρ and shrinkage $(1-l/l_o)$ are then expressed as :

$$f_l^{eff} = \frac{V_l^i - \sum\limits_{j=k+1}^{n} V_{pocket}^j}{V_s^i + V_l^i - \sum\limits_{j=k+1}^{n} V_{pocket}^j} \qquad (3)$$

$$\rho = 1 - \frac{\sum\limits_{j=n+1}^{\infty} V_p^j}{V_s^i + V_l^i + \sum\limits_{j=n+1}^{\infty} V_p^j} \qquad (4)$$

and

$$1 - \frac{l}{l_o} = 1 - \left[1 - \frac{\sum\limits_{j=k+1}^{n} V_{homo}^{j}}{l_o^3} - \frac{\sum\limits_{j=1}^{k} V_{p}^{j}}{l_o^3} \right]^{\frac{1}{3}} \tag{5}$$

Here, V_l^i is the initial volume of liquid, V_s^i the initial volume of solid, V_p^j the volume of pore j filled with liquid for $j \leq n$ or the volume of unfilled pore j for $j > n$, V_{pocket}^{j} the remaining volume of liquid pocket being homogenized, l_o the initial size of specimen, l the size of specimen at time t, and k the maximum size of completely homogenized liquid-filled pore(liquid pocket). For the parameters included in the above equations, the typical values measured in real systems were taken as : $K/G_o^3 = 0.5$ s^{-1} (for example, $G_o = 1\mu m$, $K = 5\times 10^{-19}$ m^3/sec).

3. Results and Discussion

Figure 3 shows some calculated densification and shrinkage curves of powder compacts containing pores with the size distribution shown in Fig 1. The overall shape of the curves are similar to those measured in real powder compacts(Park *et al.* 1989, Kingery and Narasimhan 1959). Since the densification occurs by pore filling, the relative density in Fig. 3 means the density measured by the water-immersion technique. The abscissa represents sintering time in seconds for the given conditions of powder compacts in terms of grain packing, pore size distribution, initial grain size/pore size ratio, K/G_o^3, wetting and dihedral angles. On varying these conditions, the densification rate changes; however, the overall shape of densification curve is unchanged.

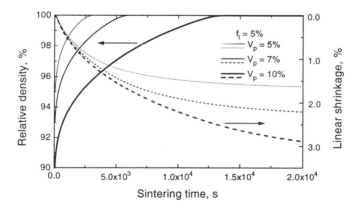

Fig. 3. Calculated densification and shrinkage curves of powder compacts with 5 vol% liquid but with various pore volume fractions.

The calculated densification curves show that for a given pore size distribution, the densification time increases considerably with increasing porosity.   In this particular case, the densification time increases more than three times when the pore volume fraction increases twice, from 5% to 10%. The densification time is therefore not proportional to the porosity.   This result arises from much reduced volume of available liquid for pore filling with increasing porosity(see Fig. 4).   The effect of pore volume fraction on shrinkage is similar to that on densification, as shown in Fig. 3 by dashed lines. The shrinkage, however, is slower than the densification.   This result is due to the fact that the densification occurs by instantaneous pore filling but the shrinkage by grain shape accommodation and change.   The shrinkage curves in the figure also imply that the microstructure homogenization, *i.e.* attainment of an equilibrium microstructure, takes long time even after the complete elimination of pores by liquid flow.

During the densification and shrinkage, the effective liquid volume fraction f_l^{eff}, which determines the liquid meniscus radius at pore and specimen surfaces, changes as shown in Fig. 4.   In general, the fraction decreases sharply at the beginning with the pore filling of liquid, but increases again and slowly reaches to the original liquid fraction with sintering time and microstructure homogenization.   Such calculated variation of f_l^{eff} with sintering time is in good agreement with the measured variation in a previous investigation(Kwon and Yoon 1981).   Figure 4 also shows that when increasing pore volume, the reduction in effective liquid volume during densification becomes higher, resulting in much longer densification time, as shown in Fig. 3.

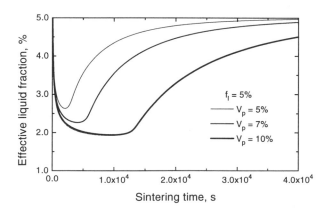

Fig. 4. Calculated variation of effective liquid volume fraction with sintering time for compacts with 5 vol% liquid but with various pore volume fractions.

Figure 5 depicts some densification and shrinkage curves of compacts with similar microstructure but with different scale.   With increasing the scale, the densification and shrinkage are retarded significantly.   Since the densification occurs by pore filling which is in turn determined by grain growth, the effect of scale on densification is expected to follow the scale effect on grain growth. The calculated densification curves confirm this expectation.   The exponent in scaling law for densification during liquid phase sintering is in fact equal to that for grain growth, 3 in our case.   The dependence of shrinkage on scale is also calculated to follow the dependence of grain growth on scale.   This result comes also from the dependence of microstructure homogenization on grain growth.

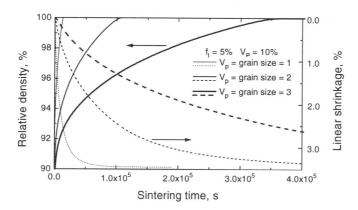

Fig. 5. Calculated densification and shrinkage curves of compacts with different scales : 1, 2 and 3.

4. Conclusion

A theoretical analysis of liquid phase sintering has been made for powder compacts with pore size distribution.   The analysis is based on microstructure observations made in previous investigations(Kang *et al.* 1985, Kwon and Yoon 1981, Kang *et al.* 1984, Park *et al.* 1989): pore filling, grain shape accommodation and microstructure homogenization.   Calculated densification curves of powder compacts with various pore volume fraction demonstrate that increase in pore volume results in considerable increase in densification time. The present theory of liquid phase sintering also shows that the effect of scale on densification is determined by grain growth.   The exponent in a scaling law for liquid phase sintering is the same as that of grain growth equation. Such evaluations on the effect of sintering parameters, which were not possible before, may provide with better understanding of the liquid phase sintering which has been thought to be much complicated.

References

S.-J. L. Kang, K.-H. Kim, and D. N. Yoon, "Densification and Shrinkage During Liquid Phase Sintering," *J. Am. Ceram. Soc.,* **74** [2] 425-27 (1991).

W. D. Kingery, "Densification During Sintering in the Presence of a Liquid Phase," *J. Appl. Phys.,* **30** [3] 301-06 (1959).

O. J. Kwon and D. N. Yoon, "The Liquid Phase Sintering of W-Ni," *Sintering Processes* (Proceedings of the 5th International Conference on Sintering and Related Phenomena, held at the University of Notre Dame, Notre Dame, Indiana, June 18-20, 1979), G. C. Kuczynski(ed.) Plenum Press, NY, pp.203-18 (1980).

H. H. Park, O. J. Kwon and D. N. Yoon, "The Critical Grain Size for Liquid Flow into Pores During Liquid Phase Sintering," *Metall. Trans. A,* **17** [11] 1915-19 (1986).

S.-J. L. Kang, W. A. Kaysser, G. Petzow and D. N. Yoon, "Growth of Mo Grains around Al_2O_3 Particles During Liquid Phase Sintering," *Acta Metall.,* **33** [10] 1919-26 (1985).

O. J. Kwon and D. N. Yoon, "Closure of Isolated Pores in Liquid Phase Sintering of W-Ni," *Int. J. Powder Metall. Powder Tech.,* **17** [2] 127-33 (1981).

S.-J. L. Kang, W. A. Kaysser, G. Petzow and D. N. Yoon, "Elimination of Pores During Liquid Phase Sintering of Mo-Ni," *Powder Metall.,* **27** [2] 97-100 (1984).

J. K. Park, S.-J. L. Kang, K. Y. Eun and D. N. Yoon, "The Microstructural Change During Liquid Phase Sintering of W-Ni-Fe alloy," *Metall. Trans. A,* **20** [5] 837-45 (1989).

H. H. Park, S.-J. L. Kang, and D. N. Yoon, "An Analysis of Surface Menisci in a Mixture of Liquid and Deformable Grains," *Metall. Trans. A,* **17** [2] 325-30 (1986).

W. D. Kingery and M. D. Narasimhan, "Densification During Sintering in the Presence of a Liquid Phase," *J. Appl. Phys.,* **30** [3] 307-10 (1959).

Relationship Between Contact Angle and Surface Tensions

Yixiong Liu and Randall M. German

Department of Engineering Science and Mechanics
P/M Lab, 118 Research West
The Pennsylvania State University
University Park, PA 16802-6809
USA

ABSTRACT
The theoretical and experimental background of Young's equation for the solid-liquid-vapor contact angle are reviewed and discussed. A recent microgravity liquid phase sintering experiment using a W-Ni alloy provides an opportunity to test the contact angle definition in terms of the surface tensions without the influence of gravity. The resulting configuration can only be described by a force balanced three-vector diagram, not by Young's equation. Microscopically, a three phase contact forms a "valley" rather than a "ridge" on an initially plane solid surface. A main problem of Young's equation is the assumption of thermodynamic equilibrium for what is a metastable system.

INTRODUCTION
Wetting behavior is of great importance during LPS [1-3]. The contact angle and surface tensions are key parameters in determining capillary force [2-4]. The angle of contact between liquid, vapor and solid phases is usually related to the surface tensions by Young's equation,

$$\gamma_{SV} - \gamma_{SL} = \gamma_{LV}\cos\theta \qquad (1)$$

where γ_{SV}, γ_{SL} and γ_{LV} are the tensions of the solid-vapor, solid-liquid, and liquid-vapor interfaces, respectively, and θ is the contact angle as shown in Figure 1 for a drop on a solid.

The validity of Young's equation was called into question in the late 1950's. Bikerman pointed out that γ_{SV} is balanced by the sum of γ_{SL} and $\gamma_{LV}\cos\theta$, leaving an unbalanced vertical component of $\gamma_{LV}\sin\theta$ [5,6]. Actually, three interface forces acting on one point in equilibrium form a three-vector diagram of balanced forces as shown in Figure 2(a), leading to the following equation:

$$\frac{\gamma_{12}}{\sin\alpha_3} = \frac{\gamma_{23}}{\sin\alpha_1} = \frac{\gamma_{13}}{\sin\alpha_2} \tag{2}$$

where γ and α denote tensions and angles, respectively, and the subscripts represent the individual phase. This equation has been verified for fluids, which can not support shear stresses. On the other hand, Bikerman [5] pointed out that Young's equation is not satisfied as long as the solid surface is treated as a plane. This is substantiated by the observation that liquid droplets cause curling of thin mica sheets [6]. Consequently, Bikerman predicted that the solid surface force $\gamma_{LV} \sin\theta$ would raise a ridge on the solid plane as illustrated in Figure 2(b), and that a solid strain energy would also act on the structure. The existence of such a ridge would invalidate Young's equation, because γ_{SV} and γ_{SL} would act along different directions.

Bikerman's argument was not totally convincing. In an air-gel-mercury experiment conducted later [7], he failed to prove the predicted bulging of the three phase contact line from $\gamma_{LV} \sin\theta$. At the same time, Pethica and Pethica [8] challenged the validity of Young's equation in a gravitational field. From numerical calculations they concluded that Young's equation is only applicable to a special case where the liquid drop is spherical with no gravitational effect. Normally, the equilibrium contact angle is not solely a function of the three surface tensions, but must include gravitational forces that are proportional to the droplet mass.

Other evidence of a gravitational effect is in the hysteresis in contact angle measurements [9]. There is a difference in contact angles for droplets that are not on a horizontal plane. This can be seen by the difference in contact angles between the top and bottom of rain droplets on a window. Likewise, advancing and receding contact angles are different. For water droplets on mineral surfaces, the advancing angle may be as much as 50° larger than the receding one, and for mercury on steel there is a difference of 154° [10]. This phenomena was studied by Lomas [11] using computer calculations of the drop profile. For an inclined drop, the solution of the Laplace equation leads to different leading and trailing contact angles. From such illustrations it is evident that gravity interferes with precise determination of surface tensions. For a solid, liquid and vapor phase equilibrium, the three surface tensions should have constant, intrinsic values, not dependent on orientation. Clearly, Young's equation is incorrect in this regard, since the contact angle is not uniquely determined by the surface tensions.

EXPERIMENTAL RESULTS

Young's equation has never been verified experimentally [10]. Additionally, a convincing experiment for disproving the equation has not been generally accepted either. Bikerman [7] failed to prove the $\gamma_{LV}\,sin\theta$ component lifted a solid ridge, due to insufficient precision in the experiment. According to Lester's calculation [12], there should be a ridge at the contact point, as shown in Figure 2(b), but only on a microscopic scale. If this is an infinitesimal ridge, then experimental confirmation may prove very difficult.

A recent microgravity sintering experiment provided an opportunity for studying the contact angle relationship. The 10 mm diameter sample was made from mixed W and Ni powders (97 wt.% W). The compact was sealed in an argon-filled container and sintered at 1550°C for 180 min during the Spacelab J mission of September 1992. After returning from space, the sample was cut and polished, and finally examined by scanning electron microscopy. Detailed experimental procedures and results are given elsewhere [13,14].

Figures 3 is a scanning electron micrograph of the solidified W-Ni alloy. According to the phase diagram [15], the system was in a two phase zone (solid plus liquid) at 1550°C. The absence of buoyancy in the microgravity condition removes the normal forces that cause the low density pores to rise and the high density solid to settle in the liquid. Thus, these samples are ideal for studying the three phase interaction. In addition, the prolonged time at high temperature allowed formation of a stable configuration. Solid motion via solution-reprecipitation contributes to significant solid mobility as evident by grain shape accommodation and grain growth [1]. The micrographs show a configuration of equilibrium at the solid-liquid-vapor contact point. Now, γ_{SV} and γ_{SL} act in different directions; this is the evidence Bikerman sought to disprove Young's equation. Moreover, there is no basis to believe that Young's equation is applicable at a microscopic level, since these microscopic results clearly show the condition sketched in Figure 2(a) other than Figure 1. Arrows in Figure 3 indicate depressions of solid at the contact points with pores. The depressions are formed to satisfy the static balance between three tensions, as required by the three-vector diagram. The ridge predicted earlier by Bikerman and Lester [5,12] is actually a valley in the solid. Also shown here is that the geometry in Figure 1 used to derive Young's equation does not correspond to an equilibrium state for solid-liquid-vapor phases in the absence of gravity. The current experiment unambiguously disproves Young's equation. By further employing serial sectioning technique on the microgravity sample, it will be possible to measure a true contact angle that is balanced solely by three surface tensions. Obtaining the true contact angle in three dimensions will be a subject of future study.

DISCUSSION

A correct description of the stable phase configuration is needed to establish the relation between contact angle and surface tension. The situation in Figure 1 is a metastable state, not an equilibrium state. There is a stress on the solid, but its rigidity prevents deformation to balance surface tension. If the solid is deformable under the stress induced by the liquid-vapor surface tension, then the solid surface will deform in response to the stress. In most situations it will take a long time to reach a balanced three-vector situation. Accordingly, Young's equation is not an equilibrium equation. Elastic strains will exist in a rigid solid that needs to be incorporated in calculating the minimum energy configuration.

Due to the great differences in deformability between solid and liquid, it is very difficult to experimentally observe an equilibrium solid-liquid-vapor configuration. This is probably why Young's equation has never been proved or disproved convincingly. Most sessile drop experiments conducted at room temperature fail to provide a condition for a solid substrate deformation since the forces imposed by surface tensions and gravity are small. Increasing the temperature in a conventional sessile experiment might accelerate solid deformation through creep or diffusion processes; however, the liquid drop might evaporate before solid deformation. Generally, an ideal liquid phase sintered alloy could be a good candidate for studying the equilibrium configuration, since the solubility of solid in liquid is relative high and a sintering temperature is only slightly higher than the melting point of the liquid phase. Thus, during liquid phase sintering the liquid component can be stable (without evaporation) while the solid deforms via diffusion. A configuration of solid-liquid-vapor contact during liquid phase sintering has been observed on powders bonded by liquid necks [2], in which the three surface tensions were in three different directions. However, these micrographs have not been used to disprove Young's equation. Such necks occur early in liquid phase sintering before equilibrium is achieved. During prolonged sintering, when equilibrium is

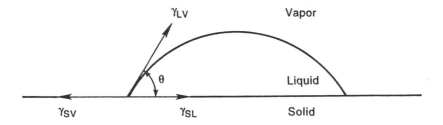

Figure 1. Alleged equilibrium of a liquid drop on a horizontal solid surface.

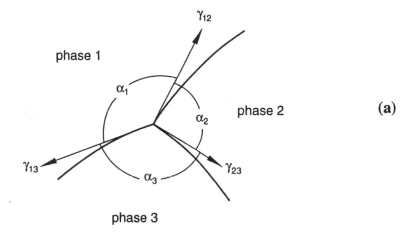

phase 1

γ_{12}

α_1

α_2

phase 2

(a)

γ_{13}

γ_{23}

α_3

phase 3

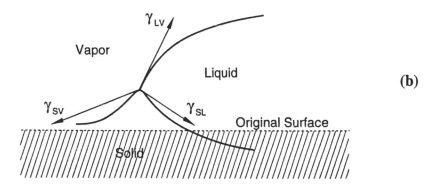

γ_{LV}

Vapor

Liquid

(b)

γ_{SV}

γ_{SL}

Original Surface

Solid

Figure 2. Schematic figures of contact configuration; (a) a three-vector diagram of balanced surface tensions; and (b) showing a ridge raised due to the vertical component of γ_{LV}.

expected the processes of solid-liquid separation or grain shape accommodation usually eliminate the desired three phase regions [1]. Under Earth-based liquid phase sintering conditions, pores are eliminated or undergo buoyant motion such that they do not remain sufficiently stationary to form a three phase equilibrium. A three phase equilibrium does exist on the outer surface of a sample, but that is often contaminated and not observed for contact equilibrium. Consequently, Earth-based samples have not been taken as good candidates for studying the contact configuration.

This microgravity sintering experiment provides an unique opportunity for solving the problem of Young's equation. Three phase equilibrium was induced by sealing the W-Ni powder compact in a low pressure (13 kPa) argon-filled container. Sintering at 1550° for 180 min increased the average W grain size by a factor of 15 (over 3000-fold volume increase), indicating significant mass transport to form an unstressed, stable configuration. The resulting solid-liquid-vapor phase equilibrium is evident in Figure 3.

On debating the validity of Young's equation, several derivations have been used to support the equation. The most widely quoted proofs in the literature are those derived by McNutt and Andes [16], Collins and Cooke [17], and Johnson [18]. Reanalysis of these derivations by the present authors have detected errors which are reported elsewhere [19], leading to further disproof of Young's equation.

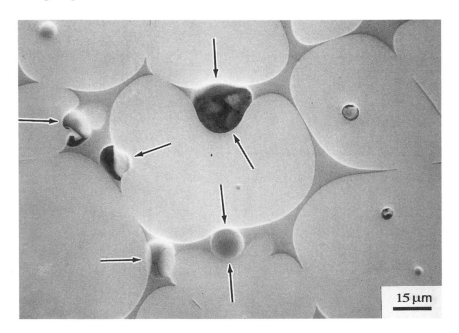

Figure 3. Scanning electron micrographs of a cross section through a 97 wt. % W - 3 wt.% Ni alloy liquid phase sintered under microgravity conditions.

CONCLUSIONS

Experimental results from a W-Ni alloy sintered under microgravity demonstrate that the solid-liquid-vapor configuration can not be described by Young's equation. Analysis of Young's equation and various derivations shows a variety of errors. A main problem in studying Young's equation is the mistreatment of a rigid, metastable system as an equilibrium configuration. The balanced three-vector diagram is the only configuration for three phases which explains the microgravity microstructure, indicating that most sessile drop experiments do not provide equilibrium contact angle and surface tension results.

ACKNOWLEDGMENT

This research is sponsored by the National Aeronautics and Space Administration (NASA) under grant NAG3-1287; the support provided by Drs. Richard DeWitt and Gilbert Santoro of the NASA Lewis Research Center is greatly appreciated. The W-Ni sample from Spacelab J was kindly provided by Prof. Kohara of Tokyo Science University.

REFERENCES

1. R. M. German, *Liquid Phase Sintering*, Plenum Press, New York, NY, pp. 13-39, (1985).
2. W. J. Huppmann and H. Riegger, "Modelling of Rearrangement Processes in Liquid Phase Sintering," *Acta Metall.*, **23**, pp. 965-971, (1975).
3. R. B. Heady and J. W. Cahn, "An Analysis of the Capillary Forces in Liquid-Phase Sintering: I. Capillary Force," *Metall. Trans.*, **1**, pp. 185-189, (1970).
4. K. S. Hwang, R. M. German, and F. V. Lenel, "Capillary Forces Between Spheres during Agglomeration and Liquid Phase Sintering," *Metall. Trans. A*, **18A**, pp. 11-17, (1987).
5. J. J. Bikerman, *Proc. 2nd Intern. Congr. Surface Activity*, London, **III**, Academic Press, New York, NY, pp. 125-130, (1957).
6. J. J. Bikerman, "Discussion", *J. Phys. Chem.*, **63**, p. 1658, (1959).
7. J. J. Bikerman, *"Contributions to the Thermodynamics of Surfaces,"* MIT Press, Cambridge, MA, p. 38, (1961).
8. B. A. Pethica and Y. J. P. Pethica, "The Contact Angle Equilibrium," *Proc. 2nd Intern. Congr. Surface Activity,* London, **III**, Academic Press, New York, NY, pp. 131-135, (1957).
9. B. K. Larkin, "Numerical Solution of the Equation of Capillary," *J. Colloid Interface Sci.*, **23**, 305, (1967).
10. A. W. Adamson, *Physical Chemistry of Surfaces*, fifth edition, Wiley, New York, NY, 1990.
11. H. Lomas, "Calculation of Advancing and Receding Contact Angles," *J. Colloid Interface Sci.*, **33**, pp. 548-553, (1970).

12. G. R. Lester, "Contact Angles on Deformable Solids," *Wetting*, SCI Monographs, No. 25, Bristol, UK, pp. 57-93, (1967).

13. S. Kohara, *"Study on Liquid Phase Sintering,"* NASDA Technical Memorandum, The Science University of Tokyo, Tokyo, Japan, May 1994.

14. Y. Liu, R. G. Iacocca, J. J. Johnson, R. M. German and S. Kohara, *"Microstructural Anomalies in a W-Ni Alloy Liquid Phase Sintered under Microgravity Conditions"*, *Metall. Mater. Trans.*, pp. 2484-2486, (1995).

15. *Binary Alloy Phase Diagram*, T. B. Massalski (ed.), ASM International, Materials Park, OH, **2**, p. 1774, (1986).

16. J. E. McNutt and G. M. Andes, "Relationship of the Contact Angle to Interfacial Energies," *J. Chem. Phys.*, **30**, pp. 1300-1303, (1959).

17. R. E. Collins and C. E. Cooke, "Fundamental Basis for the Contact Angle and Capillary Pressure," *Trans. Faraday Soc.*, **55**, 1602, (1959).

18. R. E. Johnson, "Conflicts Between Gibbsian Thermodynamics and Recent Treatments of Interfacial Energies in Solid-Liquid-Vapor Systems," *J. Phys. Chem.*, **63**, pp. 1655-1658, (1959).

19. Y. Liu and R. M. German, "Contact Angle and Solid-Liquid-Vapor Equilibrium," to be published in *Acta Metall. & Mater.* (1995).

Solid-State Sintering in the Presence of a Liquid Phase

John L. Johnson and Randall M. German

P/M Lab, Department of Engineering Science and Mechanics
The Pennsylvania State University, University Park, PA   16802-6809

Abstract
Systems which lack significant solubility of the solid in the liquid can be sintered to high densities by enhancing solid-state sintering in the presence of the liquid phase. The degree to which solid-state sintering contributes to overall densification depends on several factors including surface energies, solubilities, diffusivities, and particle sizes. Theoretical modeling is combined with experimental results to demonstrate the importance of these factors with respect to densification, microstructural evolution, and shape retention. The results are especially useful for the processing of thermal management materials such as W-Cu and Mo-Cu.

I. Introduction
Applications are increasing for composites that combine the thermal properties of two dissimilar materials. Specific examples include W-Cu and Mo-Cu heat sinks, which combine the low thermal expansion behavior of W or Mo with the high thermal conductivity of Cu to tailor the thermal properties for microelectronic packaging applications [1]. Such materials are prime candidates for liquid phase sintering due to the large difference in melting temperatures; however, the dissimilar nature of the components also results in limited inter-solubility. This lack of solubility greater hinders the solution-reprecipitation mechanism for densification [2]; however, high sintered densities can be achieved by increasing solid-state contributions to densification in the presence of the liquid phase. This work investigates the factors that affect these solid-state mechanisms.

II. Experimental Procedures
The characteristics of the W and Mo powders used for this study are summarized in Table I. Additives included a 13 μm Cu powder (Alcan 635), a 13 μm Ni powder (Inco/Novamet 4SP), a 14 μm Co powder (Alfa Products

Table I. Characteristics of the W and Mo powders.

Vendor	OSRAM	Teledyne	OSRAM	Climax	Climax
Designation	M10	C-5	M35	012293	102792
Particle size distribution					
D_{10} (μm)	0.9	1.8	2.7	1.1	2.1
D_{50} (μm)	2.1	5.1	6.2	2.5	4.1
D_{90} (μm)	7.2	16.4	14.2	4.8	7.9
BET particle size (μm)	0.23	0.49	1.08	0.47	1.52
Pycnometer density (g/cm^3)	18.4	19.0	18.4	9.62	10.1
Apparent density (g/cm^3)	1.9	2.6	2.8	2.1	3.0
Tap density (g/cm^3)	3.7	3.2	5.0	3.1	5.0

00739), and a 7 μm Fe powder (ISP CIP-R-1470). The W solubility in the liquid phase was adjusted by substituting Ni for Cu in the matrix in amounts of 0%, 1%, 2.5%, 5%, 10%, and 20%, while maintaining a constant W weight fraction of 0.9. The tungsten-based compositions were prepared by ball milling the constituent powders for 24 hours in a glass container with stainless steel balls. In all cases, milling was performed in heptane to minimize the oxidation of the powders, which were subsequently sieved and dried. The Mo powders were already milled by the vendor, so the molybdenum compositions were simply dry mixed in a Turbula mixer. For densification studies the samples were uniaxially pressed at 70 MPa into cylinders approximately 12.7 mm in diameter and 18 mm high. For structural rigidity studies the powders were uniaxially pressed at 150 MPa into cylinders approximately 12.7 mm in diameter and 18 mm high.

Sintering was performed in dry H_2 at temperatures ranging from 1300°C to 1500°C. The sintering time was one hour unless otherwise stated. In all cases, the heating rate was 10°C/minute and a one hour hold at 900°C was employed during the sintering cycle for further reduction of the powders. Following isothermal sintering, densities of sintered samples were measured by Archimedes' technique of water displacement. The samples were then sectioned, mounted, and polished to a 0.3 μm surface finish using standard metallographic procedures. Both optical microscopy and scanning electron microscopy were used for characterization of the microstructure.

III. Experimental Results

Small variations in W particle size have large effects on systems with low inter-solubility as shown in Figure 1 for the case of W-10(Cu-Ni) sintered for 1 hour at a temperature of 1.23 times that of the solidus temperature of the matrix. Even with extremely low solubility, near full density is achieved with the 0.23 μm W powder, although the sintered densities of the same compositions with the 0.49 and 1.08 μm W powders are substantially lower. As the solubility increases, this dependence on particle size decreases. Once a solubility of about 0.2 wt.% (0.06 at.%) is achieved, full density is achieved for the entire range of particle sizes examined.

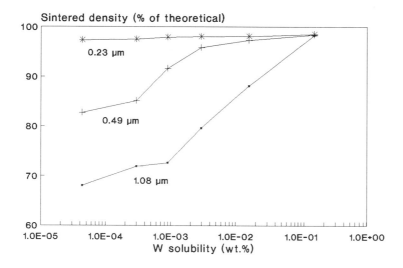

Figure 1. Effect of solubility on the sintered density of W-10(Cu-Ni) at a homologous sintering temperature of 1.23 for the three W BET particle sizes investigated in this work.

Other activators besides Ni result in high sintered densities despite nearly negligible solubility. The effects of small additions of Co, Ni, and Fe on the sintered density of W-10Cu with a 0.49 μm initial W BET particle size are plotted in Figure 2 for samples sintered at 1300°C for 1 hour in H_2. The sintered density increases nearly linearly with Ni additions, but for these quantities, the solubility is insufficient to achieve full density. Despite the low solubilities and relatively coarse W particle size, nearly full density is achieved with the addition of 0.35 wt.% Co or 0.5 wt.% Fe.

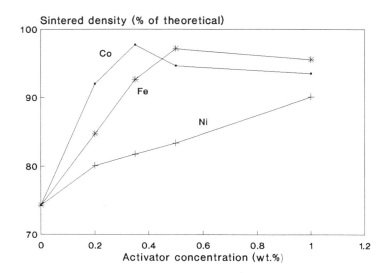

Figure 2. Effects of activator concentration on the densities of W-10Cu samples sintered at 1300°C for 1 hour in H_2.

The Mo-Cu system demonstrates a dependence of particle size on density similar to that of W-Cu. For both systems, this particle size dependency is similar to that of solid-state sintering of the base metal, as shown in Figure 3. The coarser W and Mo powders give little solid-state densification at 1400°C, while the finest powders give substantial densification, although full densities cannot be achieved. These similar trends indicate that in the case of limited inter-solubility, solid-state sintering may be an important densification mechanism despite the presence of the liquid phase.

In the case of Mo-Cu, Cu has solubility for about 1.0 wt.% (1.0 at.%) Mo at 1400°C. From the investigation into W-Cu, this solubility should be sufficient to achieve full density, but apparently additional factors influence the relative contributions of solid-state and liquid phase mechanisms to densification. Although the solubility of Mo in Cu is several orders of magnitude higher than that of W in Cu, both systems have very high dihedral angles. For W-Cu, measured dihedral angles exceed 90°, while dihedral angles greater than 70° were measured for Mo-Cu. In contrast, tungsten heavy alloys with high solubilities, including the W-8Cu-2Ni compositions investigated in this work, have dihedral angles of about 30°. The higher dihedral angles of W-Cu and Mo-Cu likely promote the formation of a rigid solid skeleton.

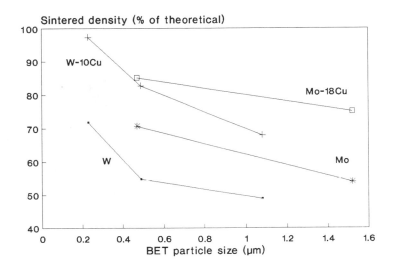

Figure 3. Effect of particle size of the base metal on the sintered densities of W, W-10Cu, Mo and Mo-18Cu. The samples were sintered at 1400°C for 1 hour in H_2.

Additional experiments were conducted with Mo-Cu and W-Cu-Ni compositions to determine the role of the microstructure on the development of a rigid skeleton. Following preliminary screening experiments, compositions of Mo-46Cu and W-25.6Cu-6.4Ni (both 50 vol.% liquid) were sintered at 1400°C for 1 hour. These compositions have similar solubilities and give similar grain sizes, but they have much different degrees of solid-solid contact as shown in Figure 4. As expected, the Mo-Cu sample with the higher dihedral angle resisted slumping despite the very high liquid volume fraction, while the W-Cu-Ni sample slumped, as shown in Figure 5. Thus, even though sufficient solubility exists for densification of Mo-Cu by liquid phase mechanisms, their contribution to densification is apparently restricted by this rigid skeleton.

IV. Discussion

A simulation has been developed to analyze solid-state contributions to densification during liquid phase sintering [3], but due to space limitations the details of this model will be omitted. This model is used to isolate both the solid-state and liquid phase contributions to densification for the compositions and conditions given in Figure 1. The sintered densities

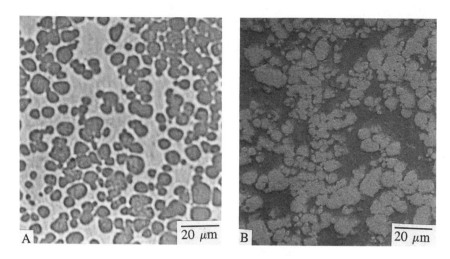

Figure 4. Microstructures of A) W-25.6Cu-6.4Ni and B) Mo-46Cu.

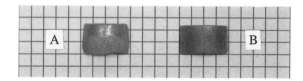

Figure 5. Slumping behavior of A) W-25.6Cu-6.4Ni and B) Mo-46Cu.

predicted from these individual components are plotted in Figure 6. For W-10Cu, the low solubility of W in the liquid phase results in domination of the densification behavior by solid-state mechanisms. In this case, the ability to achieve a high sintered density depends on the initial particle size. For the 0.23 μm W powder, liquid phase densification mechanisms can potentially contribute to densification, but solid-state sintering alone is sufficient to account for the observed sintering behavior. With increases in solubility resulting from only small amounts of Ni additions, liquid phase densification mechanisms begin to dominate the densification process, and the dependence of the density on the initial particle size decreases. Once a solubility of 0.2 wt.% is achieved, near full density is achieved for all the W powders investigated.

Due to the phase relationships between W, Cu, and Co or Fe, it is likely that these activators result in the formation of W_6Co_7 or W_6Fe_7 intermetallic phases at the surfaces of the W particles [4]. These high diffusivity phases may enhance the solid-state sintering of W in the presence

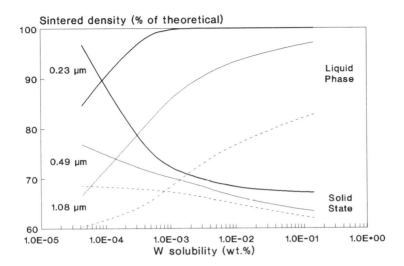

Figure 6. Predictions for the sintered density of W-10(Cu-Ni) considering only solid-state or liquid phase densification mechanisms for the three W BET particle sizes investigated in this work.

of the liquid Cu. Using the present model, this possible transport path is considered in conjunction with transport through the liquid phase. Again, analysis of densification resulting from both solid-state and liquid phase transport mechanisms identifies solid-state diffusion as the dominant densification mechanism. By increasing the solid-state diffusivity, higher densities can be achieved despite the limited inter-solubility. These results indicate the importance of considering both solid-state and liquid phase sintering mechanisms to describe the densification behavior of systems with limited inter-solubility.

Similar model simulations were conducted for Mo-Cu. Although a substantial liquid phase contribution is predicted, the actual densification behavior is closely modeled by the solid-state model alone. It is likely that the high dihedral angle of the Mo-Cu system promotes the formation of solid-state sinter-bonds, leading to the early formation of a rigid skeleton. This rigid skeleton prevents liquid penetration of the grain boundaries and densification via transport through the liquid phase. This is in agreement with the observations regarding structural rigidity. Thus, despite sufficient solubility of Mo in Cu, the densification behavior is governed by solid-state diffusion. This relationship between structural rigidity and the dominant densification mechanism also holds for the W-Cu and W-Cu-Ni systems. As the solubility increases, the dihedral angle decreases, the structural rigidity decreases and liquid phase transport events can result in higher sintered densities. Indeed, the critical solid volume fraction for structural rigidity has recently been directly linked to the dihedral angle [5]. The present work indicates that for compositions with dihedral angles that result in structural rigidity, densification is controlled by solid-state sintering of this rigid skeleton, even if sufficient solubility exists for solution-reprecipitation.

V. Conclusions

Systems with limited inter-solubility are often also characterized by a high dihedral angle, which results in a rigid solid skeleton that resists densification even if the solubility is sufficient for solution-reprecipitation. In this case, high densities can still be achieved with high solid volume fractions by increasing the magnitude of the solid-state contribution to densification. This can be done by decreasing the particle size, or by adding an activator that enhances solid-state diffusion in the presence of the liquid phase.

Acknowledgments

This research developed out of the authors' investigations into the gravitational role during sintering. Funding was provided by the National Aeronautics and Space Administration. Special thanks go to Richard DeWitt of the NASA Lewis Research Center for his active support of this research. Thanks also go to Anish Upadhyaya for his experimental assistance.

References

1. R.M. German, K.F. Hens, and J.L. Johnson, "Powder Metallurgy Processing of Thermal Management Materials for Microelectronic Applications," *Inter. J. of Powder Metall.*, **30** [2] 205-215 (1994).
2. R.M. German, *Liquid Phase Sintering*, Plenum Press, New York, (1985).
3. J.L. Johnson and R.M. German, "Solid-State Contributions to Densification during Liquid Phase Sintering," *Metall. Trans.*, submitted.
4. J.L. Johnson and R.M. German, "Phase-Equilibria Effects on the Enhanced Liquid Phase Sintering of Tungsten-Copper," *Metall. Trans.*, **24A** 2369-2377 (1993).
5. J.L. Johnson, Anish Upadhyaya, and R.M. German, "Effect of Solubility on Shape Retention during Liquid Phase Sintering," *Advances in Powder Metallurgy and Particulate Materials*, in press.

Microstructural Homogenization in PM High Strength Steels Through Sintering With Transient Liquid Phase

Herbert Danninger
Institut für Chemische Technologie anorganischer Stoffe,
Technische Universität Wien, Vienna, Austria

Abstract

The homogenization of Mo, W, and Cr in carbon steels through transient liquid phase sintering has been studied, and the resulting microstructures are described in particular with respect to alloy metal distribution and secondary pore formation. It is shown that microstructural homogeneity and secondary pores strongly influence the mechanical properties, especially the fatigue strength at high loading cycle numbers.

1. Introduction

Powder metallurgy is increasingly being used for manufacturing highly loaded precision parts e.g. for automotive engines and transmissions. The respective materials - mostly PM steels - have to exhibit superior performance compared to standard PM ferrous materials. In addition to improved manufacturing techniques, also the composition of the steels has to be optimized.

Currently, highly loaded PM steel parts are produced mainly of Fe-Cu-C or of Fe-Ni-Cu-Mo-C, for the latter material diffusion alloyed powders being commercially available. However, if the capabilities of heat treatment have to be fully exploited, the carbide-forming VIa elements have to be taken into consideration as alloy metals. In ingot metallurgy, Cr is the main alloy metal in standard tool steels, while W and Mo are present in high speed steels (Houdremont, 1956). For PM, the less oxidation sensitive Mo and W are easier to handle, but also Cr has already been used to produce tool steels.

With PM steels, various alloying techniques are possible. Pre-alloy powders offer excellent microstructural homogeneity after sintering but may be less compactible than elemental powder mixtures, and the number of pre-alloy powders available is restricted. Using elemental powder mixtures offers better compactibility and virtually unrestricted compositional flexibility, but the distribution of the alloy metals is initially inhomogeneous. Materials containing inhomogeneously distributed alloy metals may offer advantages for certain applications such as valve seat inserts, but for most structural parts fairly even distribution has to be attained, which is possible only during sintering. Formation of transient liquid phase (German 1985) has been found to be helpful here (Kaufman, 1974). Within this work, the effect of transient liquid phase on microstructure and properties of VIa alloy PM steels is described.

2. Sintering and properties of Mo steels

Mo is a very effective alloy metal, improving hardenability and resistance to heat softeneing. During heat treatment, Mo stabilizes bainite which offers better wear behaviour than martensite. Today, Fe-Mo pre-alloy powders are commercially available, but production of Mo alloy steels from powder mixtures is an interesting alternative.

Sintering tests showed that Mo homogenization is strongly affected by a transient liquid phase. Fig.1 shows microstructures of Fe-1.5%Mo-0.7%C prepared from elemental powders (Fe ASC 100.29; Mo <25 μm, natural graphite UF4) sintered in H_2 with and without liquid phase; the pronounced difference in the microstructures is clearly evident. At that carbon content, transient liquid phase is formed above approx. 1240°C, which can be clearly seen in dilatometric graphs from the abrupt expansion at that temperature (Danninger 1992). Apparently Mo homogenization through solid state sintering alone is a very slow process: even after extremely slow heating up to the sintering temperature, some Mo is still present to form liquid phase, marked by some expansion in the dilatometric graph.

During sintering, the Mo particles are transformed into carbides. At carbon contents >1%, the binary carbide Mo_2C is formed; at lower carbon activity, the binary carbide is successively transformed into the ternary η-carbide (approx. Fe_3Mo_3C), which is in excellent agreement with data given for the ternary system Fe-Mo-C (Smith and Watanabe, 1977). The liquid phase is generated in the contact area between the carbide and austenite (contact melting) and penetrates the pore channels but also the pressing contacts, solidifying there. As a consequence, swelling is observed as predictable (German 1986, Savitskij 1993). This of course implies some loss of sintered density which can be expected to affect the mechanical properties.

The formation of transient liquid phase and resulting microstructural homogenization above a certain sintering temperature decisively improves both strength and ductility, as shown for the impact energy in Fig.2. The pronounced break in the curves indicates the existence of virtually two different materials, the inhomogeneous and the homogeneous one. Despite some reduction of density as a consequence of swelling, better microstructural homogeneity due to the liquid phase sintering results in improved mechanical properties. The effect of the sintering time is significantly smaller here.

In particular for engine parts fatigue loading is frequent, and the fatigue behaviour is thus critical and requires careful investigation. Wöhler curves were taken on quenched and tempered Mo steels using a fully reversed push-

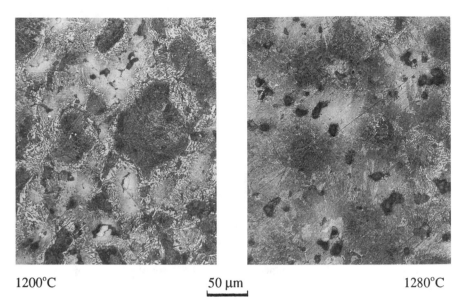

1200°C   50 μm   1280°C

Fig.1: Microstructures of Fe-1.5%Mo-0.7%C (from elemental powders) compacted 1200 MPa, sintered 60 min at different temperatures in H_2

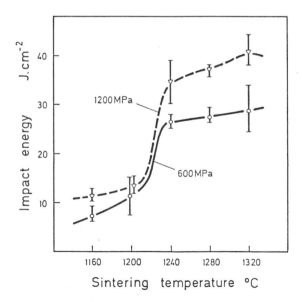

Fig.2: Impact energy of Fe-1.5%Mo-0.7%C sintered for 60 min at different temperatures in H_2

pull resonance testing system which operates at 20 kHz and enables loading cycle numbers up to 10^9. (Weiss, Stickler, and Sychra, 1990). For comparison purposes, steels prepared from pre-alloy powder (Astaloy Mo, Höganäs) were tested. The curves are shown in Fig.3; it is evident that with the specimens sintered at 1200°C the material produced from pre-alloyed powder shows much better fatigue limit than the steel prepared from elemental powder mixtures. After sintering at 1280°C however there is hardly any difference between both materials, which implies that sintering with transient liquid phase results in a microstructure well comparable to that of steel from pre-alloy powder.

One disadvantage of sintering with transient liquid phase however is the formation of secondary pores (e.g. German, 1986). In the original sites of the alloy metal particles, pores are formed that are at least the size or even larger than the particles they originate from, as visible e.g. in Fig.1b. With Mo, the liquid phase generated contains only about 13%Mo and 4%C, balance Fe, which means that a considerable amount of the iron matrix is also dissolved in the liquid. Typically, the secondary pores are about double the size of the original Mo particles.

Such defects are known to adversely affect the mechanical properties. In the case of Fe-3%Cu, both the monotonic and cyclic properties are affected (Danninger, 1987; Spoljaric and Danninger, 1993). With Fe-Mo-C, the monotonic properties are unaffected if the pores are not too large (Danninger, 1987a). The cyclic properties however are more sensitive to defects, singular defects possibly acting as crack initiators (e.g. Betz and Track, 1981). Thus, Mo alloy steels were prepared using different Mo powder fractions - which after sintering resulted in different secondary pore sizes - and fatigue tested in the resonance system. Wöhler curves were taken for each material and the effective threshold value for crack propagation was measured (Chen et al., 1991).

From above results a Kitagawa diagram (Kitagawa and Takahashi, 1976) can be plotted (Fig.4) that depicts the fatigue limit (in log scale) as a function of the defect, i.e. secondary pore, radius (in log scale). Evidently there is a critical pore radius r_{crit} below which the secondary pores do not affect the fatigue limit. Any defects with $r > r_{crit}$ are detrimental, and a singular defect as e.g. a secondary pore is sufficient to cause unexpected failure of a component. It is therefore essential to produce PM steel parts without any secondary pores. This can be attained by using fine alloy metal powders and by avoiding agglomeration of the alloy metal particles. However, as shown in Fig.3, if Mo steels prepared from powder mixtures are manufactured properly, excellent fatigue strength can be attained.

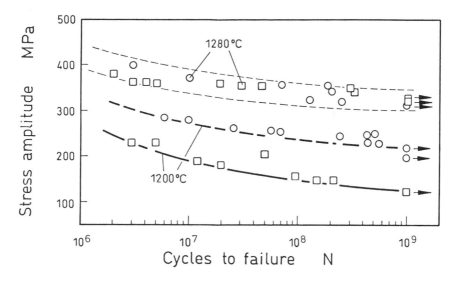

<u>Fig.3</u>: Wöhler curves (push-pull, R=-1, 20 kHz) of Fe-1.5%Mo-0.7%C from
elemental powders (□)/prealloy powders (O), sintered for 120 min in
vacuum, heat treated

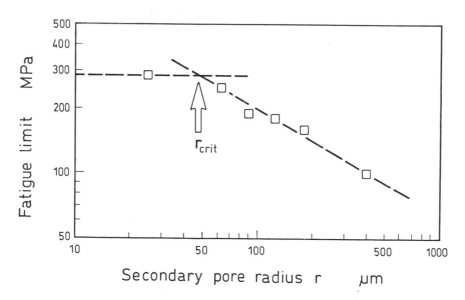

<u>Fig.4</u>: Kitagawa diagram for Fe-1.5%Mo-0.7%C containing secondary pores
of varying size. Sintered for 120 min at 1280°C in vacuum, heat treated

3. Tungsten alloy steels

Tungsten is closely related to molybdenum in its behaviour in steel and also chemically, being well reducible. Wrought tool steels, containing about 1%W, are used e.g. for drills and thread cutters. In PM, tungsten is used mainly for HSS produced by HIPing; tungsten low alloy steels are not common.

Sintering tests with Fe-W-C showed that the homogenization behaviour of W is comparable to that of Mo. Also here, carbides are formed, initially binary carbides and then η-carbides M_6C. Transient liquid phase can be generated by contact melting between carbide and austenite. The temperatures for liquid phase formation are markedly higher for W than for Mo, being near $1330°$ compared to $1240°C$ for Mo at 0.7%C. Formation of coarse secondary pores was not observed since fine W powders are readily available.

The mechanical properties of W alloy steels Fe-3%W-0.7%C, in particular the impact energy, are markedly improved by sintering with transient liquid phase. However, the effect is not as pronounced as with Mo. W alloy steels that are solid state sintered, though at high temperatures, tend to exhibit soft spots after heat treatment, apparently due to inhomogeneous W distribution. On the other hand, the high temperatures necessary for formation of transient liquid phase are near the upper limit for walking beam furnaces. Therefore, W alloy steels, if required, are produced to advantage from pre-alloy powders.

4. PM steels containing chromium

Chromium is by far the most widely employed alloy metal for wrought tool steels, while in PM its affinity to oxygen has somewhat impaired its use. Pre-alloy powders containing Cr combined with Mn and Mo (4100 series) are available e.g. in water atomized and then vacuum reduced quality. PM steels produced from these powders offer excellent as-sintered mechanical properties (Sanderow, Rodrigues, and Ruhkamp, 1985), but with pre-alloy powders the danger of oxidation during the atomization process requires special measures and increases the price of the material. With powder mixtures, on the other hand, Cr oxidation is not so much of a problem since the sintering atmosphere has to be kept oxygen-free and dry anyhow to avoid decarburization, and dissolution of Cr in the matrix reduces its chemical activity and thus the danger of oxidation. In the case of Cr steels using powder mixtures thus seems to be the simpler way.

Sintering tests showed that for Cr the contribution of solid state diffusion to homogenization is apparently much larger than for Mo and W. Thus, transient liquid phase is not as essential. If appropriately fine Cr powders are used, even heating at an average rate results in most of the Cr already being dissolved when the temperature for liquid phase formation is reached.. Only when coarse

Cr powders are used, the marked expansion well known from Mo steels is found.

This can also be seen from the mechanical properties when plotted against the sintering temperature: While in the case of Mo steels - and of Cr steels prepared from coarse powders - the liquid phase results in considerably better mechanical properties, indicated by a break in the curve, the curves for Cr steels are fairly smooth, without any break. This is shown in Fig.5 for the impact energy.

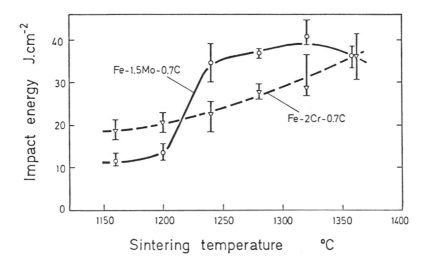

Fig.5: Impact energy of as-sintered Fe-Cr-C and Fe-Mo-C as a function of thesintering temperature. Compacted 1200 MPa, sintered 60 min

5. Summary
PM steels alloyed with VIa elements can be produced from elemental powders, which offers excellent compositional flexibility. Microstructural homogenization has to be attained during sintering to achieve highest mechanical properties. For Mo steels, a transient liquid phase generated by contact melting between Mo particles - turned into carbides - and the matrix is very beneficial here, resulting in improved mechanical properties despite some reduction in sintered density. The fatigue strength of liquid phase sintered Mo steels are comparable to that of steels prepared from pre-alloy powders. Secondary pores generated from the Mo particles during sintering affect the fatigue strength. Therefore, introduction of coarse Mo particles or Mo agglomeration during mixing should be avoided.

With W alloy steels, the homogenization of W proceeds similarly to Mo, the W particles first being transformed into carbides. However, formation of liquid phase required markedly higher temperatures, which is uneconomical, and W alloy steels are thus better produced using pre-alloy powders. With Cr, solid state diffusion plays a significant role in microstructural homogenization, and formation of transient liquid phase during sintering is not essential if sufficiently fine Cr powders are used.

6. Acknowledgment

This work was supported by the Austrian Fonds zur Förderung der wissenschaftlichen Forschung (projects 5989C and 8262TEC). Furthermore, the assistance by Prof.Dr.B.Weiss, University of Vienna, and Mr.D.Spoljaric as well as by MIBA Sintermetall AG is gratefully acknowledged.

7. References

Ed.Houdremont: Handbuch der Sonderstahlkunde (Handbook of Special Steels) 3rd Ed., Springer-Verlag, Berlin (1956)

R.M.German: Liquid Phase Sintering. Plenum (1985)

S.M.Kaufman: Modern Dev. in Powder Metall. **6** (1974) 265

H.Danninger: powder metall. int. **24** No.3 (1992) 163

Y.E.Smith, Watanabe: Modern Dev. in Powder Metall. **9** (1977) 277

R.M.German: J.Metals **1986** No.8, 26

A.P.Savitskij: Liquid Phase Sintering of the Systems with Interacting Components. Russian Academy of Sciences, Tomsk (1993)

B.Weiss, R.Stickler, H.Sychra: Metal Powder Rep. **45** (1990) 187

H.Danninger: powder metall. int. **19** (1987) 19

D.Spoljaric. H.Danninger: Proc. PTM'93 Dresden, F.Aldinger ed., DGM, Oberursel (1993) 151

H.Danninger: Powder Metall. **30** (1987) 103

W.Betz, W.Track: powder metall. int. **13** (1981) 195

D.L.Chen, B.Weiss, R.Stickler: Int. J. Fatigue **4** (1991) 327

H.Kitagawa, S.Takahashi: Proc. ICM-2 (1976) 627

H.I.Sanderow, H.A.Rodrigues, J.D.Ruhkamp: Progr. Powder Metall. **41** (1985) 283

Liquid-Phase Sintering of Iron Aluminide-Bonded Ceramics

J. H. Schneibel and C. A. Carmichael

Metals and Ceramics Division
Oak Ridge National Laboratory,
P. O. Box 2008, Oak Ridge, TN 37831-6115, U.S.A.

Abstract

Iron aluminide intermetallics exhibit excellent oxidation and sulfidation resistance and are therefore considered as the matrix in metal matrix composites, or the binder in hard metals or cermets. In this paper we discuss the processing and properties of liquid-phase sintered iron aluminide-bonded ceramics. It is found that ceramics such as TiB_2, ZrB_2, TiC, and WC may all be liquid phase-sintered. Nearly complete densification is achieved for ceramic volume fractions ranging up to 60%. Depending on the composition, room temperature three point-bend strengths and fracture toughnesses reaching 1500 MPa and 30 MPa m$^{1/2}$, respectively, have been found. Since the processing was carried out in a very simple manner, optimized processing is likely to result in further improvements.

I. Introduction

Iron aluminide intermetallics such as Fe_3Al (DO_3 structure) and FeAl (B2 structure) offer a number of potentially useful properties, which are discussed in recent conference proceedings on intermetallics (Horton et al., 1995), or, more specifically, on iron aluminides (Schneibel and Crimp, 1994). Iron aluminides are very resistant against oxidizing or sulfidizing environments, they exhibit comparatively low densities (5.6 Mg/m^3 for Fe-50 at. % Al), they have high work hardening rates suggesting high wear resistance, and consist of inexpensive raw materials. Most conventional hard metals contain significant amounts of Co or Ni, both of which are not only expensive, but also toxic when occurring as fine particulates. Since iron aluminides are presumably less toxic than Co or Ni, they may have an advantage over these elements as environmentally benign matrix materials or binders. The major drawback of iron aluminides is their low ductility in air, although appropriate processing may result in ductilities in excess of 15% (see for example Schneibel and Crimp, 1994). The low ductilities are, at least in part, due to hydrogen generated via the reduction of water vapor by aluminum. Consequently, much higher ductilites may be achieved in vacuum or dry oxygen environments. Another limitation is the poor creep strength of iron aluminides, which has, however, recently been improved (McKamey and Maziasz, 1994).

Recent thermodynamic calculations by Misra (1990) showed iron aluminides to be thermodynamically compatible with a wide range of ceramics at 1273 K. Also, since ceramics such as TiB_2 form from Fe_3Al melts containing small amounts of Ti and B (McKamey et al., 1991), it was thought possible that TiB_2

and some other ceramics might be stable in liquid iron aluminides. It was decided to investigate composites of iron aluminides with TiB_2, ZrB_2 (which has the same density as Fe-40 at. % Al), TiC [which may be successfully bonded with Ni_3Al intermetallics (Tiegs et al., 1995; Plucknett et al., 1995), and WC (which is the most common ceramic in cobalt bonded hard metals).

II. Experimental Details

Iron aluminide/ceramic composites were produced from mixtures of prealloyed Fe-40 at. % Al powder and various ceramic powders by liquid-phase sintering. Loosely packed powders were heated in a vacuum of 10^{-4} Pa to 1723 K in approximately 5 ks, held, unless otherwise stated, for 900 s at this temperature, and furnace cooled. The powder sizes, the values assumed for the densities of the powders, and the predicted theoretical densities of composites made from them are listed in Table 1.

Table 1 Densities of iron aluminide, selected ceramics, and iron aluminide composites containing 50 vol.% ceramic.

Material	Powder Size	Density (Mg/m^3)	Theoretical Density of FeAl Composite with 50 vol.% Ceramic (Mg/m^3)
Fe-40 at. %Al	-325 mesh (<45 μm)	6.06	
TiB_2	-325 mesh (<45 μm)	4.5	5.3
ZrB_2	-325 mesh (<45 μm)	6.08	6.07
TiC	3 μm	4.93	5.49
WC	5 - 8 μm	15.6	10.8

From the liquid-phase sintered samples bend specimens as well as chevron-notched specimens were fabricated by electro-discharge machining and grinding. Bend strengths and fracture toughnesses were measured in three-point bending with a span of 20 mm and a crosshead speed of 10 μm/s; fracture toughnesses K_Q were determined from the energy W absorbed during the tests, the area A swept out by the crack, and the plane strain Young's modulus E' as

$$K_Q = [(W/A) E']^{1/2}.$$

The E' values for the composites were calculated from the Young's moduli of the iron aluminide matrix, E=180 GPa, and the respective ceramics, using Ravichandran's (1994) eqn. (8) and a value ν=0.2 for Poisson's ratio.

III. Experimental Results and Discussion

Figure 1 illustrates the liquid-phase sintering process for iron aluminides. The alumina boat in this figure was initially loosely filled to the top with a

FeAl/TiB$_2$ powder mixture. The finished product illustrates the significant shrinkage occurring during liquid-phase sintering.

10 mm

Fig. 1 Liquid-phase sintered Fe-40 at. % Al + 40 vol.% TiB$_2$ composite.

Figures 2-5 show scanning electron micrographs of mechanically polished sections of FeAl liquid-phase sintered with TiB$_2$, ZrB$_2$, TiC, or WC. The micrographs reveal only little porosity and suggest the absence of significant interfacial reactions in these systems. The rounded appearance of the WC particles is quite different from the angular shapes typically found in the WC/Co system. Since the latter shapes are the result of significant dissolution of the WC in Co, the FeAl/WC microstructure may indicate a low solubility of WC in FeAl. Since iron aluminides may be alloyed with many ternary elements (Schneibel et al., 1995) it is conceivable that the solubility of the investigated ceramics in liquid iron aluminides may be improved by alloying.

Figure 6 shows the results of immersion density measurements for FeAl/TiB$_2$ containing different volume fractions of TiB$_2$. The composites containing ZrB$_2$, TiC, or WC showed qualitatively similar behavior. In the case of 40 vol.% TiB$_2$, sintering for 14.4 ks resulted in a significantly higher density as compared to sintering for 900 s. It is presently not clear why the measured density was in this case slightly higher than the theoretical density. One possible explanation might be preferential evaporation of aluminum. The processing aspects of these types of materials have so far not been addressed in great detail and will require more work.

Table 2 shows selected values for the bend strengths, the plane strain Young's moduli assumed in the fracture toughness calculations, the fracture toughnesses K_Q, and the hardnesses of the composites fabricated with TiB$_2$, ZrB$_2$, TiC, and WC. The mechanical properties of FeAl/WC are shown in more detail in Figs. 7 and 8. Figure 7 shows the room temperature bend strength of FeAl/WC for different WC volume fractions. Above 60 vol.% the bend strengths start to decrease. This finding is consistent with significant porosity found above 60 vol.% WC. The fracture toughnesses of FeAl/WC composites decrease monotonically as the WC volume fraction increases (Fig. 8). The fracture toughness of polycrystalline Fe-40 at. % Al without ceramic

Fig. 2 Liquid-phase sintered Fe-40 at. % Al + 40 vol.% TiB_2 composite.

Fig. 3 Liquid-phase sintered Fe-40 at. % Al + 40 vol.% ZrB_2 composite.

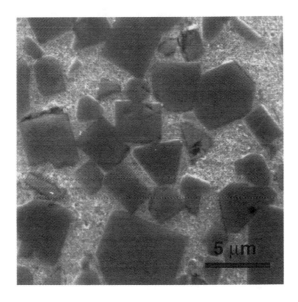

Fig. 4 Liquid-phase sintered Fe-40 at. % Al + 50 vol.% TiC composite.

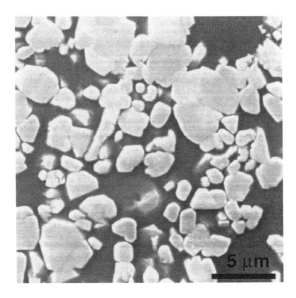

Fig. 5 Liquid-phase sintered Fe-40 at. % Al + 60 vol.% WC composite.

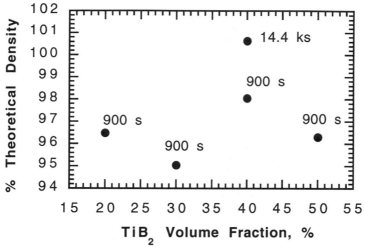

Fig. 6 Percentage of theoretical density for liquid-phase sintered Fe-40 at. % Al + TiB$_2$. The sintering times at 1723 K are indicated.

Table 2 Selected mechanical properties of iron aluminide-bonded ceramics.

Fe-40 at. % Al with	Bend Strength (MPa)	E' (GPa)	K$_Q$ (MPa m$^{1/2}$)	Hardness (Rockwell A)
40 vol.% TiB$_2$[*]	1230	276	27	75
40 vol.% ZrB$_2$[*]	1300	276	28	75
60 vol.% TiC	1050	303	15	84
50 vol.% WC	1385	333	24	77

----

[*]isothermally forged

reinforcements determined at a crack propagation velocity similar to that used in this research is approximately 30 MPa m$^{1/2}$ (Schneibel, 1993). Surprisingly, Fig. 8 shows that the incorporation of 50 vol.% WC carbide particles into iron aluminide does not degrade the fracture toughness significantly.

Although the mechanical properties of cobalt-bonded WC are significantly better than those of iron aluminide-bonded WC, the present results are encouraging. Even if it may not be possible to achieve the mechanical properties of WC/Co, some of the special properties of iron aluminides such as their oxidation and sulfidation resistance may make them of interest in those applications, where cobalt binders would degrade too quickly.

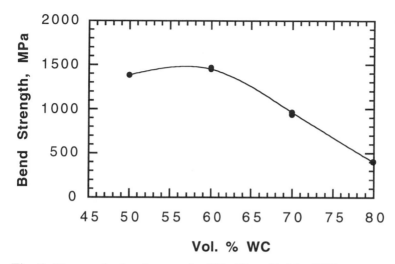

Fig. 7 Three-point bend strength of Fe-40 at. % Al + WC.

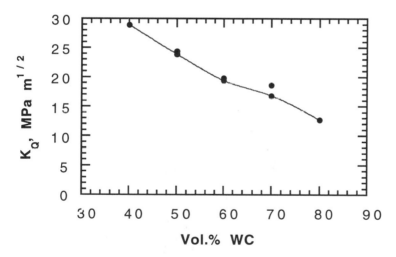

Fig. 8 Fracture toughness of Fe-40 at. % Al + WC.

Conclusions

It has been shown that ceramics such as TiB_2, ZrB_2, TiC, and WC may be liquid-phase sintered with iron aluminides. Close-to-theoretical densities have been found for ceramic volume fractions up to approximately 60 vol.%. Since the iron aluminide binder has special properties such as excellent oxidation and sulfidation resistance, composites or hard metals based on such binders may find uses in some applications. Although the processing of the composites was carried out in a very simple manner, useful mechanical properties have been found. It is anticipated that further improvements in the mechanical properties will be realized by improved processing coupled with suitable alloying modifications of the iron aluminide binder.

Acknowledgments
This research was sponsored by the Division of Materials Sciences, and the Assistant Secretary for Energy Efficiency and Renewable Energy, Office of Industrial Technologies, Advanced Industrial Materials (AIM) Program, U.S. Department of Energy, under contract DE-AC05-84OR21400 with Lockheed Martin Energy Systems, Inc.

References
J. A. Horton, I. Baker, S. Hanada, R. D. Noebe, and D. S. Schwartz (eds.), "High Temperature Ordered Intermetallic Alloys VI," Vol. 364, MRS, Boston, MA (1995).

C. G. McKamey and P. J. Maziasz, "Effect of Heat Treatment Temperature on Creep-Rupture Properties of Fe_3Al-Based Alloys," in J. H. Schneibel and M. A. Crimp (eds.), "Processing, Properties, and Applications of Iron Aluminides," TMS, Warrendale, PA, pp. 147-58 (1994).

C. G. McKamey, J. H. DeVan, P. F. Tortorelli, and V. K. Sikka, "A Review of Recent Developments in Fe_3Al-Based Alloys." J. Mater. Res. **6**, 1779-1805 (1991).

A. K. Misra, "Identification of Thermodynamically Stable Ceramic Reinforcement Materials for Iron Aluminide Matices," Metall. Trans. A **21A**, 441- 6 (1990).

K. P. Plucknett, T. N. Tiegs, K. B. Alexander, P. F. Becher, J. H. Schneibel, B. Waters, P. A. Menchhofer, "Intermetallic Bonded Ceramic Matrix Composites," to be published in "Ceramics for Structural and Tribological Applications," Vancouver, B.C., Aug. 21-23, CIM, 1995.

K. S. Ravichandran, "A simple Model of Deformation Behavior of Two Phase Composites," Acta Metall. Mater. **42**, 1113-23 (1994).

J. H. Schneibel and M. G. Jenkins, "Slow Crack Growth at Room Temperature in FeAl," Scr. Metall. Mater. **28,** 389-93 (1993).

J. H. Schneibel and M. A. Crimp (eds.), "Processing, Properties, and Applications of Iron Aluminides," TMS, Warrendale, PA (1994).

J. H. Schneibel, E. P. George, E. D. Specht, and J. A. Horton, "Strength, Ductility, and Fracture Mode in Ternary FeAl Alloys," in J. A. Horton, I. Baker, S. Hanada, R. D. Noebe, and D. S. Schwartz (eds.), "High Temperature Ordered Intermetallic Alloys VI," Vol. 364, MRS, Boston, MA, pp. 73-78 (1995).

T. N. Tiegs, P. A. Menchhofer, K. P. Plucknett, K. B. Alexander, P. F. Becher, and S. B. Waters, "Hardmetals Based on Ni_3Al as the Binder Phase," to be published in Proceedings 4th International Conference on Powder Metallurgy in Aerospace, Defense, and Demanding Applications, Anaheim, CA, May 8-10, 1995.

Alterations of Boride Particle Coarsening by an Addition of Cr to a Mo-Ni-B Alloy during Liquid Phase Sintering

Satoru Matsuo, Shinya Ozaki, Yuji Yamasaki, Masao Komai, and Ken-ichi Takagi*

Technical Research Laboratory, Toyo Kohan Co., Ltd., Kudamatsu, Yamaguchi, JAPAN, *Technical Department, Toyo Kohan Co., Ltd., Tokyo, JAPAN

Abstract

Liquid phase assisted reaction sintering of Mo-Ni-Cr-B alloys results in cermets consisting of a M_3B_2 (M: Mo, Cr, Ni) boride and metallic Ni alloy. The presence of a liquid phase drastically enhances particle coarsening kinetics as well as densification processes of green compacts during sintering. Cr addition to a Mo-Ni-B alloy influences the size and morphology of M_3B_2 particles in contact with the liquid phase. The crystal structure of M_3B_2 changes from orthorhombic to tetragonal due to the addition of 10 mass% of Cr. The paper discusses the effect of Cr on the microstructural evolution of Mo-Ni-Cr-B alloys in connection with the crystal structure transformation of the boride phase.

I. Introduction

Mo_2NiB_2 boride cermets having a Ni alloy as the metal matrix phase have been successfully produced by liquid phase sintering [1 - 3]. In order to overcome poor wetting of borides, a boronizing reaction method was exploited to improve the sinterability and form the Mo_2NiB_2 ternary boride in the cermets. This method has been labeled as reaction boronizing sintering. Extensive research has been conducted regarding the sintering mechanisms in the Mo_2NiB_2 boride cermets, and it has been revealed that the ternary boride is formed prior to liquid phase formation [3].

Effects of alloy element additions on the sintering behavior and properties of the cermets have been also studied by the authors [4 - 6]. Additions of Cr resulted in boride particle size refinement and improvement of the mechanical properties. It was also found that Mo_2NiB_2 boride having the orthorhombic structure transformed to tetragonal M_3B_2 (M: Mo, Ni, Cr) by the solid solution of Cr [5, 6].

Mechanical properties are the major concern for cermets, and hence microstructural control, namely the control of size and distribution of the boride particles during liquid phase sintering is essential. The elucidation of factors which influence the microstructural evolution of the cermets is strongly desired. Generally, a liquid phase drastically enhances densification processes during sintering and also accelerates particle coarsening kinetics due to rapid mass transport across the liquid [7]. Liquid/solid interfacial phenomena influence the shape of particles during sintering [8]. This paper describes particle coarsening and morphology change of the M_3B_2 boride during liquid phase sintering. A variation of the boride particle evolution is discussed in connection with the boride crystal structure difference between orthorhombic and tetragonal which was caused by an addition of Cr.

II. Experimental Procedure

M_3B_2 (M: Mo, Ni, Cr) borides were synthesized prior to the production of the cermets in order to differentiate the partition of Cr to the M_3B_2 phase and the metal matrix phase. Table 1 shows the compositions of two types of the boride phases synthesized, one having the Mo_2NiB_2 composition and the other containing 10% Cr in substitution for Ni with the stoichiometry of M_3B_2. The compositions of the cermets are listed in table 2. Ni metal matrices with and without an addition of 10 mass% of Cr were prepared. The mass ratio of the boride to the metal matrix is 3 for all the compositions.

Fig. 1 describes a flow chart of the experimental procedure. By blending MoB powder, carbonyl nickel powder, and electrolytic chromium powder, the compositions of the borides were adjusted so that the atomic ratio of M (Mo, Cr, Ni) to B is 3/2. The blended powders were heated to 1400 K for 3.6 ks (1 hr) to obtain the M_3B_2 borides. Carbonyl nickel powder, electrolytic chromium powder and the synthesized boride were mixed and further comminuted in acetone by a vibratory mill to an average particle size of about 1.5 μm to produce the boride cermets. The ball-milled powder was compressed at 98 MPa, and the green compacts were sintered at 1533 and 1573 K for 1.2 ks (20 min.). The boride crystal structures were identified by XRD, and microstructures of the sintered bodies were examined by SEM.

Table 1 Compositions of borides.

boride		B	Mo	Ni	Cr
Mo_2NiB_2	(mass%)	7.9	70.5	21.6	0
	(at%)	40	40	20	0
$(Mo,Cr,Ni)_3B_2$					
	(mass%)	8.0	71.4	10.6	10.0
	(at%)	40	40	9.7	10.3

Table 2 Compositions of boride cermets (mass%).

	Boride		Metal matrix
	Mo_2NiB_2	$(Mo,Cr,Ni)_3B_2$	
1.	75	0	25 (Ni)
2.	75	0	25 (Ni - 10 mass% Cr)
3.	0	75	25 (Ni)
4.	0	75	25 (Ni - 10 mass% Cr)

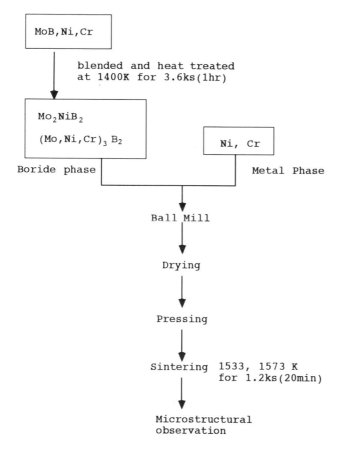

Fig. 1 Flow chart of experimental procedure.

III. Results and Discussion

Fig. 2 shows XRD patterns for the M_3B_2 borides produced by heating the premixed powders to 1400K. MoB was not detected and was thus consumed to form M_3B_2. Mo_2NiB_2 has an orthorhombic structure, while $(Mo,Ni,Cr)_3B_2$ shows a tetragonal structure with some amount of the orthorhombic structure.

Synthesized M_3B_2

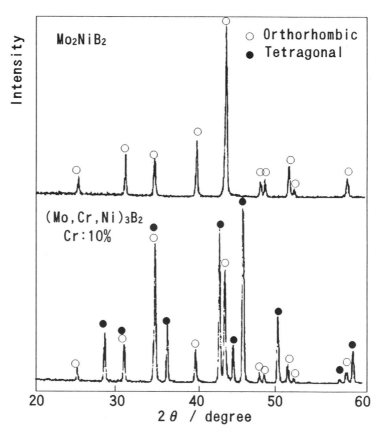

Fig. 2 XRD patterns from two types of M_3B_2 borides.

Fig. 3 shows microstructures of the sintered alloys with the combinations of Mo_2NiB_2 or $(Mo,Ni,Cr)_3B_2$ for the boride phase and Ni or Ni - 10 mass%Cr for the matrix phase as shown in Table 1, respectively. The white and gray particles in the back scattered images are the borides. The Mo_2NiB_2 boride shows extensive coarsening, while the $(Mo,Ni,Cr)_3B_2$ boride shows less coarsening tendency, in both matrix compositions. This implies that the crystal structural change of M_3B_2 associated with the addition of Cr plays a main role for particle coarsening rather than the difference in the composition of the metal matrix phase. Since the average initial particle size prior to sintering was about 1.5 µm for all the alloys, the difference in boride particle size in the cermets reflects the coarsening kinetics of the two M_3B_2.

Mo_2NiB_2 $(Mo,Ni,Cr)_3B_2$

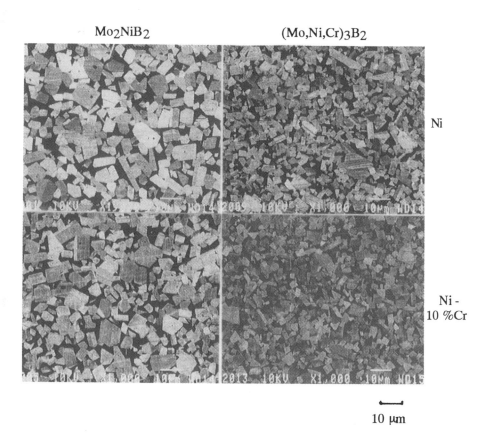

Ni

Ni - 10 %Cr

10 µm

Fig. 3 Microstructures of the sintered alloys (back scattered electron image). The compositions are the combinations of Mo_2NiB_2 or $(Mo,Ni,Cr)_3B_2$ for the boride phase and Ni or Ni - 10 mass%Cr for the matrix phase.

Fig. 4 shows magnified photographs of Mo_2NiB_2 and $(Mo,Ni,Cr)_3B_2$ particles in the sintered alloys of the compositions 1 and 4 in table 1, respectively. The Mo_2NiB_2 particles exhibit sharply faceted morphologies with sharp corners. Coalescence of some of the particles can be observed in the faceted surfaces due to different contrasts from the original grains. This indicates that the formation of low energy grain boundaries of the particles is preferable because of the interface energy balance [9]. The $(Mo,Ni,Cr)_3B_2$ particles, on the other hand, exhibit round corners with less degree of faceting and coarsening. Contacts of corners and sides of the grains are also observed.

$(Mo,Ni,Cr)_3B_2$

5 μm

Fig. 4 Morphologies of Mo_2NiB_2 and $(Mo,Ni,Cr)_3B_2$ particles.

Substances having complex crystal structures (compounds in many cases) tend to exhibit faceted growth morphologies, and this behavior stems from the atomic attachment kinetics with respect to the crystallographic plane [10]. Faceting is closely related to anisotropy of the surface energy of crystals which influences the atomic attachment kinetics. Large dependence of the attachment kinetics on crystal plane leads to uneven crystal growth. As a result, rapidly moving planes (high index planes) disappear, and the crystal becomes enveloped by slowly moving planes (low index planes).

The orthorhombic structure is more anisotropic than the tetragonal structure and resulted in more sharply faceted boride grains. It can be also assumed that rapidly moving high index planes caused faster particle growth in the orthorhombic M_3B_2 in a relatively short period.

A calculation of the equilibrium grain shape by Warren revealed that a shape change from near - spherical to perfect cubic takes place by a small decrease in the degree of crystallographic isotropy [11]. The change from the round corners to the sharp corners in M_3B_2 also corresponds to decrease in isotropy of the crystal structure and agrees with Warren's result.

The tetragonal M_3B_2 (($Mo,Ni,Cr)_3B_2$) boride provides smaller particle size and hence can be preferable to the orthorhombic M_3B_2 (Mo_2NiB_2) boride in terms of mechanical properties. The previous results that Cr additions to a Mo-Ni-B alloy resulted in improvement of the strength [2, 4] can be confirmed by the observation of the variations of M_3B_2 particle evolution in this investigation.

IV. Conclusions

The orthorhombic M_3B_2 (Mo_2NiB_2) exhibited rapid particle coarsening associated with sharp faceting and sharp angular corners, while the tetragonal M_3B_2 (($Mo,Ni,Cr)_3B_2$) exhibited slower particle coarsening with round corners. This is attributed to the degree of surface energy anisotropy of the M_3B_2 crystals. The tetragonal M_3B_2 boride provides smaller particle size and hence can be preferable to the orthorhombic M_3B_2 boride in terms of mechanical properties.

References

1. M. Komai, Y. Yamasaki and K. Takagi, "Sintering Behavior of a Reactively Sintered Ternary Boride Base Cermet", Solid State Phenomena, **25 & 26** 531 - 538 (1992).

2. M. Komai, Y. Yamasaki, K. Takagi and T. Watanabe, "Effects of Cr on the Mechanical Properties and Phase Formation of Mo_2NiB_2 Boride Base Cermet", Properties of Emerging P/M Materials, Advances in Powder Metallurgy & Particulate Materials - 1992, MPIF, Princeton, NJ, **8** pp 81 - 88.

3. M. Komai, Y. Yamasaki, S. Matsuo and K. Takagi, "Reactive Sintering Behavior of a Cr Containing Base Mo_2NiB_2 Cermet", in Powder Metallurgy World Congress PM'94 in Paris, **2** 1521 - 1524 (1994).

4. M. Komai, Y. Yamasaki and K. Takagi, "Effect of Cr Content on Properties of (Mo,Ni) Boride Base Hard Alloys", Journal of the Japan Institute of Metals, **57** [7] 813 - 820 (1993).

5. M. Komai, Y. Yamasaki, S. Ozaki and K. Takagi, "Mechanical Properties of Mo_2NiB_2 Base Hard Alloys and Crystal Structures of Boride Phases", Journal of the Japan Institute of Metals, **58** [8] 959 - 965 (1994).

6. S. Ozaki, Y. Yamasaki, M. Komai and K. Takagi, "Crystal Structural Change of Boride Observed in Boride Base Hard Alloys", pp 220 - 221 Proc. 11th Int. Symp. Boron, Borides and Related Compounds, Tsukuba, 1993, JJAP Series 10 (1994).

7. R. M. German, "Liquid Phase Sintering", pp 127 - 155, Plenum Press, NY, 1985.

8. S. Sarian and W. Weart, "Factors Affecting the Morphology of an Array of Solid particles in a Liquid Matrix", Transactions of the Metallurgical Society of AIME, **233** [11] 1990 - 1993 (1965).

9. R. M. German, "The Contiguity of Liquid Phase Sintered Microstructures", Metallurgical Transactions A, **16A** [7] 1240 - 1252 (1985).

10. W. Kurz and D. J. Fisher, "Fundamentals of Solidification", 3rd edition, pp 34 - 41, Trans Tech Publication Ltd., Switzerland, 1992.

11. R. Warren, "Microstructural Development during the Liquid-Phase Sintering of Two-Phase Alloys, with Special Reference to the NbC/Co System", Journal of Materials Science, **3** 471 - 485 (1968).

DEVELOPMENTS IN THE SINTERING OF TUNGSTEN HEAVY ALLOYS

Animesh Bose[1] and Randall M. German[2]
[1]Parmatech Corporation, Petaluma, CA 94954
[2]Penn State University, University Park, PA 16802-6809

Abstract

Tungsten heavy alloys are two phase composites that are generally processed by liquid phase sintering. The properties of heavy alloys critically depend on the sintering conditions and alloying additions. An important aspect of successful sintering is control of pores and impurities via the sintering atmosphere. Secondary factors such as the time and temperature also effect residual porosity, which in turn has a strong negative influence on mechanical properties. Alloying with refractory metals is effective in promoting higher strength via a reduction in the sintered grain sizes. This paper discusses recent developments relating to the sintering, alloying, physical metallurgy, and mechanical properties of liquid phase sintered tungsten heavy alloys.

Introduction

Tungsten heavy alloys (WHAs) are characterized by dispersed grains of almost pure tungsten embedded in a matrix that is liquid at the sintering temperature. The matrix is composed of tungsten dissolved in an alloy of usually Ni, Fe, Co, or Cu. The composite is liquid phase sintered in the 1500°C range, although solid-state sintering has been used in limited cases. The most popular WHAs are made from mixed elemental powders of W, Ni, and Fe, where the W content typically varies from 88 to 98 wt.%, the balance being the matrix phase (generally 7Ni-3Fe or 8Ni-2Fe). New compositions of matrix, such as Co-Ni-Fe, Co-Fe, Co-Fe, and Ni-Mn have been developed.

The properties of WHAs are sensitive to the sintering conditions and alloying additions. Processing conditions that result in residual porosity drastically reduce ductility, tensile strength, and toughness. This is illustrated in Figure 1 where the normalized ductility is plotted versus residual porosity for a number of WHAs (German *et al.* 1992, Ge *et al.* 1982, German and Churn 1984, Danninger *et al.* 1983, Kang et al. 1981, Churn and Yoon 1981). Clearly, one of the critical aspects of sintering is the elimination of residual porosity through proper selection of the temperature, time, heating rate, and atmosphere. This latter aspect results from a high sensitivity to residual contamination. Further, refractory alloying additives are beneficial in retarding microstructural coarsening and improving the as-sintered properties.

Sintering Atmosphere

In WHAs, retained porosity plays a critical role in determining the final mechanical properties. Incomplete densification may be the consequence of too low a sintering temperature, insufficient sintering time, solidification shrinkage due to fast cooling through the solidus temperature, entrapped gases within the pores, or pores formed due to *in situ* reaction products. This latter event is subtle, but repeatedly has been demonstrated for sintering in a reducing atmosphere. Thus, the sintering atmosphere is vital in controlling residual porosity. The mechanical properties associated with plastic flow (ultimate tensile strength, impact toughness, and fracture elongation) are strongly affected by the residual porosity.

A typical sintering and heat treatment cycle for WHAs consist of heating to a temperature in the range of 800 to 1000°C and holding for one or two hours (to reduce oxides) in dry hydrogen, followed by a ramp to the sintering temperature near 1500°C with a hold for 30 to 120 min in dry hydrogen. On completion of sintering, the compacts are slow cooled to around 1400°C (to prevent solidification shrinkage pores) and furnace cooled to room temperature. The sintered alloys are generally heat treated at 1000 to 1100° C for 1 h in vacuum, nitrogen, or inert gas to reduce hydrogen embrittlement.

There are several problems with such processing. Dry hydrogen often causes swelling during liquid phase sintering due to water vapor trapping, a problem that becomes severe with high liquid contents or longer sintering times. Pore growth can be minimized using wet hydrogen in the sintering cycle (Bose *et al.* 1986). However, wet hydrogen gives less complete oxide reduction and degraded final properties. A switch to a secondary inert atmosphere at the end of the sintering cycle is beneficial since it desaturates the embrittling hydrogen via outward diffusive flow of to the compact surface (German and Churn 1984). In spite of hydrogen desaturation, it is still beneficial to apply a post-sintering heat treatment that ends with a quench to suppress embrittlement due to impurity segregation to the grain boundaries (German *et al.* 1984).

Investigations on a 88W-8.4Ni-3.6Fe heavy alloy revealed the following (Bose and German 1988a):

1) Use of inert gas sintering atmosphere (argon) for the total sintering cycle was detrimental and resulted in a porous microstructure.

2) Elongation increased from 23% to 33% when the dew point of the hydrogen sintering atmosphere was changed from -44 to +30°C while sintering at 1500°C for 30 min.

3) Elongation decreased to 2% by changing the sintering temperature from 1500 to 1540°C when using a dry hydrogen atmosphere (dew point -44°C).

4) The tensile elongation also decreased to 0% when the sintering time was increased to 120 min at 1500°C in dry hydrogen.

These findings can be rationalized by considering Markworth's (1972) final stage sintering model. According to this model, final pore removal is governed by the gas trapped in closed pores. Due to pore shrinkage, the gas pressure inside the pores increases. Densification stops when this gas pressure is large enough to balance the compressive stress from surface energy divided by the pore radius. If the trapped gas is soluble in the matrix (such as with hydrogen), then it will slowly diffuse out of the pores allowing continued densification. Thus, hydrogen gas in the pores decreases densification as compared to vacuum sintering, but does not prevent complete densification.

The situation is different with gases such as argon or water vapor which have no solubility or diffusivity through the matrix. These remain trapped in the pores which may coalesce to grow large pores, further preventing densification. Vacuum is the best sintering atmosphere with respect to densification. However, vacuum is not recommended in practice because the vapor pressure of the non-refractory elements is high, resulting in preferential vaporization of the low melting elements. German and Churn (1984) developed a sintering model which connects the pore depressurization rate to the gas diffusivities. Calculations based on that model showed the final sintered density variation with sintering time for different atmospheres as shown in Figure 2. Fastest densification is expected with vacuum sintering. This is followed by hydrogen, where complete densification occurs in 20 min. However, this is not seen in practice due to water vapor formation.

Water vapor forms within the matrix by a reaction between dissolved hydrogen (form the reducing atmosphere) and oxygen in solution in the tungsten. Oxygen is present in tungsten as a consequence of reduction of WO_3 as an inherent aspect of powder production. The solubility is in the 3000 to 4000 ppm range, a value typical for oxygen contamination of most tungsten powder. During liquid phase sintering, solution-reprecipitation continually leads to dissolution of tungsten into the liquid matrix as a natural part of grain growth. In turn the shrinking grains release dissolved oxygen which reacts with the dissolved hydrogen to form water vapor. Similar reactions involving residual carbon and oxygen or hydrogen are possible, forming CO or CH_4 vapor products. Any condition that favor tungsten grain growth increases the tendency for internal vapor generation and pore growth. Initially, the vapor filled pores are small and will slowly exit to the surface by buoyancy driven pore migration, assuming detachment from the grain interface. Wet hydrogen inhibits the thermochemical reaction responsible for forming water vapor, allowing elimination of the water vapor filled pores. However, the use of wet hydrogen throughout the entire sintering cycle results in improper reduction of the oxides and degraded mechanical properties.

Based on the above considerations a new cycle for processing of WHAs has been reported by Bose and German (1988a). In this cycle, the mixed powder compacts are heated to a temperature of 800°C and held for 1 h followed by a ramp up to a temperature of 1300°C in dry hydrogen (dew point typically near -40°C). This reduces the oxides and is followed by a switch to wet hydrogen (dew point near +20° C) while the temperature is ramped through the liquid formation temperature. This switch (before liquid formation) reverses the kinetics of water vapor formation. The wet hydrogen atmosphere is maintained for approximately two-thirds of the sintering time of 30 min, followed by a switch to dry argon (dew point around -60°C) to suppress any embrittlement from residual hydrogen. The densification kinetics in liquid phase sintering are sufficiently fast to ensure pore closure prior to argon introduction. Thus, argon entrapment in the pores does not occur. The sintered compacts are then cooled slowly to a temperature slightly below the solidus to prevent formation of solidification shrinkage pores. After that point the compacts are furnace cooled in dry argon. To ensure maximum ductility and toughness, the sintered compacts should be heat treated using a 60 min soak near 1100°C in vacuum, Ar, or N_2 followed by water quenching. This step prevents embrittlement due to impurity segregation. Repeatedly this processing sequence has resulted in the highest mechanical properties.

Sintering Time

Sintering of WHAs for times greater than 2 h usually results in pore coalescence and pore growth, with degraded properties, especially tensile elongation. Because of the strong sensitivity to porosity, prolonged sintering is avoided in WHAs. The development of a multiple stage atmosphere cycle that minimizes pore formation, as discussed above, allowed study of the true effect of sintering time on the mechanical properties of WHAs. Extended sintering times also leads to several interesting microstructural changes. One obvious change is that there is considerable grain growth and grain shape accommodation associated with increased sintering time, with a decrease in contiguity (Kipphut *et al.* 1988, Fischmeister *et al.* 1969, Shrikanth and Upadhyaya 1986). The contiguity decrease indicated a relative increase in the solid-solid versus solid-liquid interfacial energy due to grain coalescence or grain rotation (Kipphut *et al.* 1989, Yang *et al.* 1990, German 1985). With the multiple stage sintering cycle, it was possible to sinter both 95 and 88 wt.% W heavy alloys to almost full density with sintering times as high as 600 min. For the 95W alloy, the mechanical properties were not degraded, in contrast to the degradation reported in earlier studies.

During solution-reprecipitation and coalescence controlled grain growth, the cube of the grain size varies linearly with time. The rate constant (typically near 1 to 4 $\mu m^3/s$) provides the proportionality between the two factors. One important aspect of the rate constant is a dependence on the solubility of tungsten in the liquid phase. As discussed later, refractory metal alloying additions reduce this solubility and give a smaller grain size while still allowing full densification.

Sintering WHAs for 90 min results in fully dense and generally round tungsten grains, while sintering for 600 min results in contact flattening and considerable grain shape accommodation with low levels of solid-solid contact. Based on independent observations of grain rotation (Muster and Willerscheid 1979), grains with small crystal misorientation angles progressively underwent rotation due to the grain boundary torque. This resulted in elimination of the grain boundary with grain coalescence, giving a concomitant decrease in contiguity. Grains with high angles of misorientation progressively rotated into special crystal orientations of minimized solid-solid grain boundary energy. This process increases contiguity. There will be some misorientation angles between grains where only grain growth by solution-reprecipitation is expected to occur. Thus, a schematic time versus contiguity plot starts with a near zero contiguity at the point where liquid first forms. This chemically active liquid fragments the solid skeletal structure by penetrating grain boundaries, allowing rapid grain rearrangement. This is followed by a rapid increase in the net contiguity due to neck growth. With continued liquid-phase sintering there is a drop in the net contiguity as grains coalesce and eliminate the grain boundary. Eventually the structure is stabilized into an arrangement that reduces the grain boundary energy, resulting in an increase in the contiguity.

With respect to mechanical properties, one consequence of long time sintering is a dramatic increase in ductility and decrease in strength. The yield strength of a 95W alloy sintered for different times followed the Hall-Petch behavior. The yield strength σ_y in MPa and the grain size G in μm for the 95W alloy are related as follows:

$$\sigma_Y = 497 + 507/\sqrt{G}$$

However, the ultimate tensile strength and ductility are maximum at 90 min of sintering. This correlates with a gradual increase in residual porosity during prolonged sintering.

Alloying

The properties of WHAs after optimized sintering and heat treatment cycles gives a high ductility and strength (elongation of 25% and tensile strength over 900 MPa). However, higher strengths are possible with lower ductility. Typically this trade-off has been employed via post-sintering deformation processing. Several recent investigations have attained similar property benefits via microstructure refinement using refractory metal additions (Bose and German 1988b, Bose *et al.* 1987, Bose 1994). The refractory metals such as Mo, Re, and Ta have complete solubility in tungsten and a high solubility in the matrix alloy. Additions of these alloying ingredients results in an increase in strength and hardness, a decrease in ductility. This is due to a grain size reduction in the sintered structure. The grain size results from a retarded grain growth rate. This is a direct manifestation of solubility effects that are poorly described by current theories.

The strength and hardness gains in WHAs results from two mechanisms. The refractory metals which have solubility in the tungsten and the matrix result in solid-solution strengthening. This has been verified by microhardness and other measurements (Kemp and German 1991). The second effect is the grain size reduction in full density materials, which contributes to the increase in the strength and hardness. Rhenium proves to be the most potent alloying addition [German and Bose 1991).

Acknowledgments

The authors thank Parmatech Corporation and the U. S. Army Research Office for the support provided in the preparation of this paper.

References

A. Bose, "Alloying and Powder Injection Molding of Tungsten Heavy Alloys: A Review," *Tungsten and Refractory Metals 2*, A. Bose and R. J. Dowding (eds.), MPIF, Princeton, NJ, 21-33 (1994).

A. Bose and R. M. German, "Sintering Atmosphere Effects on Tensile

Properties of Heavy Alloys," *Metall. Trans.,* **19A**, 2467-2476 (1988a).

A. Bose and R. M. German, "Microstructural Refinement of W-Ni-Fe Heavy Alloys by Alloying Additions," *Metall. Trans.,* **19A**, 3100-3100 (1988b).

A. Bose, B. H. Rabin, S. Farooq and R. M. German, "Hydrogen Atmosphere and Residual Oxygen Interactions During Liquid Phase Sintering of Heavy Alloys," *Horizons in Powder Metallurgy*, Part II, W. A. Kaysser and W. J. Huppmann (eds.), Verlag Schmid, Freiburg, Germany, 1123-1126 (1986).

A. Bose, D. M. Sims, and R. M. German, "High Strength Tungsten Heavy Alloys with Molybdenum Additions," *Prog. Powder Met.,* **43**, 79-92 (1987).

K. S. Churn and D. N. Yoon, "Pore Formation and its Effect on Mechanical Properties in W-Ni-Fe Heavy Alloy," *Powder Metall.,* **22**, 175-178 (1979).

H. Danninger, W. Pisan, G. Jangg, and B. Lux, "Veranderungen in Eigenschaften und Bruchverhalten von Wolfram-Schwermetallen Wahrend der Flussinphasensinterung," *Z. Metallkd.,* **74**, 151-155 (1983).

H. Fischmeister, A. Kannappan, L. Ho-Yi, and E. Navara, "Grain Growth During Sintering of W-Cu-Ni," *Phys. Sintering*, vol. 1, G1-G13 (1969).

C. C. Ge, X. I. Xia, and E. T. Henig, "Effects of Some Sintering Parameters and Vacuum Heat Treatment on the Mechanical Properties and Fracture Mode of Heavy Alloy 90W-7Ni-3Fe," *Proceedings P/M 82,* Associazione Italiana di Metallurgia, Milan, Italy, 709-714 (1982).

R. M. German, "The Contiguity of Liquid Phase Sintered Microstructures," *Metall. Trans.,* **16A**, 1247-1252 (1985).

R. M. German and A. Bose, "Properties of High Density Tungsten-Rhenium Alloys by Liquid Phase Sintering," *Tungsten and Tungsten Alloys: Recent Advances*, A. Crowson and E. S. Chen (eds.), TMS, Warrendale, PA, 53-59 (1991).

R. M. German, A. Bose, and S. S. Mani, "Sintering Time Atmosphere Influences on the Microstructure and Mechanical Properties of Tungsten Heavy Alloys," *Metall. Trans.,* **23A**, 211-219 (1992).

R. M. German and K. S. Churn, "Sintering Atmosphere Effects on the Ductility of W-Ni-Fe Heavy Alloys," *Metall. Trans.,* **15A**, 747-754 (1984).

R. M. German, J. E. Hanafee, and S. L. Digiallonardo, "Toughness Variation with Test Temperature and Cooling Rate for Liquid Phase Sintered W-3.5Ni-1.5Fe," *Metall. Trans.,* vol. 15A, 121-128 (1984).

T. K. Kang, E. T. Henig, W. A. Kaysser, and G. Petzow, "Effect of Cooling Rate on the Microstructure of a 90W-7Ni-3Fe Heavy Alloy," *Modern Developments in Powder Metallurgy*, vol. 14, H. H. Hausner, H. W. Antes, and G. D. Smith (eds.), MPIF, Princeton, NJ, 189-203 (1981).

P. B. Kemp and R. M. German, "Grain Growth in Liquid Phase Sintered W-Mo-Ni-Fe Alloys," *J. Less-Common Met.,* vol. 175, 353-368 (1991).

C. M. Kipphut, A. Bose, S. Farooq, and R. M. German, "Gravity and
Configurational Energy Induced Microstructural Changes in Liquid Phase
   Sintering," *Metall. Trans.*, **19A**, 1905-1913 (1988).
C. M. Kipphut, R. M. German, A. Bose, and T. Kishi, "Gravitational Effects
   on Liquid Phase Sintering," *Advances in Powder Metallurgy*, vol. 2, MPIF,
   Princeton, NJ, 415-429 (1989).
A. J. Markworth, "On the Volume Diffusion Controlled Final Stage
Densification of a Porous Solid," *Scripta Metall.*, **6**, 957-960 (1972).
W. Muster and H. Willerscheid, "Crystallographic Orientation Relationships in
   Coalescing Sintered Tungsten Spheres," *Metallog.*, vol. 12, 287-294 (1979).
V. Shrikanth and G. S. Upadhyaya, "Contiguity Variation in Tungsten
Spheroids of Sintered Heavy Alloys," *Metallog.*, **19**, 437-445 (1986).
S. C. Yang, S. S. Mani, and R. M. German, "The Effect of Contiguity on
   Growth Kinetics in Liquid Phase Sintering," *J. Met.*, **42** (5), 16-19 (1990).

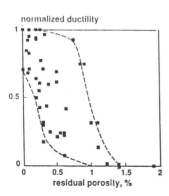

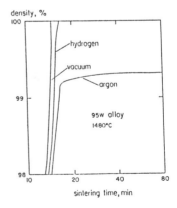

Fig. 1 The normalized ductility
versus the residual porosity for a
number of heavy alloys processed
under a variety of processing
conditions [2-6].

Fig. 2 The variation in sintered
density with sintering time for
different atmospheres.

Relationship between Sintering Atmospheres and Mechanical Properties of P/M Al-20Si-5Fe-2Ni Products

J. L. Estrada, V.M. Carreño, H. Balmori, and [1]J. Duszczyk

Dept. of Metallurgical Engineering, ESIQIE
National Polytechnic Institute, A. P. 75-872 Mexico City, Mexico 07300
[1]Delft University of Technology,The Netherlands

Abstract

Sintering, under both air and nitrogen atmospheres, of high performance air- and nitrogen-atomized Al-20Si-5Fe-2Ni P/M alloys was performed. Characterization of the powders included an evaluation of powder morphologies, determination of powder sizes and size distribution, determination of total oxygen contents and description of the microstructures. The specific surface areas were determined by the BET method and the thickness and chemical composition of the surface oxide films by Auger electron spectroscopy. Thermodifferential and thermogravimetric analyses of the powders were performed. A thermodynamical analysis, by means of the SOLGASMIX computing program, to study the influence of different sintering atmospheres is presented. Powders were consolidated by cold isostatic pressing at different pressures, sintered at different temperatures and times. The final microstructures were analyzed and the tensile properties determined.

1. Introduction

The improvement in performance of engine parts for automobile applications seems to be one of the prime targets of research for the coming years. This goal will be achieved provided that new, more temperature resistant and lighter materials than those presently applied can be developed and processed at a reasonably low cost. Among the variety of new aluminium alloys, the Al-20Si-X P/M system appears to be very suitable for high performance applications as reported by Kuroishi, Odani and Takeda (1985), and by Akechi, Odani and Kuroishi (1985). This temperature and wear resistant material, with a low coefficient of thermal expansion and good workability, has an enormous commercial potential for highly profitable applications in the automobile industry.

The well balanced combinations of mechanical properties associated with these alloys are mainly controlled by the characteristics of the powders from which these alloys are produced, and by the processing and consolidation parameters inherent in manufacturing P/M products.

2. Experimental procedure

The powders chosen for this investigation consisted of two lots of aluminium alloys based on the hypereutectic Al-Si-X system containing 20 wt% Si, 5 wt% Fe and 2 wt% Ni. Powders A1 (air-atomized) and A2 (nitrogen-atomized) were produced by Osprey Metals Ltd. (UK).

The characterization of the powders consisted of the evaluation of the powder morphologies, determination of the particle sizes and size distributions, description of the microstructures, and measurements of the specific surface areas, thicknesses of the surface oxide films, total oxygen contents, apparent densities and compressibilities.

Both powders were compacted uniaxially or cold isostatically pressed at room temperature. The compacts were sintered in air or nitrogen. The sintered products were subjected to tensile tests at room temperature.

3. Results and discussion

3.1 Characterization

The chemical compositions of both Al-20Si-5Fe-2Ni powders were about the same. The oxygen content of the nitrogen- atomized powder A2 (0.106 wt%) was approximately half of that in the air-atomized powder A1 (0.213 wt%). It is thus confirmed that powders atomized in a nitrogen atmosphere contain a lower amount of oxygen than powders of the same composition atomized in air.

X-ray diffraction patterns of the powders were similar showing that the same kind of elements and compounds: Al, Si, α-Al$_2$O$_3$, and Al$_3$Fe were present.

The melting temperature for both powders was 573^{0}C which corresponds to the melting point of the eutectic Al-12.5Si. A peak at 598^{0}C corresponding to the melting temperature of the compound FeSiAl$_5$, as reported by Mondolfo(1990), was also observed.

Powders A1 and A2 are stable under an oxidizing atmosphere at elevated temperature and long times . Isothermal analyses for both powders, (550^{0}C, 6 hours), showed that there was no weight gain for either powder.

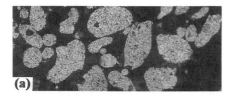
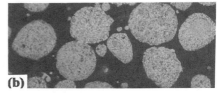

Fig. 1 Morphologies of Al-20Si-5Fe-2Ni powders: (a) A1, atomized in air, (b) A2, atomized in nitrogen, obtained by light microscopy.

Figures 1(a) and 1(b) show the morphologies of the air- and nitrogen- atomized powders A1 and A2, respectively.

Image analyses showed that air-atomized particles appear to be mainly teardrop and ligamental in shape with a mean size of 34 μm whereas the nitrogen-atomized particles are spherical in shape with a mean size of 42 μm.

Powder characteristics, such as size distribution and shape, affect their apparent density. The apparent densities of the powders A1 (1.08 gcm^{-3}) and A2 (1.38 gcm^{-3}) measured by a Hall flowmeter, were 39% and 50% of theoretical density (2.74 gcm^{-3}), respectively.

It is clear that the atomization atmosphere has an influence on the specific surface area of powders with the same composition, 0.23 m^{2}g^{-1} for the air-atomized powder A1 and 0.17 m^{2}g^{-1} for the nitrogen-atomized powder A2. A reducing atomization gas, like nitrogen, permits the production of powders with a smaller surface area.

The as-received Al-20Si-5Fe-2Ni powders were subjected to surface analysis by Auger electron spectroscopy (AES). Although information about the chemical state of the elements is not generally obtained, it was possible to distinguish silicon from silicon oxide, aluminium from aluminium oxide in the low energy range (transition KLL), according to the shape of the peaks and the energy at which they occur. Both oxides and pure elements were overlapping in the high energy range (transition LMM) exhibiting no energy shift. The average chemical compositions (at%) for the surface oxides are given in Table 1.

Depth profiles are obtained by successive ion etchings by argon ions bombardment. The surface oxide thicknesses were estimated based on the sputtering times required for a 50% drop in the oxygen profiles. For powder A1 it occurred after 3 minutes sputtering and for powder A2 after only 1 minute giving respective oxide layer thicknesses of ~36 and ~ 10 nm.

After these times, the alloying elements nickel, silicon and iron maintained constant atomic concentrations similar to their bulk

Table 1 Chemical composition of surface oxides on powders A1 and A2.

Powder	Al	Si	Fe	Ni	O
A1	46.1	1.7	1.6	0.5	50.0
A2	52.7	21.4	5.5	2.6	17.2

concentrations. After 7 minutes sputtering the proportion of oxygen in the nitrogen-atomized powder A2 was about three times lower than that in the air-atomized powder A1.

At the beginning of sputtering of the powder A1 aluminium was present only as aluminium oxide. As sputtering continued there was a mix of both metal and oxide until the latter finally disappeared after ~5 minutes sputtering. For the powder A2 before sputtering there was a mix of aluminium in aluminium oxide. After 40 seconds sputtering there was no oxide left. For the two powders silicon was present as the element. For the powder A2 the silicon peak was present from the beginning of the measurement, before etching.

Figures 2(a) and 2(b) show the microstructures of the as-received air-atomized and nitrogen-atomized Al-20Si-5Fe-2Ni powders, observed by light microscopy. Rapidly solidified Al-20Si-5Fe-2Ni powders are characterized by an extremely fine and homogeneous microstructures with intermetallic compounds and Si crystals (primary and eutectic) in the aluminium matrix. The microstructures are similar, irrespective of the atomization gas. The primary and eutectic Si crystals are finely dispersed in the background. In addition there are also primary silicon crystals block-like or rectangular in shape. The results obtained by microanalysis indicate that the intermetallic compunds seem to be $FeAl_3$, $FeSiAl_5$, or $FeSiAl_3$.

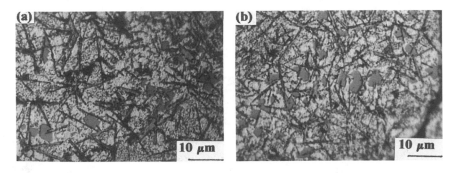

Fig. 2 Microstructures of Al-20Si-5Fe-2Ni powders: (a) A1, atomized in air, (b) A2, atomized in nitrogen, obtained by light microscopy.

3.2 Consolidation

Powders A1 and A2 exhibited approximately the same compressibility although the values for cold isostatic pressing were slightly higher than those for uniaxial compaction under the same pressure.

Microhardness Vickers measurements gave 107 and 117 HV for compacts A1 and A2, after uniaxial compaction at 440 MPa, with corresponding relative densities of 78 and 82%, and 123 and 126 HV for powders A1 and A2, after cold isostatic pressing at 408 MPa, with corresponding relative densities of 83 and 85% .

3.3 Sintering

Compacts of powders A1 and A2, obtained by CIP'ing at 408 MPa, were sintered in air at different temperatures (300 to 530°C) and times (0 to 45 min), Figs. 3(a) and 3(b).

It is apparent that both compacts behave in a similar way; during the first 15 minutes the higher the sintering temperature and time the lower the density (expansion). This expansion arise from gas trapped in the pores to some pressure; by increasing the temperature this inner pressure increases until its value is higher than the external pressure of the material producing an expansion. Thermodifferential analyses indicated that this expansion is not caused by any allotropic transformation. Sintering, in this case, was enhanced by breaking the oxide surface layer during the compaction stage. This oxide layer breaks along the contact points, which become oxide free, and during sintering atoms diffuse through these points. The surface diffusion reinforces the interparticle contacts even in cases where there is no shrinkage.

After the first 15 minutes of sintering, and at much longer sintering times (up to 300 min), the density of the compacts remained constant. At 300 and 400°C compacts A1 and A2 behaved similarly.

Compacts of powders A1 and A2, obtained by CIP'ing at 408 MPa, were sintered in nitrogen at 550°C where the highest densities (89 and 91%) appeared after sintering for 60 minutes. From 60 to 120 minutes there was only a slight increase in density. However, the densities of compacts sintered under this atmosphere were higher than compacts sintered in air at the same pressure and temperature (Fig.4). Figures 5 and 6 show the corresponding microstructures from which it is apparent that the longer the sintering time the coarser the microstructure.

A thermodynamical analysis, by means of the SOLGASMIX computing program, was performed. The influence of different atmospheres during sintering showed that sintering powders A1 and A2 in air under ideal conditions (1 atm pressure) the oxygen in the compacts and

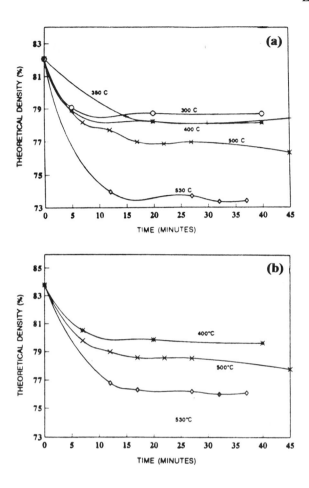

Fig. 3 Air-sintering curves of Al-20Si-5Fe-2Ni powders: (a) A1, atomized in air, (b) A2, atomized in nitrogen, at different temperatures and times (CIP'ed at 408 MPa).

the oxygen in the air react with the metal forming aluminium oxide. This aluminium oxide hinders the amount of Al atoms diffusing through the oxide layer which increases the compact volume. In this case, sintering is produced only by diffusion through the interparticle contact points. On the other hand, sintering in nitrogen avoids the formation of aluminium oxide permitting more Al atoms to diffuse through the oxide layer.

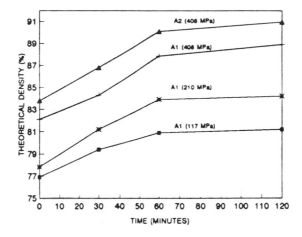

Fig. 4 Nitrogen-sintering curves of Al-20Si-5Fe-2Ni powders: A1, atomized in air, and A2, atomized in nitrogen, at different times and 550⁰C (CIP'ed at 408 MPa).

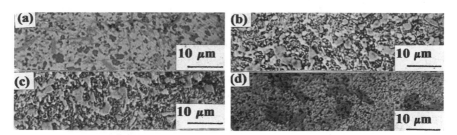

Fig. 5 Microstructures of Al-20Si-5Fe-2Ni compacts A1 after air-sintering for (a) 0, (b) 30, (c) 60, and (d) 120 minutes, at 550⁰C .

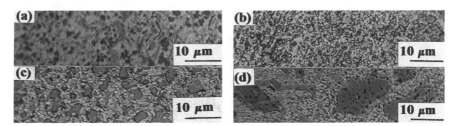

Fig. 6 Microstructures of Al-20Si-5Fe-2Ni compacts A2 after air-sintering for (a) 0, (b) 30, (c) 60, and (d) 120 minutes, at 550⁰C .

3.4 Tensile properties

The tensile values, at room temperature, for samples obtained from compacts A1 and A2 cold isostatically pressed at 408 MPa and sintered in air at 550⁰C during 60 minutes showed that the material A2 (nitrogen-atomized) has better properties than A1. A2 had an ultimate tensile strength σ_{UTS} = 360 MPa and elongation ϵ = 0.82% and the material A1 (air-atomized) had an ultimate tensile strength σ_{UTS} = 346 MPa and elongation ϵ = 0.78%.

4. Conclusions

1. Microhardness values gave 107 and 117 HV for powders A1 and A2, uniaxially compacted at 440 MPa, with relative densities of 78 and 82%, and 123 and 126 HV for powders A1 and A2, CIP'ed at 408 MPa, with relative densities of 83 and 85%, respectively .

2. Compacts A1 and A2 showed no significant difference in density after being sintered in air at different times (10 to 300 minutes) at the same temperature.

3. Compacts A1 and A2 sintered in nitrogen showed higher densities than those sintered in air at the same pressure and temperature.

4. The highest densities, 89% and 91% for compacts A1 and A2, respectively, were obtained after sintering in nitrogen for 60 minutes.

5. The tensile values, at room temperature, for samples obtained from compacts A1 and A2 CIP'ed at 408 MPa and sintered in air at 550⁰C for 60 minutes showed that the material A2 has better properties, σ_{UTS} = 360 MPa and ϵ = 0.82% than those corresponding to the material A1, σ_{UTS} = 346 MPa and ϵ = 0.78%.

Acknowledgments

The authors thank to the National Polytechnic Institute (DEPI Projects), COFAA (Becarios) and SNI (National System of Researchers) for financial support. The powders were supplied by Osprey Metals Ltd. (UK).

References

N. Kuroishi, Y. Odani and Y. Takeda, Int. J. of Powder Metallurgy, 3642(1985).

K. Akechi, Y. Odani, N. Kuroishi, Sumitomo Electric Tech. Rev., 24191(1985).

L.F. Mondolfo, Aluminium alloy structure and properties, pp.759-899, edited by William Clowes & Sons Ltd., 1990.

A NEW METHOD FOR DETERMINING THE WULFF SHAPE OF ALUMINA

M. Kitayama and A. M. Glaeser
Department of Materials Science and Mineral Engineering,
University of California
&
Center for Advanced Materials, Lawrence Berkeley National Laboratory,
Berkeley, CA 94720-1760, USA

ABSTRACT

Internal arrays of micron-sized pores were produced in both undoped and Ti-doped sapphire by combining microfabrication methods and hot pressing. Bonded sapphire/sapphire assemblies were annealed in air at elevated temperature. The equilibration rate of pores in Ti-doped sapphire was much higher than that in undoped sapphire, apparently consistent with findings of sintering studies. The shapes of equilibrated pores in Ti-doped sapphire were examined using an AFM (Atomic Force Microscope). Pores introduced into the $(11\bar{2}0)$ plane of Ti-doped sapphire were bounded by $(11\bar{2}3)$-type planes, suggesting that these represent the stable, low-energy planes in Ti^{4+} doped sapphire. The equilibration rates of pores etched into the $(11\bar{2}0)$, (0001), $(10\bar{1}2)$, and $(10\bar{1}0)$ planes of undoped sapphire were compared. Only pores etched into the $(10\bar{1}0)$ plane equilibrated rapidly, and the equilibrium shape of these pores was examined using AFM. Although indexing has not been completed, the potential of the method has been demonstrated. Once the work is completed, an assessment of the effect of Ti-doping on the surface energetics of sapphire will be possible.

1. INTRODUCTION

The Wulff shape provides a graphical display of the surface energy anisotropy of a material, and can be used to assess the relative surface energies of different crystallographic planes (Wulff (1901) and Herring (1951)). Dopants, particularly those that segregate anisotropically, can radically alter the Wulff shape. The changes in Wulff shape will, in turn, affect the driving force for morphological and microstructural change. Thus, understanding the effect of dopants on surface energetics is essential to a meaningful interpretation of kinetic data.

In metal systems, shapes of particles annealed for prolonged periods at elevated temperature have been determined by Martin (1976). Such measurements assume that the interactions between the particle and annealing environment (inert gas or vacuum) are either unimportant or known, and that the particle can be approximated as a closed system, *i.e.,* vaporization in a vacuum does not change the equilibrium shape. As the melting temperature of the material is increased, the diffusivities at fixed temperature tend to decrease, and thus, either finer scale features (smaller particles) must be examined to reduce the transport distance, or the anneal temperatures must be increased to achieve comparable transport rates, or both. Higher anneal temperatures increase the risk of contamination, and finer scale features make the shape changes more difficult to monitor. For high melting temperature ceramics, the experimental challenges are particularly severe. Despite its fundamental importance, there are relatively few determinations of the Wulff shape of ceramics (Okamoto (1989) and Wang (1986)), and particularly of high melting temperature oxide materials available in the literature.

Measurements of pore shapes provide an alternative means of assessing the Wulff shape. Such measurements have the advantage (and disadvantage) of shifting the experimental focus to internal features. Pores, particularly when situated within a large grain, are isolated to at least some degree from the furnace environment. Such pores often develop during densification, but can also be developed by crack healing processes (Rödel (1990)). One difficulty with conventional crack healing approaches is that they do not offer control over the size and location of the features that develop. If one examines large pores, there is concern over whether an equilibrium shape has in fact been reached. If the finest features are examined, they can be too fine for optical study, and TEM images of sections taken along high symmetry directions must be examined.

Our efforts have focussed on developing alternative methods that provide some experimental advantages in dealing with this difficult problem. Such considerations led to the combined use of microfabrication techniques, hot pressing, annealing, controlled fracture (or polishing), and most importantly, the study of the pore morphology with an atomic force microscope (AFM).

2. EXPERIMENTAL

The method applied is an extension of a method suggested earlier for studying pore elimination and coarsening (Rödel (1988)) and uses procedures described in detail by Rödel (1991). Using microfabrication methods and ion beam etching, it is simple to introduce large arrays containing geometrically and crystallographically identical micron-sized surface cavities into the surface plane of a substrate. In the present experiments, 250×250 pore arrays were generated, yielding a total of 62500 pores per sample. These arrays consisted of pores initially $16 \, \mu m \times 16 \, \mu m$, etched to a depth of $\approx 0.45 \, \mu m$.

To compare pore evolution characteristics in undoped and Ti-doped ($\approx$300 ppm) sapphire, identical arrays of pores were etched into the $(11\bar{2}0)$ surface of both wafers. By bonding an etched and unetched wafer of identical orientation with a slight twist (5-10°) using hot pressing, these surface patterns were transferred to an internal interface, yielding identical arrays of internal pores. Both undoped and Ti-doped bonded sapphire/sapphire assemblies were annealed in air for up to 160 h at 1650°C, and the pore shape evolution was monitored nondestructively after various annealing times using optical microscopy.

To assess the effect of initial pore crystallography on pore equilibration in undoped sapphire, pore arrays identical to those just described were introduced into wafers of varying orientation. Since the pores are initially wide and shallow, the change in wafer orientation changes the dominant bounding plane of the as-etched pores. Pores were introduced into the $(11\bar{2}0)$, (0001), $(10\bar{1}0)$ and $(10\bar{1}2)$ surfaces of sapphire, and after hot pressing, the bonded samples were annealed in vacuum for periods of up to 16 h at 1900°C.

After combinations of anneal time and temperature were reached that yielded an apparently stable pore shape, samples were either fractured (Ti-doped $(11\bar{2}0)$ plane) or polished (undoped $(10\bar{1}0)$ plane) to re-expose the pores. These pores were then probed using an AFM, allowing the pore shape to be determined and recorded both as an image and as a data file. Using software provided with the instrument, a broad range of pore cross sections or traverses could then be examined, and from a knowledge of the bond plane and directions within the bond plane, the facet planes, if observed, could be indexed.

3. RESULTS AND DISCUSSION

3.1. Ti-doping effect on the equilibration rate

Optical microscopy of specimens after identical annealing times revealed that the equilibration rate in Ti-doped sapphire was much higher than that in undoped sapphire, Figure 1. In the Ti-doped sample, pores quickly changed their shape. A near-final shape was reached in the first 5 h; increasing the anneal time to 160 h at 1650°C leads to only relatively minor change in pore morphology. In the undoped sample, evolution were very sluggish; pores are not equilibrated even after 160 h. This result is qualitatively consistent with the findings of prior sintering studies by Bagley (1970), wherein the higher initial stage sintering rate is attributed to an increased cation vacancy concentration due to Ti^{4+} doping. Similar increases in diffusivity should also increase the rate at which pore shapes change. Since there is no reason to believe a glassy phase is affecting the observed results, a true solid-state effect is indicated. However, it is also possible that the driving force for pore shape changes in these systems is dramatically altered by the effect of Ti^{4+} doping on the stability of the $(11\bar{2}0)$ surface that dominates the initial, "as-bonded" pore shape. This possibility warrants consideration.

3.2. Equilibrium pore shape of Ti-doped sapphire

In Ti-doped sapphire it was possible to propagate a crack along the original bonded interface. This exposed the equilibrated pores, and their shapes were subsequently examined using an AFM. Figure 2 shows the AFM-derived image of the surface topography (top), and the information taken from three traces across the pore (bottom), suggesting a shape that combines $(11\bar{2}3)$-type facets and smoothly curved sections. Prior experiments investigating pore channel instability in Ti-implanted sapphire by Powers (1993) have indicated that $(11\bar{2}3)$-type facets do develop and appear to stabilize channels oriented along the $[1\bar{1}00]$ direction of sapphire. In the present experiment, the pore shape might be affected by the existence of a low angle twist boundary that intersects the pores. Although further work addressing this issue is clearly needed, the results do show the potential of the technique to provide detailed information on the surface topography of a large number of pores in one sample.

3.3. Equilibration rates of undoped sapphire with various orientations

Due to the sluggish equilibration in $(11\bar{2}0)$-oriented undoped sapphire annealed in air at 1650°C, a second sample was annealed in vacuum at 1900°C. Figure 3 shows optical micrographs at annealing time of 0, 4 and 16 h along with the results from samples of other orientations, *i.e.*, (0001), $(10\bar{1}0)$ and $(10\bar{1}2)$. Pores etched parallel to the (0001) plane equilibrated slowly, with slight edge regression evident at certain orientations. Pores etched parallel to the $(11\bar{2}0)$ and $(10\bar{1}2)$ planes showed inhomogeneous equilibration. For these three surface orientations, the most pronounced changes in pore morphology occurred during the first 4 h anneal, and further annealing resulted in negligible shape change. In contrast, for pores etched into the $(10\bar{1}0)$ plane, equilibration appears complete after 4 h of annealing. Further annealing (up to 24 h) resulted in no obvious change in pore morphology, suggesting that pores were fully equilibrated. Clearly the differences in evolution behavior are striking. The relatively greater stability of nonequilibrium shape pores bounded by $(11\bar{2}0)$, (0001), $(10\bar{1}2)$ planes may indicate that these planes have relatively lower surface energies than the $(10\bar{1}0)$ and may be part of the Wulff shape. If the original pore shape is dominated by a plane that is part of the Wulff shape, the development of inclined planes may be difficult. In contrast, if the $(10\bar{1}0)$ plane is not part of the Wulff shape, nucleation of inclined planes should occur easily (Herring (1951), and equilibration may be accelerated. Relating these results to the observations in Ti-doped sapphire, the results suggest that dopant induced changes in surface stability could dramatically impact evolution rates.

3.4. Equilibrium pore shape of undoped sapphire

Efforts to cleave undoped sapphire to expose the pores introduced into the $(10\bar{1}0)$ plane were unsuccessful. Thus, the sample was polished at a slight angle to the original bond plane. This ultimately exposed a ribbon of pores, *e.g.*, 4 pores wide by 250 pores long that could then be examined using the AFM. The AFM images and traverses indicate what may appear to be several facets. There is a facet that is properly traced by the AFM tip, and it can be indexed as a $(10\bar{1}2)$ plane. There are also artifacts in the image that are due to the interaction between the pore edges and the pyramidal shaped AFM tip (5×5 µm square, 5 µm height). Use of a high aspect ratio AFM tip will be necessary to accurately resolve the orientation of the other facets that are indicated by optical and scanning electron microscopy. Once the characterization of the Wulff shape of undoped alumina is completed, it should be possible to assess the effect of Ti^{4+} doping on relative surface energies in sapphire. A full accounting of the facet structure will be reported elsewhere.

ACKNOWLEDGMENTS

Research comparing evolution rates in undoped and Ti-doped sapphire was supported by the Director, the Office of Energy Research, Office of Basic Energy Sciences, Materials Sciences Division of the U.S. Department of Energy under Contract No. DE-AC03-76SF00098. The work examining Wulff shapes is supported by the National Science Foundation under Grant No. DMR-8821238. We also acknowledge an NSF Equipment Grant No. DMR-9119460 which allowed the acquisition of hot pressing equipment critical to this work.

REFERENCES

G. Wulff, "XXV. Zur Frage der Geschwindigkeit des Wachsthums und der Auflösung der Krystallflächen," *Z. Kristallgr.*, **34**, 449-530 (1901).

C. Herring, "Some Theorems on the Free Energies of Crystal Surfaces," *Phys. Rev.*, **82**, [1], 87-93 (1951).

J. W. Martin and R. D. Doherty, STABILITY OF MICROSTRUCTURE IN METALLIC SYSTEMS, Cambridge University Press, (1976).

A. Okamoto, "Effect of Surface Free Energy on the Microstructure of Sintered Ferrites," *Am. Ceram. Soc. Bull.*, **68**, [4], 888-90 (1989).

Z. Y. Wang, M. P. Harmer, and Y. T. Chou, "Pore-Grain Boundary Configurations in Lithium Fluoride," *J. Am. Ceram. Soc.*, **69**, [10], 735-40 (1986).

J. Rödel and A. M. Glaeser, "High-Temperature Healing of Lithographically Introduced Cracks in Sapphire," *ibid.*, **73**, [3], 592-601 (1990).

J. Rödel and A. M. Glaeser, "A Technique for Investigating the Elimination and Coarsening of Model Pore Arrays," *Mater. Lett.*, **6**, [10], 351-55, (1988).

J. Rödel and A. M. Glaeser, "Microdesigned interfaces: New Opportunities for Materials Science," *J. Ceram. Soc. Japan*, **99**, [4], 251-65 (1991).

R. D. Bagley, I. B. Cutler, and D. L. Johnson, "Effect of TiO_2 on Initial Sintering of Al_2O_3," *J. Am. Ceram. Soc.*, **53**, [3], 136-41 (1970).

J. D. Powers and A. M. Glaeser, "High Temperature Healing of Cracklike Flaws in Titanium Ion-Implanted Sapphire," *ibid.*, **76**, [9], 2225-34 (1993).

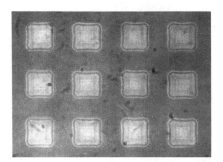

Initial pore structure, 16 μm × 16 μm pores

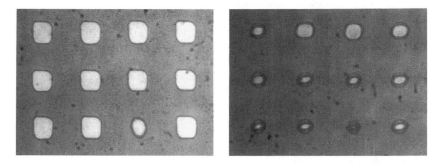

Pore structures in undoped (left) and Ti-doped (right) sapphire after 5 h at 1650°C in air.

Pore structures in undoped (left) and Ti-doped (right) sapphire after 160 h at 1650°C in air.

Figure 1   Optical micrographs of pore shape evolution in undoped and Ti-doped sapphire.

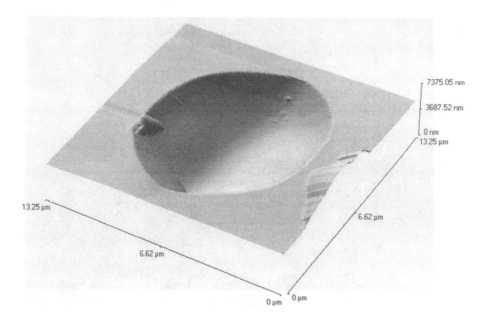

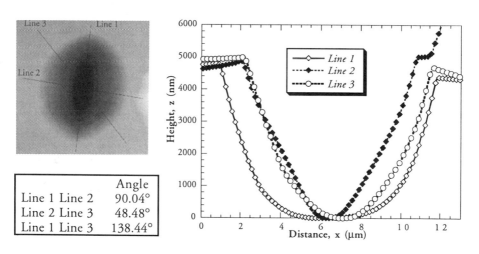

		Angle
Line 1	Line 2	90.04°
Line 2	Line 3	48.48°
Line 1	Line 3	138.44°

Figure 2   Above) AFM image of pore shape for Ti-doped sapphire, showing a shape that combines $\{11\bar{2}3\}$ facets and smoothly curved sections.

Below) Top view of AFM image (left) and line traces (right).

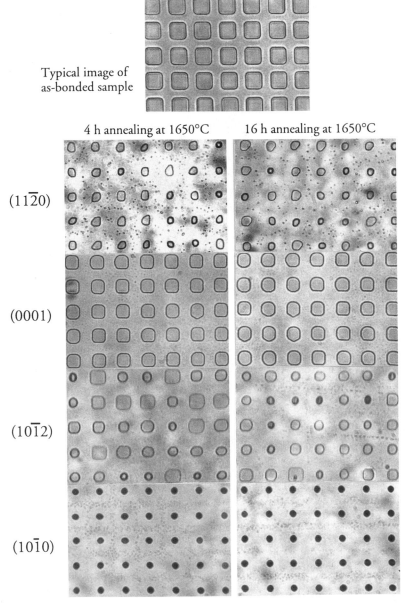

Figure 3   Optical micrographs of various sapphire planes after annealing time of 0, 4 and 16 h. Note that (0001) plane shows very sluggish equilibration, ($11\bar{2}0$) and ($10\bar{1}2$) planes show inhomogeneous equilibration, and the ($10\bar{1}0$) shows quick and complete equilibration.

Processing and Sintering of a High Purity α-Al$_2$O$_3$ Powder

G. Cisneros-Gonzalez, J.L. Estrada-Haén,
D. Jaramillo-Vigueras and H. Balmori-Ramírez

Department of Metallurgical Engineering
National Polytechnic Institute, A.P. 75-872, Mexico City, 07300 Mexico

Abstract

An α-Al$_2$O$_3$ powder with fine grain size and high purity was processed by cold isostatic pressing and slip casting and sintered in a dilatometer. The influence of different green microstructures on the sintering behavior was analyzed. Increasing the green density leads to higher sintered densities and faster sintering because the pore size and its coordination number decrease. The sintering path for particles packed uniformly or randomly is different. When the green density is > 57%, it is possible to sinter to almost full density at 1300°C.

1. Introduction

The exploitation of the good characteristics of α-Al$_2$O$_3$ ceramic articles for modern applications requires that these pieces have full density, controlled grain size and excellent reliability. Sintering submicrometer size particles allows the attainment of full density at reasonably low temperatures without the addition of sintering additives (Yeh and Sacks, 1988; Hay et al., 1989). However, the sintering behavior and the sintered microstructure depend on the consolidation method (Roosen and Bowen, 1988) and particle packing characteristics (Lange, 1989; Hirata et al., 1992).

The purpose of the present paper is to analyse the influence of the green microstructure on the sintering behavior of a commercial α-Al$_2$O$_3$ powder with fine particle size and high purity. The powder was consolidated by cold isostatic pressing (CIP) and slip casting to produce uniform green microstructures.

2. Experimental Procedure

The α-Al$_2$O$_3$ powder employed (Taimicron TMA-10) had an average and a maximum particle size of 0.25 and 0.70 μm, respectively, and a specific surface area of 12.52 m^2/g. As received, the powder consisted of irregular agglomerates of 10 to 20 μm. According to the producer, its purity was higher than 99.99%.

The powder was consolidated by CIP and slip casting. For CIP, the powder, as received, was placed in flexible molds and pressed at 100, 200, 300 and 400 MPa to produce cylindrical samples of 10 mm diameter and 50 mm length. The molds were de-aired before pressing.

For slip casting, 100 ml of suspension containing 20 and 30 vol % solids were prepared at pH=3.5. To deagglomerate the powder, it was mixed with deionised water and milled for 24 h in a polypropylene jar with 100 g of zirconia milling media. The suspensions were outgassed before casting into plaster of Paris molds to produce rectangular plates of 50 x 30 x 7 mm^3 or discs of 25 mm diameter x 10 mm height.

The cakes were calcined for 1 h at 800°C and density was measured by the Archimedes method using water as the immersing medium. Pore size distribution was determined by mercury intrusion (Micromeritics).

Sintering was performed in a dilatometer (Setaram TMA92) employing samples of 10 mm heigth cut from the calcined samples, heating at 10°C/min from 25°C to 1600°C. The equipment thermal dilation was subtracted so that only the sample shrinkage was considered. Comparative isothermal experiments at 1300°C were made in a programable furnace (Carbolite, RHF 17/3E).

The sintered microstructures were observed on surfaces polished with diamond paste and etched at 1400°C for 15 to 60 min. The cast samples required longer etching times. The grain size was measured by the linear intercept method.

3. Results and Discussion

3.1 Powder Consolidation

The density and median pore size of the green samples are given in Table 1. The density of the pressed samples increases with the applied pressure. The slip cast samples are denser and not affected by the amount of solids in suspension, but the samples cast at 30% solids are less homogeneous.

The porosity of the green samples is shown in Figure 1. The curve for the sample pressed at 100 MPa is not presented to avoid confusion, but it is similar to the sample pressed at 200 MPa. The median pore size lowers as density increases. The porosity of the isopressed samples displays a monomodal character. Increasing the compaction pressure provoked the disappearance of big pores. The similarities of their porosity curves indicate that compression produced some particle rearrangement without breaking down the agglomerates, so that the green microstructures of the pressed samples are similar. The porosity of the cast samples presents some differences with respect to the pressed samples. Their pore size distribution is narrower, starts at a smaller pore size, and has a big maximum at 0.0365 μm and a smaller one at 0.03 μm.

Table 1.- Characteristics of the green samples

Consolidation Method	Density@ (% of theoretical[*])	P_o (%)	d_p (μm)	Vol % of pores >0.1 μm[**]
CIP, 100MPa	56.08±0.11	43.64	0.0551	0.67
CIP, 200MPa	56.84±0.08	42.85	0.0564	1.17
CIP, 300MPa	57.83±0.05	---	---	---
CIP, 400MPa	57.80±0.27	41.91	0.0521	0.44
Casting, 20%sol	63.50±0.58	36.15	0.0378	1.83
Casting, 30%sol	63.55±2.84	35.87	0.0375	0.43

P_o: open porosity; d_p:average pore size; @ Average of 4 to 5 samples;
* Theoretical density=3.99g•cm⁻³;** Referred to the total pore volume.

All the porosity curves are cut at pore sizes near 0.02 μm, so that this is probably the smallest pore size that can be obtained for the powder used. All the samples analyzed by mercury intrusion contained a small fraction of pores > 0.1 μm (Table 1). These pores play an important role during sintering and determine the mechanical reliability of ceramics.

Colloidal treatment of ceramic powders gives regular particle packings (Lange, 1989). Employing an α-Al₂O₃ powder of 0.35 μm, Hirata et al. (1992) obtained cakes with a density of 68% and a median pore size of 0.13 μm. The density differences with respect to the samples prepared for this work may be a consequence of the greater difficulties encountered when processing fine particles by slip casting. The difference in pore size results from the different particle diameters. So, it is possible that the particles in the current samples are packed

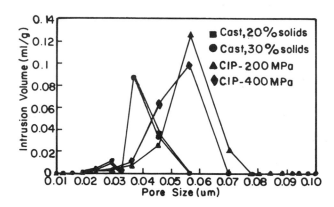

Figure 1.- Porosity of samples calcined for 1 h at 800°C.

in a regular arrangement. This is a difference with respect to the microstructures of the isopressed samples, which must have inherited some characeristcs of the more random particle packing of the agglomerated as-received powder.

As shown in Table 1, the pore size decreases as density increases. Lange (1989) and Hirata et al. (1990) proposed that the coordination number is higher for big pores, either in a regular or random particle packing. So, the samples analyzed here represent a broad spectrum of green microstructures.

3.2 Sintering.

The dilatometric curves and the shrinkage rate vs. temperature of the isopressed and cast samples are presented in Figure 2. Table 2 contains some characteristics of the sintered samples. The shrinkage behavior of all the isopressed samples was very similar, so that for the sake of clarity only the curves corresponding to the samples isopressed at 200 and 400 MPa were drawn.

The influence of green density on sintering can be inferred from the results of Figure 2. As the green density increases, the final density increases and the total shrinkage decreases. The samples with green densities > 57% attain sintered densities > 99.6%, so that this can be a condition necessary to reach full density. The sintered microstructures of the samples isopressed at 400 MPa and cast at 20 volume % solids are shown in Figure 3. Both samples are almost fully dense, retaining some porosity at triple junctions. Intergranular porosity is almost completely absent. The sintered grain size of most of the samples is almost equal (Table 2). However, based on the results for the samples pressed at <200 MPa, it seems that the less dense green samples require more grain growth to reach a density as high as the denser green samples.

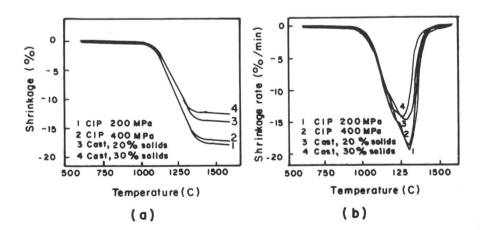

Figure 2.- (a) Shrinkage and (b) shrinkage rate as a function of temperature.

Table 2.- Characteristics of the samples after sintering from
room temperature to 1600°C at 10°C/min.

Consolidation Method	Shrinkage (%)	Density (% of theoretical)	Grain Size (μm)
CIP, 100MPa	17.39	97.87	3.18
CIP, 200MPa	17.30	99.47	4.26
CIP, 300MPa	16.90	100.23	---
CIP, 400MPa	16.84	99.81	3.19
Casting, 20%sol	13.47	99.69	3.28
Casting, 30%sol	12.28	98.71	3.32

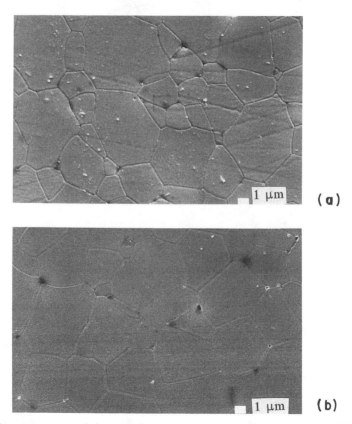

(a)

(b)

Figure 3.- Microstructures of the samples sintered in the dilatometer to 1600°C
at 10°C/min. (a) CIP at 400 MPa, (b) Cast at 20 volume % solids.

The cast samples densify less because they contained crack like defects (Figure 4) located within 1 mm of the surface in contact with the mold walls during casting, so that they formed during the cake growth. To determine their role during sintering, they were eliminated from the blocks cast at 20% solids by grinding the green samples and sintering at 1300°C. The results for the as cast sample, the ground sample and the isopressed sample at 400 MPa are shown in Figure 5. The ground sample sintered faster and attained a density slightly higher than the isopressed sample even after sintering for 8 h, while the as cast sample was always less dense.

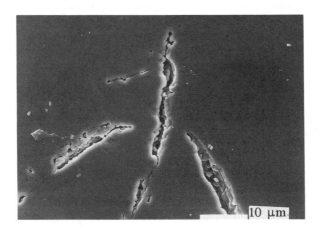

Figure 4.- Crack like voids observed in the cast and sintered samples.

Other characteristics related to a high green density are: (1) The third stage of sintering starts at lower temperatures and sintering ends faster. (2) The maximum shrinkage rate decreases. (3) The temperature for maximum shrinkage rate is lower.

Three possible reasons for the above behavior are: (1) The pores of the denser green samples are smaller, so that less mass transport is required to close them. (2) The less dense samples contain pores which are bigger than the critical pore size for pore closure, as predicted by Kingery and François (1976). Actually, the samples with green densities < 57% had pores > 0.07 μm that were not present in the denser samples. (3) The pore coordination number of the samples that densified better was smaller than the critical value required for a pore to disappear, as proposed by Lange (1989). This argument is also consistent with the results presented in the previous section, since the coordination number increases as the pore size increases for equal size particles.

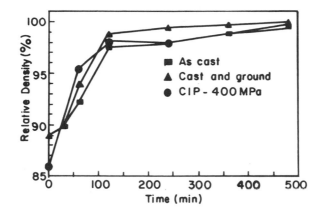

Figure 5.- Density variation at 1300°C as a function of time.

Influence of Particle Arrangement.- The influence of particle arrangement can be analyzed by comparing the shrinkage curves of the samples isopressed and cast. In Figure 2, the sintering path is different for the two groups of samples. This difference can be attributed to the different particle packing that results from different consolidation techniques. Although this observation requires further analysis, one explanation was given by De Jonghe et al. (1989) who suggested that the sintering path depends on a sintering stress that is determined by the green microstructure. Lange (1989) also suggested that a difference in the pore coordination number induces a different sintering behavior.

The set of results described here are essentially identical to the influence of green density on the sintering of BaTiO₃ (Chen and Ring, 1993), so that the behavior observed may be general for ceramic powders.

3.3 General Comments
The results discussed above demonstrate that the α-Al₂O₃ powder employed here can be sintered to a high density when the green density is > 57%. This condition can be achieved either by cold isostatic pressing with pressures above 300 MPa and by slip casting. However, both methods have advantages and disadvantages. CIP produces homogeneous green samples, but sintering to full density is accompanied by a larger shrinkage than for the slip cast samples. The main disadvantage of slip casting is the formation of crack like pores in the sintered body. These defects quite probably developed from particle packing heterogeneities formed when the cake was growing in the plaster mold. Some possible ways to avoid their formation is increasing the solids charge in the slurry and pressure casting.

Conclusions

1.- CIP and slip casting produce different green microstructures. Green density of slip cast samples is higher, with smaller pores and pores coordination number.

2.- The green microstructure exerts a great influence on the sintering behavior. Decreasing the pore size < 0.07 μm and the corresponding pore coordination number leads to almost full density and faster sintering.

3.- The sintering path for particles packed uniformly or randomly is different.

4.- Increasing the green density above 57% relative density leads to almost full density at $1300°$C.

Acknowledgements.- The authors thank JICA, IPN-DEPI and CONACYT for their financial support. G.C. thanks CONACYT for granting a fellowship. The porosity measurements were made by M. Garcia, from CENAM. The suggestions and material donation by M. T. Iga, from NIRIN, Japan, is specially acknowledged. This paper is dedicated to the memory of Dr. K. Kawakami.

References

T.S. Yeh and M.D. Sachs, "Low Temperature Sintering of Aluminum Oxide", *J. Am. Ceram. Soc.* **71**[10]841-844(1988).

R.A. Hay, W.C. Moffatt and H.K. Bowen, "Sintering Behavior of Uniform-Sized α-Al_2O_3 Powder", *Mats. Sci. and Eng.* **A108**(213-219)1989.

A. Roosen and H.K. Bowen, "Influence of Various Consolidation Techniques on the Green Microstructure and Sintering Behavior of Alumina Powders", *J. Am. Ceram. Soc.* **71**[11]970-977(1988).

F.F. Lange, "Powder Processing Science and Technology for Increased Reliability", *J. Am. Ceram. Soc.* **72**[1]3-15(1989).

Y. Hirata, A. Nishimoto and Y. Ishihara, "Effects of Addition of Polyacrylic Ammonium on Colloidal Processing of α-Alumina", *J. Ceram. Soc. Japan* **100**[8]983-990(1992).

Y. Hirata, I.A. Aksay and R. Kikuchi, "Quantitative Analysis of Hierarchical Pores in Powder Compact", *J. Ceram. Soc. Japan* **98**[2]126-135(1990).

W.D. Kingery and B. François, "The Sintering of Crystalline Oxides", pp. 471, *Sintering and Related Phenomena*, edited by G.C. Kuczinsky, N.A. Hooton and C. Gibbon, Gordon and Breach Sci., NY, (1976).

L.C. De Jonghe, M.Y. Chu and M.K.F. Lin, "Pore Size Distribution, Grain Growth and the Sintering Stress", *J. Mat. Sci.* **34**(4403-4408)1989.

Z.Ch. Chen and T.A. Ring, "Sintering of $BaTiO_3$", pp. 275-284, *Dielectric Ceramics: Processing, Properties and Applications*, edited by K.M. Nair, J.P. Guha and A. Okamoto, The American Ceramic Society, Inc., Columbus, (1993).

Characterization of the Sintering Behavior of Commercial Alumina Powder with a Neural Network

Heinrich Hofmann

Laboratoire de Technologie des Poudres
Département de Matériaux
Ecole Polytechnique Fédéral de Lausanne, CH-1015 Lausanne

Abstract

The sintering behavior of 31 commercially available alumina powders has been investigated. Typical properties of the powders, such as particle size distribution, specific surface area, agglomerate factor and chemical composition were determined and their influence on the sintered density and the microstructure was investigated. Since none of the known sintering theories could describe the whole range of properties of the commercial powders, the 'neural network' method was used for the description of the relationship between the powder characteristics and the microstructure of the sintered samples. Following the analysis, a general rule for the sintering behavior of commercial alumina could be formulated which will be presented. Also the possible use of the 'neural network' theory for the characterisation of processes which do not have well defined parameters, like the sintering of commercial powders, will be shown.

1. Introduction

The influence of the properties of commercial alumina powder on the sintering behavior as well as on the final microstructure (e.g. grain size, porosity) is still not understood in detail. Neither the well-known sintering equation for the first sintering step [1-3] nor the relation between the density and grain size at the final sintering step [e.g. 4,5] can describe the densification behavior of commercial alumina powder. The reasons for this difference between theory and practice are that the sintering process as well as the final development of the microstructure are both complex processes which are very difficult to observe and model. Therefore sintering models are developed from experiments with well defined powders (pure, monosized, spherical , without agglomerates, ideally random packed) or theoretically based on relatively simple assumptions [1-5]. Also the more global approaches to understand the densification phenomena [6-8] are very attractive because of their realism, but have proven to be limited in their ability to make specific kinetic predictions.

The aim of this work is to develop a better understanding of the relation between the powder properties and the densification behavior of commercial alumina powder. Another aim of this paper is to evaluate if modeling or prediction of the final density and microstructure based only on the known powder characteristics is possible.

2. Experimental Procedure

31 alumina powders from different producers were characterized regarding the chemical composition (Na_2O, SiO_2, TiO_2, Fe_2O_3, MgO), the grain size distribution (Malvern Mastersizer), specific surface area(N_2 adsorption, Micromeretics), α-alumina content (X-ray diffraction), loss on ignition and the particle shape (SEM). All powders were spray dried in a Niro spray dryer with a drying capacity of 20 kg powders per hour. For the spray drying, aqueous slurries starting from 50 kg of powder were prepared by wet ball milling with 0.05 weight % MgO and the commercial additives (Zusoplast, Optapix and Dolapix). The powders were cold isostatically pressed (2000 bar) into cylinders of 2 cm diameter and a length of 10 cm. The sintering was carried out in gas fired oven at temperatures between 1500 and 1750 °C for 1h. The heating rate was 1.35 °C/min whereas the cooling rate was only 0.35 °C/min. For the investigation of high purity alumina powder, we used only 2 kg powder per batch and the spray drying was carried out in a Lab spray dryer (Büchi). With this procedure, we had an industrial like manufacturing process with well defined production parameters. Additionally, characteristic powder properties such as mean grain size in μm calculated from the specific surface area (d_{BET}):

$$d_{BET} = 6/ 3.98 \text{ BET} \tag{1}$$

where BET = specific surface area in m^2/g measured by nitrogen adsorption. or the agglomerate factor (AGF)

$$AGF = d_{v50}/d_{BET} \tag{2}$$

where d_{v50} = median particle size by volume were also calculated.

The samples were characterized with respect to their density (after cold pressing as well as after the sintering step), linear shrinkage, open porosity, microstructure and some mechanical properties such as hardness and fracture toughness.

As neural network approach was used to treat all 31 investigated powders with 10 typically powder properties as an input (SiO_2, Na_2O, Fe_2O_3, TiO_2, d_{v90}, d_{v50}, d_{v10} BET, AGF α content) and 6 properties of the sintered part as the output [green density, sintered density (1500°C/h), sintered density (1600°C/h), mean grain size (1750°C/h), max. grain size (1750°C/h), aspect ratio (1750°C/h)] were used in a second step. A detailed description of these neural networks can be found elsewhere [9-12].

3. Results

The density of the samples after sintering at different temperatures are shown in fig. 1. There exists at least 5 groups of powders:

I.   These powders sinter at low temperatures and the optimal sintering temperature is 1600 °C. Higher sintering temperatures leads to the well-known phenomenon of "over sintering" of the samples accompanied with a desintering; (16 powders).

II. These powders can be also nearly 100 % densified but the sintering temperature is higher than 1700 °C; (5 powders)

III. These powders show a very strange behavior. At lower temperatures, they sintered very fast, but the maximum density of these powders is <93 % of the theoretical density; (6 powders).

IV. These two powders show a behavior like the powders of group II at low temperature, but the optimal sintering temperature is 1650 °C (2 powders).

V. Powders with a very low densification rates (2 powders)

Fig 2 shows a more schematic description of the sintering behavior of the commercial alumina powders.

The analysis of the experimental results with a neural network gives us a complete description of the sintering behavior of commercial alumina powder. Because the presentation of all results of the calculation with the neural network is not possible only examples based on the purified Bayer-alumina of Reynolds RCHPDM will be given. We calculated predictions of final densities and grain sizes based on the neural network approach using the data of the 31 alumina powders. Figs 3a and 3b show the influence of d_{BET} and d_{v50} (with this choice the influence of AGF is also considered, see eqn (2)) and of Na_2O and SiO_2, respectively, on the density of the sintered samples. The results regarding the influence of the particle size on the density show clearly that the density increases with increasing specific surface area as well as with decreasing grain size. The influence of the agglomerate factor can be demonstrated with a powder with a constant specific surface area but with increasing mean diameter. The density of such ceramics decrease with increasing AGF. The influence of the impurities is shown in fig 3b. The influence of Na_2O is relatively small, only a small reduction in the final density can be observed which depends slightly on the SiO_2-content. The SiO_2 -content seems to have a very strong influence at low concentrations whereas at concentrations > 200 ppm the influence is negligible.

4. Discussion

The well-known equations for the description of the sintering and grain growth behavior of ceramic can not be used for the explanation of the sintering behavior of commercial powders. The equations developed by Brook[13] and very successfully used by Harmer [5] for the explanation of the role of MgO in pure and theoretically monosized and agglomerate free powder do not consider the role of the grain size distribution and agglomerates. Both studies determined the pore size distribution and consequntly predict the sintering behavior as Roosen and Bowen have also shown in their work using a relatively pure alumina[14]. Similar results were found by Allemann et al. [15] who investigated the sintering behavior of different TZP powders. The sintering behavior was mostly determined by the pore size distribution which could be correlated to the specific surface area and the agglomerate factor of the investigated powders.

In this work, the experimental results show that the density, e.g. after sintering at 1600 °C for 1h, depends on the agglomerate factor and on the quantity of the major impurities. Fig 4 gives a schematic view of the microstructure of the green compacts. In the upper left corner, we have a fine powder with a mean grain size of 0.6 μm which is completely deagglomerated. In the lower right corner, we will have a very fine powder (mean diameter of the primary particles of 0.075 μm) but very strongly agglomerated. Based on this picture, we can easily explain the sintering behavior of the commercial alumina powder. With coarse but unagglomerated powders the fabrication of green compacts with a nearly uniform pore size distribution with a relatively coarse mean pore size is possible. Therefore we will observe a densification rate which is not very high but uniform (powders of type II in Fig 2b). The worst case is located in the lower left corner. These powders are coarse and in addition agglomerated. Therefore, the pore size distribution will be bimodal with an additional peak at larger diameter rising to pores between the agglomerates (powders of type V in fig. 2b). The best powder is located in the upper right corner. The green bodies produced from these powders show a lower green density, but the pore size distribution will be very narrow and the mean pore diameter is very small. These powders are very reactive and the optimal sintering temperature can be reduced (powders of type I in Fig 2b). Powders with properties as represented in the lower right corner are also very reactive, but the strong agglomeration enhance the sintering of the agglomerates themselves. Therefore, after a first densification the densification rate decrease and the sintering behavior will be similar to that of a very coarse one (powders of type III (coarse agglomerates) and IV (finer agglomerates)).

The influence of the impurities on the densification rate is not very important for the commercial aluminas studied under the aspect of densification. In agreement with Bae and Baik [16,17], the results from the neural network calculations shows that in the range from 0 to 200 ppm SiO_2 has a positive effect on the density. At higher concentrations, Bae and Baik observed the occurrence of abnormal grain growth which lead to a lower final density. This latter observation is in agreement with the results from Sumita and Bowen [18] who also observed a negative influence of SiO_2 on the densification of alumina. In the commercial aluminas, we can not see this negative influence. Perhaps the negative influence is compensated by other impurities which may have a positive influence (18). The results regarding Na_2O is in agreement with the results reported in the literature (18). A small negative influence on the density can be observed at Na_2O-contents higher than 200 ppm.

5. Conclusions

This study of the sintering behavior of commercial alumina powders covering high purity as well as Bayer alumina has shown that the density of the sintered samples as well as the final grain size depends mainly on the structure of the green body. This indicates that the pore size distribution in the green body ,which is determined by the particle size and their distribution and by the quantity and structure of any agglomerates structure, is the key factor for the understanding of the sintering behavior of commercial powders. Regarding the impurities, in the investigated range of commercial powders, their role is not as important as the role of the granulometry of the powders. Only in high purity powders, a grain growth as well an enhanced densification with increasing impurity content is observed. These results also show that besides the powder quality the powder processing, especially the dispersion and granulation (e.g., spray drying), is one of the most important factors influencing the quality of a sintered alumina ceramic.

This study also shows that the treatment of the experimental data with an optimized neural network is a very useful method for the modeling of the consolidation behavior of commercial alumina.

Acknowledgments
The author is grateful to Alusuisse-Lonza and Martinswerk for the carrying out the experimental work and L.Silvestri and T. Cornu from the Department of Computer Science from the Swiss Federal Institute of Technology , Lausanne, for the calculations with the neural network and Paul Bowen from the Powder technology Laboratory of the Swiss Federal Institute of Technology for helpful discussions during the preparation of the manuscript.

References:
1) Ja. E. Greguzin, "Physik des Sinterns", edited by VEB Deutscher Verlag für Grundstoffindustrie, Leipzig, (1973).
2) W. Schatt, "Sintervorgänge, Grundlagen", edited by VDI Verlag, Düsseldorf, (1992).
3) D. Bernache-Assollant, "Chimie-physique du frittage", edited by Hermes, Paris, (1993).
4) R.J. Brook, "Processing Technology for High Performance Ceramics"*Mater.Sci.Res.* 71 305-312 (1985)
5. M.P. Harmer, "Use of Solid-Solution Additives in Ceramic Processing" *Structure and Properties of MgO and Al$_2$O$_3$,* Vol. 10, W.D. Kingery (ed.), pp 679-696, (1983).
6. R.T. Dettoff, *Mat. Sci. Res.* 16, 23 (1984).
7. G.C. Kuczyinski, *Mat. Sci. Res.* 10, 325 (1975).
8. M.P. Anderson, D.J. Strolovitz, G.S. Grest and P.S. Sahni, "Computer Simulation of grain Growth -I Kinetics", *Acta Metall.* 30, 783 (1984).
9. H. Hofmann, "Alumina Powder: Relation between Powder Properties and Microstructure", pp 641-645, Ceramic Transactions, Vol. 51, *Ceramic Processing Science and Technology,* edited by H. Hausner, G.L. Messing and S. Hirano, The American Ceramic Society, Westerville, (1995).
10. G. Kateman and J.R.M. Smith, *Analytica Chimica Acta,* 277, 179 (1983).
11. S. Haykin, "Neural Network, a Comprehensive Foundation", edited by Macmillian, New York, (1994).
12. S. Leonardo, "Comportement des poudres céramiques par réseaux de neurones", rapport de projet de semestre, DMI-LAMI-EPFL, (1995).
13. R.J. Brook, "Controlled Grain Growth" , pp 331-364, Treatise on Materials Science and Technology, Vol. 9, *Ceramic Fabrication Processes,* edited by F.F.Y. Wang, Academic Press, New York, (1976).
14. A. Roosen and H.K. Bowen, "Influence of Various Consolidation Techniques on the Green Microstructure and Sintering Behaviour of Alumina Powders", J. Am. Ceram. Soc. 71 [11] 970-977 (1988).
15. J. Alleman, H. Hofmann and L. Gauckler, "Sintering Behaviour of Tetragonal Zirconia Polycrystalline Powders", *Berichte der Deutschen Keramischen Gesellschaft* 67 434-442 (1990).
16. S.I. Bae and S. Baik, "Critical Concentration of MgO for the Prevention of Abnormal Grain Growth in Alumina", *J. Am. Ceram. Soc.* 77 [10] 2499-2504 (1994).

17. S.I. Bae and S. Baik, "Determination of Critical Concentration of Silica and /or Calcium for Abnormal Grain Growth in Alumina", J. Am. Ceram. Soc. 76 [4] 1065-1067 (1993).
18. S. Sunrita and H.K. Bowen, "Effects of Foreign Oxides on Grain Growth and Densification of Sintered Al_2O_3", pp 840-847, Ceramic Transaction, Vol. 1, *Ceramic Powder Science II*, edited by G.L. Messing, E.R. Fuller and H. Hausner, The American Ceramic Society, Westerville, (1987).

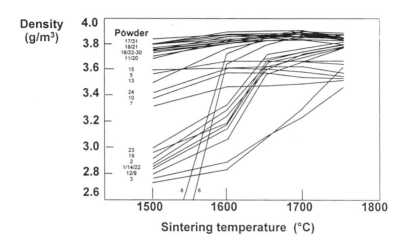

Figure 1: Density of alumina samples after 1 h sintering at temperatures between 1500 and 1750°C.

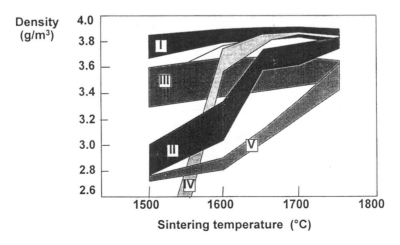

Figure 2: Density of alumina pieces after 1h sintering at different temperatures, schematic view.

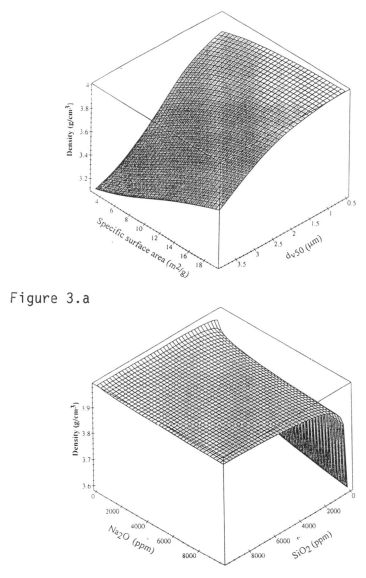

Figure 3.a

Figure 3.b

Figure 3: Predicted density using the optimized neural network in samples fabricated from a RCHPDM type powder(Sintered at 1600°C during 1h)
a) Influence of the mean particle size and specific surface area.
b) Influence of the impurities.

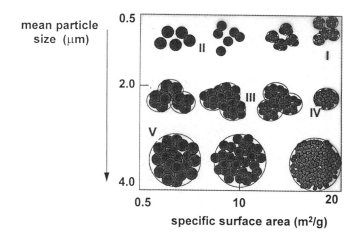

Figure 4: Agglomerate structure (schematic) and type of sintering behavior of commercial alumina powder.

Direct Observation of Mg and Ca Segregation in Sintered Al_2O_3

A. M. Thompson*, M. P. Harmer, and D. B. Williams
Materials Research Center, 5 E. Packer Avenue
Lehigh University, Bethlehem PA 18015

K. K. Soni, J. M. Chabala, and R. Levi-Setti
The Enrico Fermi Institute
The University of Chicago, Chicago, IL 60637

*Currently at GE Corporate Research and Development, Schenectady, NY 12301

Abstract

Distributions of Mg and Ca in 250 ppm MgO-doped Al_2O_3 are examined using a scanning ion microprobe in combination with secondary ion mass spectrometry (SIMS). The high spatial resolution and analytical capabilities of this technique permit compositional mapping of these elements with unprecedented detail. Segregation of Mg to the Al_2O_3 grain boundaries is observed, providing strong evidence that the primary role of MgO is to reduce grain boundary mobility through a solute-drag mechanism. Surfaces of internal pores are also found to be enriched with Mg. Ca, a background impurity in the Al_2O_3 is seen to co-segregate to grain boundaries and pore surfaces. Implications of these Mg and Ca distributions to sintering are discussed.

1. Introduction

Since Coble[1] first demonstrated that MgO additions enable Al_2O_3 to densify to a translucent material, significant progress has been made in understanding the role of this classic sintering additive[2]. However, the underlying atomistic mechanisms through which MgO operates remain unresolved primarily because the distribution of Mg solute is unknown. For example, critical kinetic experiments have established that the overriding influence of MgO is to reduce the grain boundary mobility[3,4]. Although the atomistic mechanism for this reduction in boundary mobility remains debatable, the most commonly accepted hypothesis is Cahn's solute drag model[5] in which the larger magnesium solute ions segregate preferentially to Al_2O_3 grain boundaries and inhibit boundary migration. *However, magnesium has never unambiguously been shown to segregate in polycrystalline, single-phase Al_2O_3.*

MgO additions also promote stable microstructural development. Even in low-purity Al_2O_3, MgO prevents abnormal grain growth, a process that is commonly associated with the presence of non-uniformly distributed glass-forming impurities[6]. The minimum Mg-doping level required to stabilize the grain growth is governed by the content of impurities such as Ca and Si[7]. Atomistic theories that explain the stabilizing role of MgO have consequently hypothesized about the interaction between Mg and cation impurities.

In order to resolve the atomistic roles of MgO, it is therefore important to know both the distribution of Mg within the microstructure and its distribution relative to other cations such as Ca and Si. In previous microanalytical studies, unambiguous verification of Mg segregation has proved elusive, primarily because of the limited sensitivity of these techniques to magnesium at the nanometer level. Further complications have arisen from Mg-rich grain-boundary phases [8,9] and Mg segregation induced by slow cooling rates[10,11].

In this study, ultra-high purity MgO-doped Al_2O_3 has been examined with a scanning ion microprobe utilizing secondary ion mass spectrometry (SIMS). This instrument developed at The University of Chicago provides both the analytical sensitivity and high spatial resolution necessary to study grain boundary chemistry (for a recent review see ref. [12]). A recent communication describes the successful application of this imaging-SIMS instrument in detecting grain boundary segregation of Mg in Al_2O_3[13]. The purpose of this paper is to examine the Mg segregation in more detail, focus on its distribution relative to Ca, and discuss the implications of these observations to sintering.

2. Experimental Details

Samples were prepared using an ultra-high purity (99.995%) monosized Al_2O_3 powder mixed with a suitable aliquot of magnesium nitrate solution (99.999% purity) to yield a magnesia doping level of 250 ppm (Mg/Al). This doping level is below the solubility limit at the sintering temperature [14]. As in previous kinetic studies, all powder processing was carried out using precleaned Teflon ware under clean-room conditions. Dried powders were cold-pressed to form cylindrical pellets which were subsequently isostatically pressed at 200 MPa.

In order to remove possible carbon and sulfur contamination, the green pellets were calcined in air at 1000°C for 24 h while surrounded by powder of the same composition. Two sintering schedules were employed: In the first schedule, pellets were sintered for 18 h at 1800°C in a graphite-lined furnace under an atmosphere of flowing nitrogen, and cooled at 10°C/min. The second sintering schedule was conducted in order to test the influence of cooling rate on solute segregation. Pellets were pressureless sintered in a cold-walled, tungsten-resistance-heated furnace for 1 h at 1800°C in a hydrogen atmosphere, and cooled at 100°C/min. Sintered samples were polished to a 1 μm finish and sputter-coated with Au to provide sufficient conductivity during SIMS analysis.

3. Results

Early SIMS maps of the MgO-doped Al_2O_3 cooled at 10°C/min revealed Mg enrichment at the surfaces of isolated grain boundary pores (see Fig. 1(A)). Although Ca was only present as a background impurity in the starting powder (<10 ppmw) the imaging-SIMS was sufficiently sensitive to detect Ca segregation at the same pore surfaces (Fig 1(B)). Subsequent examination of fracture surfaces confirmed that this surface enrichment of Mg and Ca was a genuine observation and not an artefact introduced during polishing. Also observed in Fig 1(B) is a faint network indicating grain boundary segregation of Ca. A corresponding network of Mg segregation is not evident in Fig 1(A).

Recently, the scanning ion microprobe was interfaced to a magnetic sector mass spectrometer, which enhanced the ion collection and the transmission optics. This modification has resulted in a dramatic improvement in analytical sensitivity (40-100X relative to the previous RF quadrupole system). The spectrum shown in Fig 2. was acquired using this new instrument to analyze the MgO-doped Al_2O_3 cooled at 10°C/min. In addition to the Ga^+ ions arising from the incident ion beam, this spectrum displays peaks corresponding to the main components, the dopant Mg, and impurities, Ca, Na and K. Silicon was not detected in this sample. SIMS maps acquired from this polished section using the magnetic sector mass spectrometer are shown in Fig. 3. Segregation of Mg to the grain boundaries can be clearly discerned in Mg^+ map (Fig. 3(A)): Boundaries are uniformly enriched with Mg in comparison to the grain interiors. The accompanying Ca map (Fig. 3(B)) indicates that Ca also segregates uniformly to the same grain boundaries.

The extent of segregation to the grain boundaries was quantified by determining the enrichment factor, the ratio of Mg concentration at the grain boundary to that within the grains. Assuming matrix effects to be uniform across the grain and grain boundary regions, the enrichment factor is given by the ratio of corresponding signal intensities per unit area. Corrected values of solute signal emitted per unit boundary area were calculated assuming a boundary width of 2 monolayers (1 nm) and a probe size equal to the pixel size (39 nm at this magnification). Solute concentrations corresponding to the grain interiors were determined by averaging the signal intensity within each grain. The grain boundary enrichment factors determined from the SIMS maps in Fig 3 are 100 and 60 for Mg and Ca, respectively.

A Mg^+ map of the fast-cooled (100°C/min) MgO-doped sample is shown in Fig. 4. Again, boundaries are enriched in Mg relative to the grains, in this case by a factor of approximately 75. Therefore, although Mg segregation is slightly enhanced by slower cooling, the fast-cooled sample confirms strong segregation of Mg at the sintering temperature.

4. Discussion

These SIMS maps form a consistent picture of the Mg distribution in single-phase, polycrystalline MgO-doped Al_2O_3: Mg segregates to *all* grain boundaries and pore surfaces. Even in a sample cooled at 100°C/min, magnesium enrichment at the grain boundaries is clearly seen, indicating that these observations represent the true Mg distribution at the sintering temperature. While Mg surface enrichment has been reported previously for internal pores and single crystals[15,16], it is believed that these results constitute the first unambiguous evidence for grain boundary segregation of Mg in a single-phase, polycrystalline Al_2O_3.

The observed Ca distribution is in agreement with previous microanalytical studies. Baik *et al.*[17] have indicated that Ca segregation is isotropic in the presence of a sufficient quantity of Mg and SIMS maps of the polished section in Fig. 3(B) support these observations: Ca is distributed uniformly along each boundary and between different grain boundaries. However, the co-segregation of Mg and Ca to pore surfaces is in disagreement with another study by Baik *et al.* which indicated that Mg suppressed surface enrichment of Ca in single crystal Al_2O_3 [15].

Implications to Sintering

The current observations of Mg grain boundary segregation are consistent with the solute drag model, the most widely accepted hypothesis for the reduction in grain boundary mobility. It should be noted that these SIMS observations alone do not directly prove solute drag to be operating in this system. However, in combination with the previous kinetic studies, this long-awaited result presents strong evidence in favor of the solute-drag model.

Kinetic experiments have indicated that a secondary beneficial effect of MgO is the enhancement of surface diffusivity [18]. This, in combination with the reduction in boundary mobility, enables pores to remain attached to migrating grain boundaries. It has been suggested that this increase in surface diffusivity arises from a Mg-induced reconstruction of the surface[18]. An alternative hypothesis contends that Mg enrichment in the near-surface region enhances diffusion of the rate-limiting ionic species through an increase in the defect concentration[16]. Both theories require that Mg ions segregate to the surface. Consequently, the atomic mechanism cannot strictly be deduced from these observations of Mg-enrichment at pore surfaces.

Interpretations of the atomistic role of MgO in the densification process have also varied widely[2]. While it is generally agreed that the incorporation of Mg into Al_2O_3 generates compensating defects which alter the diffusion rate of the rate-limiting ionic species, opinions differ as to the type of compensating defect, the role of background impurities, and the diffusion mechanism responsible for densification. The observations of Mg grain boundary segregation support kinetic studies in which MgO alters the grain boundary diffusivity.

Knowledge of both the Mg and Ca distributions sheds light on magnesia's unique ability to stabilize microstructural evolution in Al_2O_3. It has been suggested that Mg counteracts the anisotropic segregation of Ca by preferentially segregating to boundaries deficient in solute ions[3]. The combination of Mg and Ca solute drag produces isotropic grain growth. In contrast, Baik and Moon[17] proposed that Mg segregates to all boundaries and reduces both the magnitude and the anisotropy of Ca segregation. From the results shown in Figs. 3, it is clear that the former hypothesis is unlikely: Ca and Mg are uniformly distributed along all boundaries. It is also evident that Mg does not completely inhibit Ca segregation either to pores or to grain boundaries. These co-segregation observations, however, are insufficient to comment on MgO's influence on the anisotropic nature of Ca segregation. Consequently, work is now underway to examine the Mg and Ca distribution in undoped and co-doped Al_2O_3, and will be the subject of a future communication.

5. Conclusions

(i) Grain boundary segregation of Mg is detected in a single-phase, polycrystalline MgO-doped Al_2O_3 — strong evidence in favor of a solute drag mechanism. Grain boundaries are uniformly enriched to a level 75-100 times greater than that within the grains.
(ii) Ca and Mg are observed to co-segregate to pore surfaces and grain boundaries.
(iii) The ability to detect magnesium, where other methods have failed, underscores the power of this imaging-SIMS technique.

Acknowledgments

This material is based on work supported by the National Science Foundation through the Materials Research Center at Lehigh University and the Materials Research Science and Engineering Center at The University of Chicago.

References

1. R. L. Coble, "Transparent Alumina and Method of Preparation," U.S. Patent # 3,026,210 March (1962).
2. S. J. Bennison and M. P. Harmer, "A History of the Role of MgO in the Sintering of α-Al$_2$O$_3$," *Ceram. Trans.*, **7**, 13-49 (1990).
3. S. J. Bennison, and M. P. Harmer, Effect of MgO Solute on the Kinetics of Grain Growth in Al$_2$O$_3$," *J. Am. Ceram. Soc*, **66**[5], C90-C91 (1983).
4. S. J. Bennison and M. P. Harmer, "Grain Growth Kinetics for Alumina in the Absence of a Liquid Phase," *J. Am. Ceram. Soc*, **68**[1], C22-C24 (1985).
5. J. W. Cahn, "The Impurity-Drag Effect in Grain Boundary Motion," *Acta Metall.* **10**, 789-98 (1962).
6. S. I. Bae and S. Baik, "Determinations of Critical Concentrations of Silica and/or Calcia for Abnormal Grain Growth in Alumina," *J. Am. Ceram. Soc.*, **76**[4], 1065-67 (1993).
7. W. C. Johnson, D. F. Stein, "Additive and Impurity Distribution at Grain Boundaries in Sintered Alumina," *J. Am. Ceram. Soc.*, **58**, 485-488 (1975).
8. S. S. C. Tong and J. P. Williams, "Chemical Analysis of Grain Boundary Impurities in Polycrystalline Ceramic Materials by Spark Source Mass Spectrometry," *J. Am. Cer. Soc.*, **53**, 58-59 (1969).
9. H. L. Marcus, M. E. Fine, "Grain-Boundary Segregation in MgO-Doped Al$_2$O$_3$," *J. Am. Ceram. Soc.*, **55**, 568-570 (1972).
10. R. I. Taylor, J. P. Coad, R. J. Brook, "Grain Boundary Segregation in Al$_2$O$_3$," *J. Am. Ceram. Soc.*, **57**, 539-540 (1974).
11. W.C. Johnson, "Mg Distributions at Grain Boundaries in Sintered Alumina containing MgAl$_2$O$_4$ Precipitates," *J. Am. Ceram. Soc.*, **61**[5-6], 234-37 (1978).
12. J. M. Chabala, K. K. Soni, J. Li, K. L. Gavrilov, and R. Levi-Setti, "High-Resolution Chemical Imaging with Scanning Ion Microprobe SIMS," *Inter. J. Mass Spectrometry and Ion Processes*, **143**, 191-212 (1995).
13. K. K. Soni, A. M. Thompson, M. P. Harmer, D. B. Williams, J. M. Chabala, and R. Levi-Setti, "Solute Segregation to Grain Boundaries in MgO-doped Alumina," *Appl. Phys. Lett.*, **66**, 2795-97 (1995).
14. S. K. Roy and R. L. Coble, "Solubilities of Magnesia, Titania, and Magnesium Titanate in Aluminum Oxide," *J. Am Ceram. Soc.*, **51**[1], 1-6 (1968).
15. S. Baik, D. E. Fowler, J. M. Blakely, R. Raj, "Segregation of Mg to the (0001) Surface of Doped Sapphire," *J. Am. Ceram. Soc.*, **68**, 281-286 (1985).
16. C. Sung, G. C. Wei, K. J. Ostreicher, and W. H. Rhodes, "Segregation of Magnesium to the Internal Surface of Residual Pores in Translucent Polycrystalline Alumina," *J. Am. Ceram. Soc.*, **75**, 1796-1800 (1992).
17. S. Baik and J. H. Moon, "Effects of Magnesium Oxide on Grain-Boundary Segregation of Calcium During Sintering of Alumina," *J. Am. Ceram. Soc.*, **74**[4], 819-822, (1991).
18. S. J. Bennison and M. P. Harmer, "Effect of Magnesia Solute on Surface Diffusion in Sapphire and the Role of Magnesia in the Sintering of Alumina," *J. Am. Ceram. Soc.*, **73**[4], 833-37 (1990).

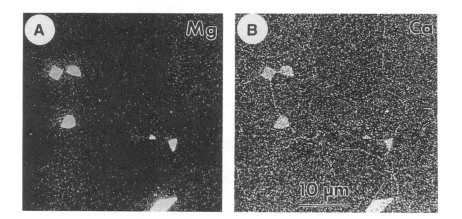

Fig 1. (A) Mg$^+$ and (B) Ca$^+$ maps of 250 ppm MgO-doped Al$_2$O$_3$ cooled at 10 C/min. Image size = 40 μm x 40 μm (512 x 512 pixels).

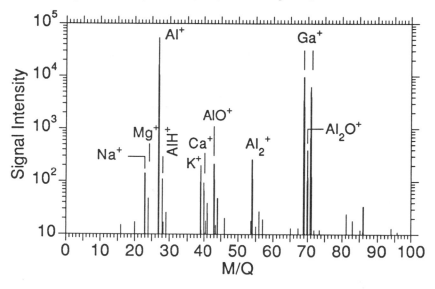

Fig. 2. A SIMS mass spectrum obtained using magnetic sector mass spectrometer from 250 ppm MgO-doped Al$_2$O$_3$ cooled at 10 C/min .

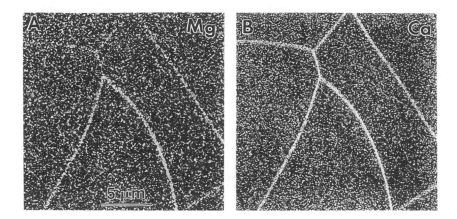

Fig 3. (A) Mg$^+$ and (B) Ca$^+$ SIMS maps of 250 ppm MgO-doped Al$_2$O$_3$ cooled at 10 C/min. Image size = 20 μm x 20 μm (512 x 512 pixels).

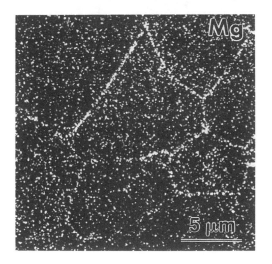

Fig 4. Mg$^+$ SIMS map of 250 ppm MgO-doped Al$_2$O$_3$ cooled at 100 C/min. Image size = 20 μm x 20 μm (512 x 512 pixels).

Sintering Behavior of Ultra-High Purity Al₂O₃ Doped with Y and La

Jianxin Fang, A. Mark Thompson, Martin P. Harmer and Helen M. Chan

Department of Materials Science and Engineering and Materials Research Center
Lehigh University, Bethlehem, Pennsylvania 18015

Abstract

Final-stage sintering has been investigated in ultra-high pure alumina and alumina doped individually with 1000 ppm Y and 1000 ppm La. In all undoped and doped materials, the dominant densification mechanism was grain-boundary diffusion. Doping with Y and La decreased the densification rate (grain-boundary diffusion coefficient) by a factor of about 11 and 21, respectively. In addition, Y and La doping decreased grain growth during sintering. In the undoped alumina surface-diffusion-controlled pore drag governed grain growth. In the doped materials, grain growth followed a simple cubic law, suggesting that a solute-drag mechanism was operative. Overall, Y and La decreased the coarsening rate relative to the densification rate and, hence, shifted the grain size/density trajectory to higher density for a given grain size. It is believed that the role of the additives is linked strongly to their segregation to the alumina grain boundaries.

I. Introduction

Recent studies have shown that doping of 500~1500 ppm yttrium can improve both tensile and compressive creep resistance of Al₂O₃ (Lartigue, Carry and Priester 1992, French *et al.* 1994). For instance, French *et al.* (1994) have found that addition of 1000 ppm Y to ultra-pure alumina decreased the creep rate from $10^{-5}S^{-1}$ to $10^{-6}S^{-1}$ at 1350°C under the applied stress ranging from 35 MPa to 75 MPa. French *et al.* attributed the beneficial effect of yttrium to the segregation of yttrium to alumina grain boundaries, which reduces the rate of ion transport along grain boundaries. However, it was not clear whether yttrium decreased the creep rate by directly lowering the grain boundary diffusivity or by inhibiting other processes such as grain boundary sliding.

Some processes, which occur during creep, such as grain boundary sliding, do not operate in sintering. Therefore, a major objective of the present work was to clarify the effect of Y-doping on the creep of Al_2O_3 by studying its influence on sintering kinetics under identical conditions. In addition, 1000 ppm La-doped alumina was examined to investigate the effect of ion size on sintering kinetics of alumina and, hence, to assess the potential of La dopant for enhancing creep. La was selected because it has the same valence as Al and Y, but a much larger ionic radius (ionic radii are 0.51, 0.893 and 1.061 for Al^{3+}, Y^{3+} and La^{3+}, respectively (Lide 1991).

II. Experimental Procedure

Sintering experiments were performed on undoped Al_2O_3, Y-doped Al_2O_3 and La-doped Al_2O_3. Cylindrical pellets of 14 mm in diameter x 5 mm were prepared by mixing, drying, crushing and pressing. Initial powder was an ultra-high purity alumina powder (Sumitomo AKP-53) with 99.995% purity and a mean particle size of 0.2 μm. Yttrium ($Y/Al = 1000 \times 10^{-6}$) or lanthanum ($La/Al = 1000 \times 10^{-6}$) doping was achieved by the addition of a high-purity $Y(NO_3)_3$ or $La(NO_3)_3$ solution to a batch of alumina powder. All processing was performed using pre-cleaned Teflon ware in an ultra-clean environment.

The pellets were calcined at 1000°C and sintered at 1350°C in air. Densities were measured using Archimedes method with deionized water as the immersion medium. Grain boundaries were revealed by thermally etching polished sections at 1150~1250°C in air. Grain sizes were determined from SEM micrographs using the linear intercept technique. Further details of the experimental procedure are described elsewhere (Fang *et al.* 1995).

III. Results and Discussion

3.1. Effect of dopants on densification kinetics

Figure 1 shows the density as a function of sintering time at 1350°C for undoped Al_2O_3, $Y-Al_2O_3$ and La-doped Al_2O_3. Both Y and La doping inhibited densification; Y showing a stronger overall effect. It should be pointed out that this stronger overall effect does not necessarily indicate that yttrium doping has more effectively inhibited the diffusion process for densification because the influence of the grain size has not been taken into account.

During final stage sintering, the densification rate, $d\rho/dt$, can be expressed in the following form (Coble 1962, Zhao and Harmer 1991, Zhao and Harmer 1992):

$$\frac{d\rho}{dt} = \frac{C\gamma_s DN_g}{G^n} \tag{1}$$

where ρ is the density, C the constant, γ_s the solid/gas surface energy, D the diffusion coefficient, N_g the number of pores per grain, and G the grain size. Here, we assume that N_g is constant during densification. The grain size exponent, n, is 3 for lattice diffusion control and 4 for grain boundary diffusion control. Fig. 2 shows a fit of the experimental data to Eq. (1).

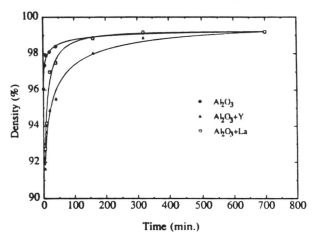

Fig. 1. Density as a function of sintering time at 1350°C for undoped, Y-doped and La-doped Al₂O₃.

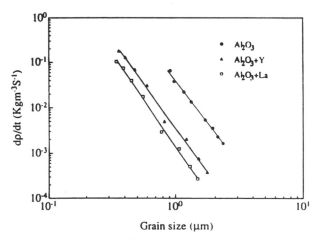

Fig. 2. Densification rate vs. grain size for undoped and doped Al₂O₃.

The grain-size exponent was found to be 3.8 for undoped alumina, 3.9 for Y-doped alumina and 4.1 for La-doped alumina. The fact that the grain-size exponents are close to 4.0 for all three materials strongly suggests that grain boundary diffusion was the predominant transport mechanism during densification process, consistent with previous studies of sintering kinetics for model final-stage microstructures of Al_2O_3 (Zhao and Harmer 1991).

It is evident from Eq. (1) that a reduction in densification rate may result either directly from a decrease of the diffusion coefficient (D) or indirectly from an increase in the grain size (G). To evaluate the direct effect of a dopant on the diffusion mechanism responsible for densification, the indirect effect of the grain size has to be considered. Comparing densification rates at constant grain size in Fig. 2. reveals that Y and La doping decreased the densification rate directly by factors of 11 and 21, respectively. Usually, upon doping, the change in the surface energy (γ_s) is small in comparison to the change in diffusivity. Thus, the reduction in the densification rates for the doped materials are postulated to be caused by a reduction in the grain boundary diffusion coefficient.

It is hard to explain the lower diffusion coefficients for the doped alumina in terms of defect chemistry since Al, Y and La have the same charge valence. Y and La can substitute for Al ions without creating point defects. The more likely cause for the reduction in Db is segregation of dopants to grain boundaries. Earlier investigations have well established that yttrium strongly segregated to the grain boundaries in Al_2O_3 due to the large size misfit (Nani, Stoddart and Hondros 1976, Bender, Williams and Notis 1980, Li and Kinegery 1984, Gruffel and Carry 1994). It is conceivable that the large yttrium cations at grain boundaries block the diffusion of ions along grain boundaries, leading to reduced grain boundary diffusivity.

The factor of 11 decrease in densification rate due to Y-doping compares favorably well with the factor of 10 decrease in creep rate. This result implies that the beneficial effect of yttrium in creep stems from a lowered grain boundary diffusivity. La should, therefore, also have a beneficial influence on the creep of Al_2O_3.

3.2. Effect of dopants on grain growth

Fig. 3 shows the dependence of the average grain size on the sintering time at 1350°C for undoped, Y-doped and La-doped Al_2O_3. Clearly, both Y and La doping retarded the grain growth during sintering, La doping having a stronger effect.

Previous studies (Zhao and Harmer 1991, Zhao and Harmer 1992) on pure alumina have shown that grain growth during the final stage of sintering is governed by surface-diffusion-controlled pore drag, for which the grain growth models predict

$$\frac{dG}{dt} = \frac{CD_s N_g}{G^3 (1-\rho)^{4/3}} \qquad (2)$$

where C is a constant and D_S the diffusion coefficient. The present data for the undoped alumina gave an excellent fit to such a model of surface-diffusion-controlled pore drag as shown in Fig. 4.

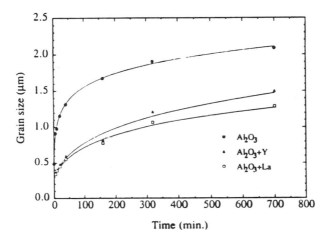

Fig. 3. Grain size as a function of sintering time at 1350°C for undoped, Y-doped and La-doped Al₂O₃.

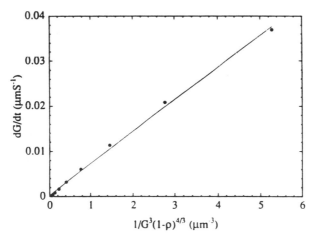

Fig. 4. Grain growth rate plotted according to the model for surface diffusion-controlled pore drag for undoped Al₂O₃.

The grain size data for the doped materials, however, did not fit any of the pore-drag models. Fig. 5 shows that the grain growth obeys a simple cubic kinetic relationship:

$$G^3 - G_0^3 = Kt \qquad\qquad (3)$$

where G is the grain size after time t, G_0 is the initial grain size and K is a temperature-dependent grain growth rate constant. The growth rate constant (K) was determined to be 4.2×10^{-3} and 2.0×10^{-3} $\mu m^3 s^{-3}$ for Y-doped and La-doped samples, respectively.

One possible mechanism for grain growth inhibition is particle pinning. However, given the small volume fraction of precipitates present (<0.5%), this influence is likely to be small. A more plausible explanation is that grain growth is controlled by solute-drag, consistent with the observed cubic kinetics and grain boundary segregation.

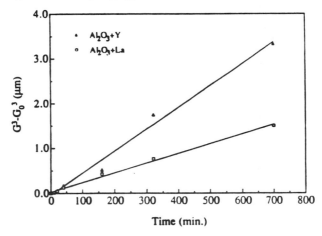

Fig. 5. Grain growth kinetics during sintering for Y-doped and La-doped Al_2O_3.

3.3. Effect of the dopants on Microstructure Development

Fig. 6 shows the microstructure of the undoped and Y-doped Al_2O_3 sintered for 160 minutes. The majority of pores are located within the interior of the grains in the undoped sample (Fig. 6 (a)), but situated at the grain boundaries and triple grain junctions in Y-doped material (Fig. 6 (b)). These observations indicate that in pure Al_2O_3 pore-boundary separation occurred after about 160 minutes of sintering, whereas in the doped Al_2O_3, the pores remained attached to the grain boundaries.

A figure of merit to evaluate the pore-boundary resistance is (Berry and Harmer 1986)

$$(\frac{1}{G^2})\rho \frac{D_s}{M_b} \qquad\qquad (4)$$

It is postulated that Y and La doping increased this figure of merit by increasing $(1/G_2)_\rho$ and decreasing M_b (grain boundary mobility), resulting in the improved resistance to pore-boundary separation.

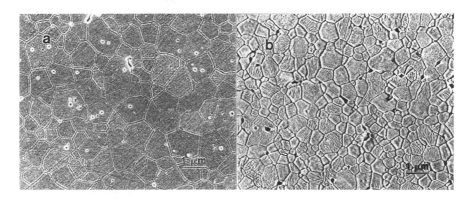

Fig. 6. Microstructure of undoped (a) and Y-doped (b) Al₂O₃ sintered for 160 minutes at 1350°C.

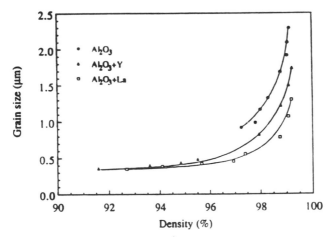

Fig. 7. Grain size-density trajectory for undoped, Y-doped and La-doped Al₂O₃ sintered at 1350°C.

Fig. 7 shows grain size/density trajectories for undoped Al₂O₃, Al₂O₃+Y and Al₂O₃+La. The grain size/density trajectory is a function of the relative ratio of densification rate to grain growth rate. Y and La doping flattened the G-ρ trajectories, indicating that the doping enhanced the densification rate relative to the grain growth rate.

IV. Conclusions

In undoped and doped materials, densification was controlled by grain-boundary diffusion. Doping with Y and La decreased the densification rate directly by a factor of about 11 and 21, respectively, through a reduction in the grain boundary diffusion rate. In undoped alumina grain growth was governed by surface-diffusion controlled-pore drag, whereas in the doped alumina the grain growth was more likely controlled by solute-drag. Doping with Y and La inhibited grain growth during sintering. The overall effect of Y and La was to enhance the densification rate relative to the grain growth rate, thereby displacing the grain size-density trajectory to higher densities for a given grain size. The role of the dopants was discussed in terms of grain boundary segregation due to the ionic size misfit between the dopant and the host.

References

Bender, B., Williams, D. B., and Notis, M. R., "Investigation of Grain-Boundary Segregation in Ceramic Oxides by Analytical Scanning Transmission Electron Microscopy", *J. Am. Ceram. Soc.*, **63** [9] 542-46 (1980).

Coble, R. L., "Sintering Crystalline Solids. I. Intermediate and Final Stage Diffusion Models", *J. Appl. Phys.*, **32** 787-92 (1962).

Fang, J., Thompson, A. M., Chan, H. M., and Harmer, M. P., "Sintering Behavior of Ultra-High-Purity Alumina Doped with Y and La", submitted to *J. Am. Ceram. Soc.* (1995).

French, J. D., Zhao, J., Harmer, M. P., Chan, H. M., and Miller, G. A., "Creep of Duplex Microstructures",*J. Am. Ceram. Soc.*, **77** [11] 2857-65 (1994).

Gruffel, P. and Carry, C., "Effect of Grain Size on Yttrium Grain Boundary Segregation in Fine Grained Alumina", submitted to *J. Eur. Ceram. Soc.*, (1994).

Lartigue, S., Carry, C., and Priester, L., "Grain Boundaries in High Temperature Deformation of Yttria and Magnesia Co-Doped Alumina", *Coll. Phys.* **C1**, 51 [1] 985-90 (1990).

Li, C-W., and Kingery, W. D., "Solute Segregation at Grain Boundaries in Polycrystalline Al_2O_3", *Structure and Properties of MgO and Al_2O_3 Ceramics*, W. D. Kingery, American Ceramic Society, Columbus, OH, pp. 368-378, (1984).

Lide, D. R., *CRC Handbook of Chemistry and Physics*, 71st Edition, CRC Press Inc., p. 12-1 (1991).

Nanni, P., Stoddart, C. T. H., and Hondros, E. D., "Grain Boundary Segregation and Sintering in Alumina", *Materials Chemistry*, **1** 297-320 (1976).

Zhao, J., and Harmer, M. P., "Sintering Kinetics for a Model Final-Stage Microstructure: A Study of Al2O3", *Phil. Mag. Lett.*, **63** [1] 7-14 (1991); "Effect of Pore Distribution on Microstructure Development: III, Model Experiments", *J. Am. Ceram. Soc.*, **74** [4] 830-43 (1992).

Grain Boundary Pinning in Al$_2$O$_3$-SiC

Laura C. Stearns and Martin P. Harmer

Materials Research Center
Lehigh University, Bethlehem, PA 18015

Abstract

An experimental study was conducted to examine the influence of ultrafine SiC particles on grain growth in Al$_2$O$_3$ as a function of annealing time, particle volume fraction, and annealing temperature. The SiC particles reduced the grain growth rate of Al$_2$O$_3$ by over three orders of magnitude, resulting in final grain sizes which decreased with increasing particle volume fraction. It was observed that ϕ (fraction of particles on grain boundaries) and G (grain size) were strongly correlated, revealing a path for microstructural evolution in this system, which involved significant particle-boundary breakaway. This correlation was used to modify Zener's expression for the dependence of equilibrium grain size on particle volume fraction, and the resulting expression was used to describe the experimental data.

1. Introduction

It has long been known that inert second phase particles can inhibit grain growth, and can lead to a pinned microstructure with a limiting grain size. Many analytical models and experimental studies exist; however, the underlying mechanisms of particle pinning remain to be understood. There are basically two types of models to predict the equilibrium grain size dependence on particle volume fraction; those based on the Zener model [Zener,1948; e.g, Gladman, 1966; Haroun et al, 1968; Louat, 1983] which assumes a random correlation between particles and grain boundaries, and the topological pinning model [Srolovitz et al., 1984; Doherty et al, 1987] which incorporates a strong correlation between particles and boundaries. Experimental data from studies of grain boundary pinning have not confirmed either model [Olgaard & Evans, 1986]. Here we present some details of an experimental study in Al$_2$O$_3$-SiC, in which characterization of the microstructural evolution provides insight into the true role of pinning particles during grain growth.

2. Experimental Details

The α-Al_2O_3 powder used in this study (Sumitomo AKP-53, Sumitomo Chemical Company, Japan) was 99.995% pure with a mean particle diameter of 0.2 μm. The β-SiC powder (Performance Ceramics Company, Peninsula, OH) had a median particle diameter of 0.15 μm. Composite powders were prepared by mixing the Al_2O_3 and SiC powders, with volume fractions ranging over two orders of magnitude, from 0.2 to 20 vol% SiC. These powders were calcined before hot-pressing; Al_2O_3 at 850°C, and the composite powders at 600°C, for 10 hours in air. Hot-pressing was conducted in a graphite-lined, graphite element, hot press at 50 MPa for 15 minutes under vacuum. Hot-pressing temperatures ranged from 1400°C to 1650°C, increasing with SiC content. Fully dense specimens of each volume fraction were then annealed in the hot press in flowing nitrogen for grain growth. Annealing temperatures ranged from 1500°C to 1800°C, and times ranged from 0.1 to 100 hrs. The samples were polished and thermally etched to reveal the grain boundaries and matrix-particle interfaces. The Al_2O_3 grain size was determined using a linear intercept technique with a total of 400-500 intercepts counted for each measurement. The fraction of particles on grain boundaries, ϕ, was determined using the same micrographs used for the grain size measurement, with a total of 400-800 particles counted for each measurement. Further details of the processing methods in this system have been published elsewhere [Stearns et al., 1992, Stearns et al., 1995a].

3. Results and Discussion

Figure 1 shows the grain size as a function of annealing time at 1700°C for fully dense samples with volume fractions ranging from 0 to 20 vol% SiC. It is apparent that the SiC particles severely inhibit grain growth. In fact, the Al_2O_3 grain growth rate was decreased by over three orders of magnitude: At 100 hrs the rate for Al_2O_3 was about 5 μm/hr, while the rate for the 20 vol% SiC material at 100 hrs was about 1 nm/hr.

The 100-hr grain size as a function of particle volume fraction is given in Figure 2 so that comparisons can be made with the two main types of equilibrium models. A power law fit to the data yields a slope of -0.64. For comparison, lines are drawn on the plot with the predicted dependencies from the Zener model (slope -1) and the topological model (slope = -0.33). Clearly, the experimental data are not consistent with either model's prediction.

The fraction of particles on grain boundaries, ϕ, was measured for volume fractions 1, 2, 5, 10, and 20 vol% SiC, and for the as-pressed, 0-hr., and 100-hr. samples. Figure 3 shows ϕ as a function of time in the annealing treatment. As the anneal proceeded, ϕ decreased with annealing time, and the decrease was more pronounced for the smaller volume fractions.

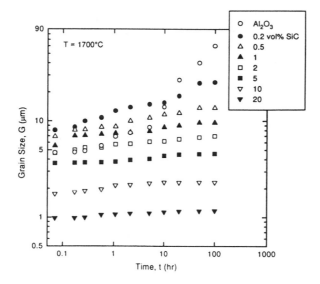

Figure 1. Grain size as a function of annealing time at 1700°C for all compositions.

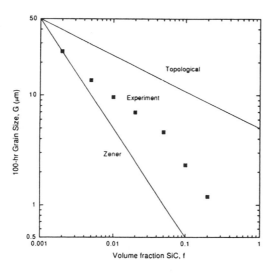

Figure 2. 100-hr. grain size as a function of particle volume fraction, including the predicted dependencies for the Zener and topological pinning models along with the experimental data.

This observation of decreasing ϕ indicates that a substantial degree of particle-boundary breakaway occurred during grain growth, inconsistent with the assumptions of the topological model.

Both the extent of grain growth and particle-boundary breakaway at 100 hrs were enhanced by an increase in annealing temperature [Stearns et al., 1995a, b], and this enhancement was far less pronounced in the larger volume fraction materials. Further, comparison of all of the G and ϕ data revealed that the suppression of thermally activated breakaway in the larger volume fraction materials is strongly correlated with a suppression of the thermally activated process of grain growth.

Figure 4 illustrates the correlation between ϕ and G, showing all the data for ϕ and G generated in this study, including all annealing times, all volume fractions, and all annealing temperatures. The fraction of particles located on grain boundaries decreased with increasing grain size for all materials. The line through the data represents a power law fit of the form:

$$\phi = \varepsilon G^{-x} \qquad (1)$$

where $\varepsilon = 1.88 \times 10^{-4} \, \mathrm{m}^{0.6}$ and the slope, $x = 0.60 \; +/- \; 0.03$.

The universal behavior exhibited in Figure 4 suggests that there is a single path for microstructural evolution in this system that couples ϕ and G. Once either of these parameters is set initially by the processing parameters, then the other is set, and any microstructural change from this point (by time, volume fraction or temperature) is determined by the general relation of Equation 1.

This relationship between ϕ and G is a key finding, as it provides insight into how the microstructure evolves, but also because it can be used directly in the analysis of the grain growth behavior. Specifically, it will allow direct comparison of the experimental results with predictions from the Zener model, and the development of an equilibrium expression by modification of the Zener force balance, as will be shown in the following paragraphs.

Zener assumed that the number of particles per unit grain boundary area, N_{ba}, was constant and equal to the number of particles per unit area of a random section through a random microstructure, N_{ua}. Using the definition of ϕ, a general expression for N_{ba} can be developed [Stearns et al., 1995a, c]:

$$N_{ba} = \frac{N_v \phi G}{3.6} \qquad (2)$$

where N_v is the number of particles per unit volume. Using the experimental values of N_v, ϕ, and G, N_{ba} was calculated, and was found to increase with increasing grain size (for a given volume fraction), inconsistent with Zener's assumption. This observed increase indicates (from Equation 2) that although ϕ was decreasing, it was not decreasing *enough* to account for the increasing

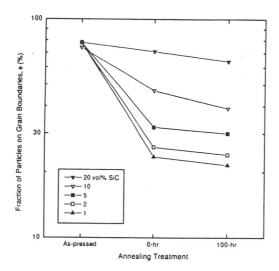

Figure 3. Fraction of particles on grain boundaries as a function of time in the annealing treatment at 1700°C for volume fractions 1,2,5,10, and 20 vol% SiC.

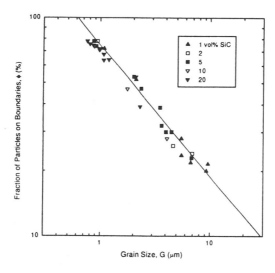

Figure 4. Fraction of particles on grain boundaries as a function of grain size for all annealing times, all volume fractions, and all annealing temperatures.

grain size so that N_{ba} did not remain constant (as Zener would have predicted), but increased.

Although the experimental results were not consistent with Zener's model, modifications can be made to this model to account for the observed non-random particle-boundary interactions. (It is assumed that it is merely Zener's choice of N_{ba}, rather than the force balance approach, that is incorrect).

The crux of the Zener model is that at equilibrium the driving force for grain boundary migration, F_b, is equal to the resistive force of the pinning particles, F_p:

$$F_b = F_p \tag{3}$$

Here F_b is taken to have its conventional form:

$$F_b = \frac{\beta \gamma_b}{G} \tag{4}$$

where γ_b is the grain boundary energy per unit area and β is a constant which relates radius of curvature to grain size. The pinning force is assumed to be the general form of that given by Zener:

$$F_p = N_{ba} \pi r \gamma_b \tag{5}$$

where N_{ba} is now taken to be the quantity calculated from experimental measureables, as given in Equation 2. Hence, combining Equations 1-5, a relation for the limiting grain size in terms of particle volume fraction is:

$$G_L = \left(\frac{4.8\beta r^2}{f\varepsilon}\right)^{1/(2-x)} \tag{6}$$

and thus for this system, the dependence of G_L on f is:

$$G_L \propto f^{-0.71} \tag{7}$$

Figure 5 plots this relationship as the solid line. The dotted and dashed lines are best fits to the experimental data at the temperatures indicated. As can be seen, the slope of the line characterizing the G_{100}-f dependence increased with increasing anneal temperature, *approaching* the predicted equilibrium behavior.

4. Conclusions

A comprehensive set of experiments in a pinning system has shown that the assumptions of the two dominant models for equilibrium grain size are too restrictive to describe a real system: Pinning particles are not distributed as randomly as the Zener model assumes, nor are they correlated as strongly with grain boundaries as the strong pinning model assumes. By incorporating the details of microstrucutral evolution - in particular the fraction of pinning particles, ϕ - into a modified Zener force balance, relationships were developed that described the dependence of grain size on particle volume fraction.

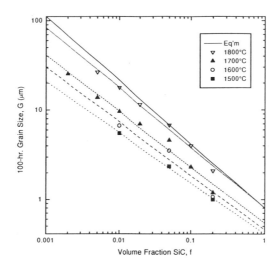

Figure 5. Grain size as a function of particle volume fraction for the experimental results at 1500°C, 1600°C, 1700°C, and 1800°C, and for the precited dependence from the equilibrium model.

Acknowledgments

The authors gratefully acknowledge the National Science Foundation for funding this research and IBM for providing support through a joint study program for LCS.

References

C. Zener, as communicated by C. S. Smith, "Grains, Phases, and Interfaces: An Interpretation of Microstructure," *Trans AIME*, **175** 15-51 (1948).

T. Gladman, "On the Theory of the Effect of Precipitate Particles on Grain Growth in Metals," *Proc. Roy. Soc. A.*, **294** 298-309 (1966).

N. A. Haroun and D. W. Budworth, "Modifications to the Zener Formula for Limitation of Grain Size," *J. Mater. Sci.*, **3** 326-328 (1968).

N. Louat, "The Inhibition of Grain Boundary Motion by a Dispersion of Particles," *Phil Mag. A.*, **47** [6] 903-912 (1983).

D. J. Srolovitz, M. P. Anderson, G. S. Grest, and P. S. Sahni, "Computer Simulation of Grain Growth - III, Influence of a Particle Dispersion," *Acta Metall.*, **32** [9] 1429-1438 (1984).

R. D. Doherty, D. J. Srolovitz, A. D. Rollett, and M. P. Anderson, "On the Volume Fraction Dependence of Particle-Inhibited Grain Growth," *Scripta Metall.*, **23** 753-758 (1987).

D. L. Olgaard and B. Evans, "Effect of Second Phase Particles on Grain Growth in Calcite," *J. Am. Ceram. Soc.*, **69** [11] C-272-C-277 (1986).

L. C. Stearns, J. Zhao, and M. P. Harmer, "Processing and Microstructure Development in Al_2O_3-SiC 'Nanocomposites'," *J. Eur. Ceram. Soc.*, [10] (1992).

L. C. Stearns, "Particle-Inhibited Grain Growth in Al_2O_3-SiC" Ph. D. Thesis, Lehigh University (1995a).

L. C. Stearns and M. P. Harmer, "Particle-Inhibited Grain Growth in Al_2O_3-SiC: I, Experimental Results," submitted (1995b).

L. C. Stearns and M. P. Harmer, "Particle-Inhibited Grain Growth in Al_2O_3-SiC: II, Equilibrium and Kinetic Analyses," submitted (1995c)

TITANIUM EFFECTS ON SINTERING AND GRAIN GROWTH OF ALUMINA

J. D. Powers and A. M. Glaeser
Department of Materials Science and Mineral Engineering
University of California, Berkeley, CA 94720-1760, USA

ABSTRACT

High-purity undoped and 500 ppma Ti-doped alumina compacts were prepared and sintered at temperatures from 1350°C to 1550°C in vacuum. Hot isostatic pressing was used to fully densify the compacts. The growth of matrix grains in these dense compacts during anneals in vacuum and in air, and the growth of single crystal seeds into the dense alumina were investigated. Microstructural evolution was observed by SEM. Ti doping increases the sintering rate during early stages of sintering, but does not allow sintering to full density. Ti-doped compacts develop faceted microstructures, with the faceting most pronounced in the sample center, where Ti^{4+} is the dominant impurity. During grain growth, faceted microstructures become more pronounced. Preliminary seed growth results suggest that the mobility of $(11\bar{2}0)$ seeds growing into Ti-doped material is >3× that previously reported for growth into nominally undoped alumina.

1. INTRODUCTION

Microstructural development during high-temperature processing of alumina has long been known to be highly sensitive to the presence of impurities. Recent work by Bae and Baik (1994) has clearly demonstrated that the presence of extremely small amounts of dopant (in the range of 10-100 ppm) can lead to substantial changes in the final microstructure and density. This dominance of impurities is not necessarily a problem; while Ca and Si are known to be highly detrimental to effective sintering of alumina, Mg is well known as an essential sintering aid. It would seem that an understanding of the behavior and influence of individual impurities could lead to methods by which these impurities are intentionally used to create a novel and desirable microstructure. However, with the notable exception of Mg, there has been little work to isolate the effects of individual dopants on microstructural development, and even less on understanding the mechanisms by which these dopants exert this control.

Several investigations of titanium effects on alumina sintering have been conducted. For years, Ti has been known as a sintering aid; Bagley *et al.* (1970) demonstrated that it increases the rate of initial-stage sintering. More recently, Ikegami *et al.* (1987) similarly found an increase in initial-stage sintering rate due to titanium, but also reported that Ti does not help to increase final density of the sintered compact.

Some recent work has focused on the surface energetics and microstructure of Ti-doped alumina. The present authors observed that Ti strongly stabilizes certain planes in sapphire relative to undoped material (Powers and Glaeser, 1993). A recent analysis of the Wulff shapes of undoped and Ti-doped sapphire suggests that Ti-doping has a dramatic effect on relative surface energies and therefore on the anisotropy (Kitayama and Glaeser, 1995). Horn and Messing (1995) have observed grain growth in fine-scale dense alumina compacts, and find that Ti-doping leads to extremely faceted microstructures, which could provide opportunities to tailor materials with improved fracture toughness.

The previous sintering and grain growth work on Ti-doped alumina has involved materials with relatively high dopant levels: 1% Ti or more. Further, the overall purity of the starting material is in question in many of these studies; it has been proposed that the beneficial effects of Ti are due to the presence of an intergranular liquid phase (Morgan, 1985). In the current work, we attempt to isolate the effects of small amounts of Ti on sintering behavior and the anisotropy in grain boundary motion. Low levels of Ti dopant ($\leq$500 ppm) were incorporated into extremely clean alumina powders (99.997%). Sintering and grain growth behavior was observed as a function of dopant level, processing temperature, and atmosphere. Further, controlled studies of grain-boundary mobility and pore drag are underway to quantify the anisotropies.

2. EXPERIMENTAL PROCEDURES

The starting material was Showa Denko UA-5105 alumina powder, with a total impurity content of about 30 ppm[♠] and an average particle size of 0.4 μm. Hard agglomerates were removed from the starting powder by centrifuging a dispersion of the starting powder in ethanol with 0.1% by weight 4-aminobenzoic acid added as a dispersant. The supernatant was retained and dried in a rotary evaporator, then crushed through a size 60 mesh sieve. Ti-doped powders were prepared by hydrolysis of 99.999% pure titanium isopropoxide, added after the centrifugation. The amount of organic precursor was chosen to give a concentration of 500 ppma Ti in the final compact.

Compacts were formed via cold isostatic pressing at 230 MPa. After this step, the organics were burnt out with a 2 h air anneal at 600°C. In this step, as in all firings, samples were packed in UA-5105 powder and contained in a commercial 99.8% purity alumina crucible.

[♠] Levels of major impurities in starting powder: 1 ppm Mg, 2 ppm Ca, 6 ppm Si, 15 ppm Na.

At this point, sintering studies were performed on green compacts, both to study sintering behavior of the aluminas and to find ideal processing conditions for preparing fully dense compacts. Sintering was performed in vacuum and in air at temperatures between 1350° and 1550°C. Samples were again packed in powder, as described previously, to minimize contamination. After sintering, pieces were polished to a 1 μm finish and microstructures were examined in the SEM.

Hot isostatic pressing (HIP) was used to obtain fully dense samples (1400°C, 180 MPa). The purity of compacts at this stage of processing has previously been determined by wavelength-dispersive x-ray flourescence, which claims a resolution limit of <1 ppm; no loss of purity could be observed when compared to the starting powder.

Two experiments were performed on fully dense compacts. First, grain growth studies were performed. Anneals were performed in air and in vacuum, at 1600°. Samples were loaded in the furnace as described above for sintering anneals, and were polished and microstructures observed in the SEM.

Grain-boundary migration studies were performed by diffusion bonding a fully dense polished compact to an oriented sapphire crystal via hot pressing (1 h, 1350°C, 10 MPa). After bonding, samples were annealed in vacuum as per the grain growth anneals. Samples were cut perpendicular to the interface, polished, and the extent and uniformity of interface migration was observed in the SEM.

3. RESULTS AND DISCUSSION

3.1 Sintering Studies

Figure 1 shows density as a function of temperature for constant-time anneals for undoped and Ti-doped (500 ppma) aluminas. Results for Mg-doped material are plotted on the same graph for comparison. In all cases, the material sinters readily, as can be seen by the high final densities achieved at relatively low sintering temperatures. This is an effect of the removal of hard agglomerates, a step which significantly increases the sinterability of the powder.

The presence of Ti clearly increases the sintering rate in the low-temperature anneals, suggesting that it has a strong effect during initial and intermediate stage sintering. However, Ti does not appear to increase the final compact density; at higher temperatures, where final densities approach the theoretical value, there is no significant difference between the undoped and Ti-doped compacts, while Mg-doped material readily sinters to theoretical density. This result is consistent with the work of Ikegami *et al.*, who observed that Ti-doping enhanced sintering only in the initial stage.

Ti is a multivalent impurity in alumina; the most common valences are 3+ (Ti_2O_3) and 4+ (TiO_2). In previous work, only TiO_2 has been observed to enhance sintering. In this work, Ti dopant was incorporated as TiO_2; however, it was necessary to sinter in vacuum to avoid contamination from the environment, which would favor the reduction of Ti from 4+ to 3+. There is qualitative evidence that such a reduction occurs during sintering of these compacts. The outer shell (~1mm thick) of each compact acquires a pink color, indicative of the

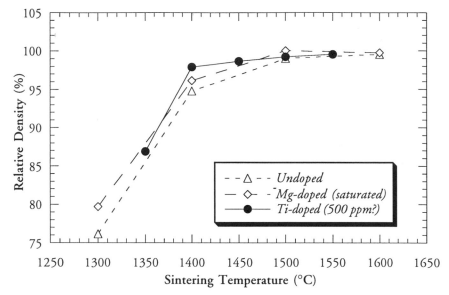

Figure 1:   Sintered density of undoped and doped alumina compacts as a function
of temperature.

presence of Ti_2O_3; the inside remains white, however, suggesting no significant
reduction at the compact center. At this time, no quantitative determination of
valence state in the sintered compacts has been made. Microstructural observations
were made at both the center and edge of each sample; observed differences in
evolution were attributed to differences in Ti valence.

Microstructures of Ti-doped sintered compacts can be seen in Figure 2.
Clear differences can be observed between compacts sintered at 1400°C and
1550°C, as well as between the sample center (TiO_2 dominated) and edge (Ti_2O_3
present) at each temperature.

In undoped compacts, there was no significant difference in microstructure
between the edges and center of the sample. The edge regions had very slightly
higher density and larger grain size. Grains remained equiaxed at all sintering
temperatures, and no abnormal or anisotropic grain growth was observed.

The microstructure of Ti-doped compacts sintered at 1400°C was similarly
equiaxed; however, the grain size varied significantly between the compact center
and edge. The TiO_2 dominated center region (representing the bulk of the
sample) exhibited a much larger grain size than the edge region, where the grain
size was comparable to undoped compacts. Such an increase in grain size can
result from an increase in density from the edge to the center of the compact.
Since Ti^{4+} is known to enhance densification while Ti^{3+} does not (Bagley,
1970), it is possible that the valence difference between the edge and center of the
sample results in the development of a density gradient within the sample during
sintering, which causes the observed grain size difference.

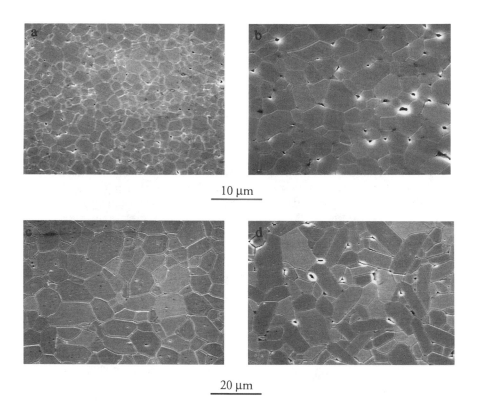

10 μm

20 μm

Figure 2: Ti-doped alumina, sintered 2 h at 1400°C (a, b) and 1550°C (c, d) in vacuum. Micrographs are taken from the sample edges (a, c) and sample center (b, d).

The microstructures which develop during sintering at 1550°C are somewhat more striking. Here, the grain size difference between the compact center and edge has essentially disappeared. However, while the grain structure in the edge region is still equiaxed and compares favorably to the microstructure of undoped compacts, the center region exhibits a faceted microstructure. A large number (>50%) of the grains have grown anisotropically. Such faceted microstructures have been observed by Horn and Messing (1995) in dense samples containing larger amounts of Ti (≈1000 ppma) and annealed in air. In the current work, we find that these structures can develop during sintering in samples containing smaller levels of dopant and annealed in vacuum.

The difference in microstructure between sample center and edge, driven by impurity valence differences, has interesting implications for materials design. It is possible to develop a graded microstructure with a starting material of uniform composition, by proper choice of impurity and sintering atmosphere.

3.2   Grain Growth Studies

Samples used for grain growth studies were all sintered for 2 h at 1450°C, and pressed to full density in the HIP, as described in the Experimental Procedures. Microstructures of fully dense compacts can be seen in Figure 3; both undoped and Ti-doped samples exhibit equiaxed microstructures. Grain sizes are similar at ≈4 μm, with the exception of the center of the Ti-doped specimen, where the average grain size is ≈8 μm.

After 8 h annealing in air at 1600°C, both undoped and Ti-doped samples exhibited abnormal, highly faceted grain growth. We assume that the compacts incorporate a large amount of impurity from the environment during this process, as Bae and Baik have shown that these annealing conditions are not sufficient to maintain purity and avoid development of a liquid phase which controls boundary motion (Bae and Baik, 1994). However, the effects of Ti can still be observed, as the Ti-doped samples have a much greater grain size.

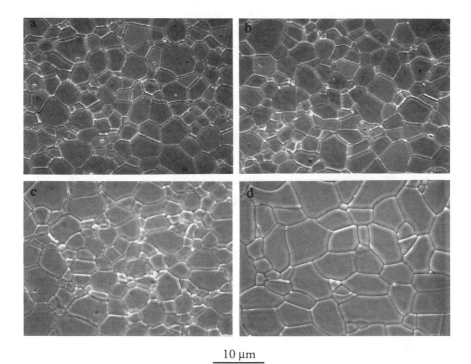

10 μm

Figure 3:   Fully dense undoped (a, b) and Ti-doped (c, d) compacts after hot isostatic pressing. Micrographs are taken from the sample edges (a, c) and center (b, d).

Anneals performed in vacuum yield very different microstructures. Micrographs of undoped alumina annealed for 2 and 8 h can be seen in Figure 4. The grains remain relatively equiaxed even after 8 h at 1600°C, with no indication of the abnormal grain growth observed during air anneals. This result suggests that the annealing environment is effective in keeping the samples free from contamination.

Figure 5 shows the microstructure of Ti-doped material annealed for 4 h in vacuum. Here, faceting of the microstructure is pronounced, especially in the sample center. Most of the sample exhibits this anisotropic growth, but some edge regions remain relatively equiaxed. This result is surprising, given that after samples are pressed to full density in the HIP, they are pink throughout, indicating that much of the Ti^{4+} has been converted to Ti^{3+}. It is possible that the tendency for grain faceting had already developed during sintering, but the grains

30 μm

Figure 4: Dense undoped alumina, annealed for 2 h (a) and 8 h (b) at 1600°C in vacuum.

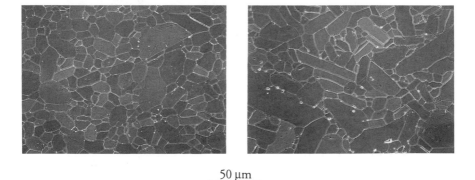

50 μm

Figure 5: Dense Ti-doped alumina, annealed for 4 h at 1600°C in vacuum. Micrographs taken at sample edge (a) and center (b).

were not large enough to observe the anisotropy until further grain growth. Long-time sintering experiments are underway to study the onset of faceting.

3.3. Boundary Mobilities

The elongated faceted structures which develop during sintering and grain growth indicate a significant anisotropy in boundary mobilities. By hot-pressing an oriented sapphire crystal to a polished, fully dense compact, it is possible to observe the growth of a large single grain of known orientation and thus study mobility as a function of orientation. This work is in the preliminary stages; growth of [11$\bar{2}$0]-oriented Ti-doped sapphire into a Ti-doped dense compact has been observed. From our initial results in this study, we estimate a boundary mobility of $\geq 100 \times 10^{-15}$ m^3/N·s, which is >3× values estimated from previous work studying the growth of [11$\bar{2}$0]-oriented sapphire into a nominally *undoped* alumina (Rödel, 1990). The alumina used in the previous work was of lower purity than in the present study, and was processed differently. In continuing work, we are repeating this study with the processing methods described here, and examining growth of basal plane sapphire into undoped and doped aluminas.

ACKNOWLEDGMENTS

This work is supported by the National Science Foundation under Grant No. DMR-8821238. We also acknowledge an NSF Equipment Grant No. DMR-9119460 which allowed the acquisition of hot pressing equipment used in this work.

REFERENCES

S. I. Bae and S. Baik, "Critical Concentration of MgO for the Prevention of Abnormal Grain Growth in Alumina," *J. Am. Ceram. Soc.*, 77 [10] 2499-504 (1994).

R. D. Bagley, I. B. Cutler, and D. L. Johnson, "Effect of TiO$_2$ on Initial Sintering of Al$_2$O$_3$," *J. Am. Ceram. Soc.*, 53 [3] 136-41 (1970).

T. Ikegami, K. Kotani, and K. Eguchi, "Some Roles of Mgo and TiO$_2$ in Densification of a Sinterable Alumina," *J. Am. Ceram. Soc.*, 70 [12] 885-90 (1987).

J. D. Powers and A. M. Glaeser, "High-Temperature Healing of Cracklike Flaws in Titanium Ion-Implanted Sapphire," *J. Am. Ceram. Soc.*, 76 [9] 2225-34 (1993).

M. Kitayama and A. M. Glaeser, "A New Method for Determining the Wulff Shape of Alumina," *in this proceedings.*

D. S. Horn and G. L. Messing, "Anisotropic Grain Growth in TiO$_2$-doped Alumina," *Mat. Sci. and Eng. A*, A195 169-78 (1995).

P. E. D. Morgan and M. S. Koutsoutis, "Phase Studies Concerning Sintering in Aluminas Doped with Ti^{4+}," *J. Am. Ceram. Soc.*, 68 [6] C-156-C-158 (1985).

J. Rödel and A. M. Glaeser, "Anisotropy of Grain Growth in Alumina," *J. Am. Ceram. Soc.*, 73 [11] 3292-301 (1990).

Anisotropic Grain Growth in Alumina Ceramics

Matthew Seabaugh, Debra Horn, Ingrid Kerscht, Seong Hyeon Hong and Gary L. Messing, Department of Materials Science and Engineering, The Pennsylvania State University, University Park, PA 16802 (814) 863-9706.

Abstract

Anisotropic grain growth is an important but poorly-understood phenomenon in ceramic materials. By viewing anisotropic growth as the growth of a single crystal in a polycrystalline matrix, it is possible to divide criteria into two groups; those intrinsic to the single crystal (crystallography, thermodynamic behavior) and those extrinsic to the single crystal (driving force due to matrix grain size, transport paths, and defect chemistry due to dopants). Al_2O_3 is presented as a model system and experimental results from this system are used to propose criteria for ceramic materials in general.

1. Introduction

An equiaxed grain microstructure yields the best combination of structural, electrical, thermal and other properties for most ceramics. However, some ceramic systems, like Si_3N_4, SiC and YBC, benefit from needle or platelet-shaped grains interspersed in the microstructure. Anisometric grains are not simply abnormal or exaggerated grains; they grow as well faceted, shaped grains in a polycrystalline matrix.

In contrast to normal grain growth (Atkinson, 1988), there is little information about the fundamental criteria and process factors leading to anisotropic grain growth in a polycrystalline matrix. Such knowledge would allow improvements of existing systems that rely on anisometric grain microstructures, the development of systems with new microstructure-property relations, and the design of processes with unique microstructures.

Anisotropic grain growth can be viewed as the growth of a single crystal (i.e., the anisometric grain) in a polycrystalline matrix. Thus, both intrinsic and extrinsic criteria are important. Intrinsic criteria are crystallographically and thermodynamically inherent to the crystal, while extrinsic criteria include the kinetic and driving force which arise from the microstructure, composition, and temperature of the sample. In an attempt to identify these criteria we present examples from our work on alumina-based ceramics to establish the influence of experimental parameters on anisotropic grain growth.

2. Experimental Methods

Anisotropic grain growth was studied in α-Al_2O_3 doped with TiO_2, Fe_2O_3, or liquid phase sintered with a calcium aluminosilicate glass. Samples were prepared from either alkoxide-derived (Tartaj and Messing, 1996, Horn and Messing, 1995), or colloidal boehmite (Sabol et al., 1994, Kerscht, et al., 1995). The boehmite was doped with Ti-isobutoxide, $Fe(NO_3) \cdot 6H_2O$ or SiO_2

plus $Ca(NO_3)_2$. To ensure fine initial grain size, sols were seeded with 1.5 wt% $\alpha-Al_2O_3$ particles ~0.1 μm in diameter. In Fe_2O_3 doped samples, $\alpha-Fe_2O_3$, particles were used as seeds. Seed particles nucleate the $\alpha-Al_2O_3$ transformation and result in a 99% dense microstructure of ~0.4 μm when sintered at <1250°C. Samples were formed by spin coating, fiber drawing, or tape casting, then dried, and fired in air between 1300 and 1600°C for 6 minutes to 16 hours.

2.1 TiO₂-doped Alumina

The equiaxed microstructure in Figure 1a is typical of all samples prior to the onset of anisotropic grain growth. The microstructure in Figure 1b is characteristic of samples that undergo anisotropic grain growth in TiO_2-doped samples. Anisotropic grain growth initiates after a period of normal grain growth with the "incubation" time decreasing with increasing temperature. The microstructure of the 0.15 wt% TiO_2-doped sample consists of equiaxed grains of ~0.8 μm average diameter at the onset of anisotropic grain growth. The 0.4 wt% sample undergoes normal grain growth to an average size of ~1.0 μm prior to the onset of anisotropic grain growth. In both cases, the intercept lengths ranged from ~0.5 to 1.5 μm, suggesting the onset or 'nucleation' of anisotropic grain growth is a function of matrix grain size and grain size distribution.

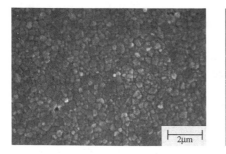 

Figure 1 Initial dense alumina microstructure (a) before and (b) after anisotropic grain growth.

To observe the initiation of anisotropic grains, a 0.4 wt% TiO_2 sample was heated intermittently in a furnace at 1500°C and the same grains were observed by SEM. A few large grains of ~2 μm were first observed after 2.5 min. (Figure 2) In Figure 2a anisotropic grains were observed to form in a matrix of significantly smaller equiaxed grains after 5 min at 1500°C. After 9 minutes (Fig. 2b) only a few grains have grown extensively. Grain **a** has developed into an anisotropic grain and consumed a number of finer grains including anisotropic grains several times larger than the average matrix grain size. Grain **a** increased from 5.5 to 8 μm in length and from 1.5 to 2 μm in width between 5 and 9 min.. The straight boundary bordering grain **b** on the right has not changed since it has too little driving force for movement relative to more curved boundaries. After 9 min. no new anisotropic grains develop.

Figure 2. Sequence of anisotropic grain development in 0.4 wt% TiO$_2$-doped alumina at 1500°C after a) 5 min., b) 9 min. (From Horn and Messing, 1995, by permission)

The radial and thickening growth kinetics of anisotropic grains in the 0.15 wt% TiO$_2$ sample are plotted in Figure 3. Some anisotropic grains undergo very rapid growth, or growth 'spurts', over a short period of time. These kinetics were obtained for the same grains at each time and thus are truly representative of the initial growth process. Radial growth kinetics are significantly faster than the rate of grain thickening. Once anisotropic grain growth begins, the grain growth exponent changes from 2 to 3 (Horn and Messing, 1995).

The growth kinetics first increase rapidly, then slow, with the growth exponent reaching values >10. The kinetics are reduced by anisometric grain impingement and the associated loss of driving force. Because the grains are so large and the basal surfaces straight, there is little driving force for grain thickening and the aspect ratio remains relatively constant at ~10 over a range of

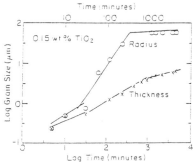

Figure 3. Radial and thickening growth kinetics in 0.15 wt% TiO$_2$-doped alumina at 1500°C. (From Horn and Messing, 1995, by permission)

processing conditions. The volume fraction of anisotropic grains maximizes of ~60 vol% in the 0.15 wt% TiO_2 doped sample after 5 h at 1450°C.

The boundary mobility for radial growth and thickening can be calculated from $M = K(n\gamma_{gb}G_a^{n-2})^{-1}$ where K is the growth constant, n is the grain growth exponent, and G_a is the average grain size (Yan, et al., 1976). Unfortunately, the effect of TiO_2 on grain boundary energies is not known. At low temperatures the mobilities for normal grain growth are similar to values reported by Handwerker et al. for undoped alumina (1989). The boundary mobility for radial growth is almost 2 orders of magnitude greater than for thickening. With increasing temperature the mobility significantly decreases as a result of the increased grain size. The mobilities in the thickening direction, however, are much lower because of the flatness of these boundaries. This result agrees with Rödel and Glaeser's (1990) observation that prismatic surfaces of sapphire single crystals grew faster into a polycrystalline matrix than basal surfaces.

2.2 Fe_2O_3-doped Alumina

To observe whether anisotropic grain growth could be 'nucleated', acicular hematite particles were used to seed a boehmite sol. Fibers were drawn from the sol to orient the acicular particles. Figure 4 shows the effect of seed orientation on microstructure development. In the random acicular seed case the α-Al_2O_3 grains were not oriented. When the seeds are oriented during forming, the seed particles nucleate an α-Al_2O_3 grain adjacent to the seed When densified, the oriented grains catalyze the growth of an oriented, microstructure.

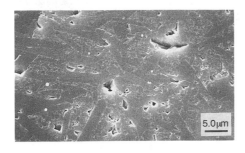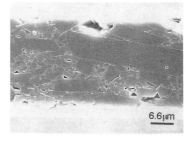

Figure 4. Acicular Fe_2O_3 seeded Al_2O_3, a) random seeds (1400°C, 30 min); b) oriented seeds (1450°C, 2h)

For samples without the hematite seed particles, the presence of Fe^{3+} appears to have little effect on microstructure development below 1450°C. However, at temperatures >1450°C anisometric grains begin to form when the Fe^{2+} concentration approaches 2% of the total iron concentration. It is suggested that

Fe^{2+} segregates to the grain boundaries, particularly the high energy prismatic planes, and enhances transport and growth in the prismatic direction.

2.3 CaO/SiO$_2$ doped Alumina

To observe the effect of a prevalent liquid phase on anisotropic grain growth, calcium-aluminosilicate glass formers were added to a seeded boehmite precursor sol and sintered. The resulting microstructure showed both anisotropic grain growth and faceted grains. Adding α-alumina platelet particles to the precursor mixture and then tape casting the slurry allowed samples to develop an extraordinary degree of texture (Figure 5).

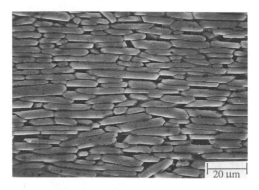

Figure 5. Liquid Phase Sintered Al$_2$O$_3$, with Al$_2$O$_3$ Platelet Addition

3. Intrinsic Criteria for Anisotropic Grain Growth

The most fundamental criterion for anisotropic grain growth is the crystallography of the material. A material with an isotropic crystal structure is less likely to grow anisotropically. Indeed, most ceramics which display anisotropic grain growth have an asymmetric unit cell or a crystal structure consisting of chains or layers of polyhedra. According to Hartman and Perdok (1955,1956), an uninterrupted series of strong bonds is required for crystal growth in a particular direction. This statement is supported by anisotropic grain growth in superconductor and β-aluminas ceramics, where large tabular grains grow parallel to the strong bonded directions of their perovskite and spinel layers. Whisker growth in mullite and Si$_3$N$_4$ can also be related to the strong chain bonding of polyhedra.

Planes with the close atomic packing have lower surface energy and are more thermodynamically stable. Maximizing the area of the lowest energy plane is thermodynamically favorable, so growth occurs parallel to these planes, and they become facets in crystals grown from melts or vapors.

Surface energy anisotropy has been shown to be a sufficient driving force for anisotropic grain growth. Kunaver and Kolar (1993) showed grains with anisotropic surface energies grow anisotropically in a matrix of isotropic grains.

The anisotropic grains demonstrate faceting, provided the lower anisotropic surface energy is less than that of the matrix. More fundamentally, Yang, et al (1995) derived the surface energy of all grains from a hypothetical Wulff plot, and demonstrated that faceted, anisometric grains develop from an initially equiaxed matrix, if the surface energies are anisotropic.

4. Extrinsic Criteria for Anisotropic Grain Growth

Intrinsic factors predispose a crystalline material toward anisotropic grain growth, but the development of an anisometric, polycrystalline microstructure is also dependent upon extrinsic factors. Extrinsic criteria include the nucleation rate and density of anisotropic grains, diffusion paths and rates in the matrix, and driving force for grain growth.

A large grain is unlikely to become anisometric late in its development, as the benefit of achieving an equilibrium anisotropic morphology at large grain size is small (Herring, 1952). However, if anisometric grains nucleate or form early in the microstructure development, their anisotropy may be preserved, if material transport to the grain surface is sufficient for anisotropic grain growth. Such growth is obvious in the model of Kunaver and Kolar which also shows the initial number of anisometric grains strongly affects the final aspect ratio.

No particular nucleation mechanism has been conclusively established in unseeded systems. In Si_3N_4, the growth of anisometric grains is initiated by the nucleation and growth of rod-like β-Si_3N_4 crystals, and the addition of β-Si_3N_4 seed particles accelerates the growth of β-Si_3N_4. In α-Al_2O_3, the nucleation of anisotropic grains was concurrent with the twinning of the basal boundary (Handwerker et al 1989, Rödel and Glaeser 1990, Kaplan, et al., 1995), but no conclusive connection was demonstrated. Similar to Si_3N_4, seed crystals can be used to nucleate anisotropic grain growth in Al_2O_3.

The growth of anisotropic grains is dependent upon a continuous supply of material to the growing grain. Anisotropic grain growth must occur faster than the normal grain growth or the driving force for the anisometric grains will be lost. To account for the obvious differences in surface curvature and their effect on boundary motion, the boundary mobility should be calculated. Faceted surfaces must have the slowest mobility since they have virtually no radius of curvature. Therefore, it is difficult to distinguish whether anisotropic growth is due to local differences in diffusion, local curvature differences, or changes in grain boundary energies.

Nevertheless, the increased rate of diffusion associated with grain boundary liquids appears to have a dominant effect in systems in which a liquid phase is present, such as Si_3N_4, and YBCCO superconductors. In liquid assisted systems, such as SiC, Si_3N_4, and ZnO anisotropic grain growth is attributed to dissolution-precipitation. The solubility of matrix grains in the liquid phase and the viscosity of the liquid phase are important controlling factors. In contrast,

Fe_2O_3- and TiO_2-doped alumina and GdBaCuO superconductors, form highly anisometric grains, but do not have a liquid present (Shin, et al 1991, 1992).

The role of dopants remains unclear and varies from system to system. Dopants may modify the intrinsic surface energy anisotropy, enhance diffusion by segregating to certain grain boundaries, change the defect chemistry, or create a liquid phase at the grain boundaries. In alumina, Bae and Baik (1993) demonstrated calcium segregation depends on the crystallographic orientation, and is responsible for 'abnormal' grain growth. In Al_2O_3 doped SiC, Al segregates to the tip of anisotropic grains, where it forms a catalytic region for grain growth (Shinozaki, 1993).

The driving force for densification, normal grain growth, and anisotropic grain growth is the decrease of interface free energy. The three processes compete, and if normal grain growth of the matrix continues, anisotropic grain growth will be limited by impingement with large matrix grains, in a manner similar to that described by Hillert (1965) for abnormal grain growth. According to Hillert, abnormal grain growth requires a microstructure in which normal grain growth has halted, the average grain size is small, and at least one grain is much larger than the average. Assuming anisometric grains develop similarly, the nuclei for anisotropic grain growth may simply be those grains that grew to a size larger than the average grain size.

5. Summary

Anisotropic grain growth is observed in many oxide and non-oxide systems. Using theoretical calculations, numerical simulations and experimental data, a two-tiered set of criteria were identified which must be met for a microstructure to experience anisotropic grain growth. Intrinsic criteria are fixed for a material system, and determine whether it is prone to develop an anisotropic microstructure in a polycrystalline sample. Extrinsic criteria determine whether the desired microstructure develops. Two factors available to tailor the extrinsic criteria are the initial microstructure and the system composition. Smaller initial grain sizes increase the energy for anisotropic grain growth. A dense, fine grained matrix presents the ideal growth environment for anisotropic grain growth; matrix grains have stopped growing, no pores exist on the boundaries to inhibit anisotropic grain growth, and a large driving force is available due to the large grain boundary area and the subsequent high grain boundary free energy.

Acknowledgments: The authors gratefully acknowledge the support of ONR Grant N00014-94-1-0007. M. Seabaugh gratefully acknowledges the receipt of a National Science Foundation Graduate Research Fellowship.

References:

H.V. Atkinson, "Theories of Normal Grain Growth in Pure Single Phase Systems", *Acta Metall.* **36** [36] 469-491 (1988).

D.S. Horn, *Grain Growth Anisotropy in Titania-Doped Submicrometer Alumina*, Ph D. Thesis, The Pennsylvania State University (1990).

J. Tartaj and G.L. Messing, "Anisotropic Grain Growth in a-Fe_2O_3-doped Alumina", submitted for publication in *J.Eur.Ceram.Soc.* (1996).

D.S. Horn and G.L. Messing, "Anisotropic Grain Growth in TiO_2-doped Alumina", *Mater.Sci.Eng.*, **A195** 169-178 (1995).

I.H. Kerscht, M.M.Seabaugh, and G.L. Messing, "Templated Grain Growth for Development of Highly Textured α–Al_2O_3 Ceramics," submitted to *J. Am. Ceram. Soc.* (1995).

M.F. Yan, R.M. Cannon, and H.K. Bowen, "Grain Boundary Migration in Ceramics", pp. 276-307 in *Ceramic Microstructures '76*. Edited by R. M. Fulrath and J.A. Pask. Westview Press, Boulder, CO (1976).

C.A. Handwerker, P.A. Morris, and R.L. Coble, "Effects of Chemical Inhomogeneities on Grain Growth and Microstructure in Al_2O_3", *J.Am.Ceram.Soc.*, **72** [1] 130-136 (1989).

J. Rödel and A.M. Glaeser, "Anisotropy of Grain Growth in Alumina", *J. Am. Ceram. Soc.*, **73** [11] 3292-301 (1990).

P. Hartman and W.G. Perdok, "On the Relations Between Structure and Morphology of Crystals", *Acta Crystallogr.* **8** 49-52, 521-524, 525-529 (1955).

U. Kunaver and D. Kolar, "Computer Simulation of Anisotropic Grain Growth in Ceramics", *Acta Metall. Mater.*, **41** [8] 2255-2263, (1993).

W. Yang, G.L Messing, and L-Q. Chen, "Computer Simulation of Anisotropic Grain Growth", *Mater.Sci.Eng.*, **A195** (1995).

C. Herring, "The Use of Classical Macroscopic Concepts in Surface Energy Problems", *Structure and Properties of Solid Surfaces,* R. Gomer and C.S. Smith, (eds.) National Research Council, Lake Geneva, WI, pp. 5- 72, (1952)

W.D. Kaplan, P.R. Kenway, and D.G. Brandon, "Polymorphic Basal Twin Boundaries and Anisotropic Growth in α-Al_2O_3", *Acta metall.mater.* **43** [2] 835-848 (1995).

M.W. Shin, T.M. Hare, A.I. Kingon, and C.C. Koch, "Grain Growth Kinetics and Microstructure in the High T_c $YBa_2Cu_3O_{7-\delta}$ Superconductor", *J.Mater.Res.*, **6** [10] 2026-2034 (1991).

M.W. Shin, T.M. Hare, A.I. Kingon, and C.C. Koch, "Grain Growth Kinetics and Microstructure in the High T_c $GdBa_2Cu_3O_{7-\delta}$ Superconductor", *J.Mater.Res.*, **7** [12] 3194-3201 (1992).

S. Bae and S. Baik, "Determination of Critical Concentrations of Silica and/or Calcia for Abnormal Grain Growth in Alumina", *J.Am.Ceram.Soc.* **76** [4] 1065-1067 (1993).

S.S. Shinozaki, J. Hagas, K.R. Cardunuer, M.J.Rokosz, K. Suzuki, and N. Shinohara, "Correlation Between Microstructure and Mechanical Properties in Silicon Carbide with Alumina Addition", *J.Mater.Res.* **8** [7] 1635-1643 (1993)

M. Hillert, "On the Theory of Normal and Abnormal Grain Growth", *Acta metall.* **13**, 227-238, (1965).

Modeling Grain Growth and Pore Shrinkage During Solid-state Sintering

Péter Varkoly and Gerhard Tomandl

Institute of Ceramic Materials, TU Bergakademie Freiberg
Gustav-Zeuner-Str. 3., 09596 Freiberg, Germany

Introduction

The aim of our work is to set up a 2D sintering model for ceramic materials and its computer simulation and to expand it into 3D. A computer program was written, modeling the sintering processes using materials constants and the distribution functions of grains and pores. The sintering model is started with a 2D and 3D microstructure respectively. The sintering process is described by interactions between the microstructural elements (grains, pores). The model shall enable us to calculate the processes during sintering and to compare this with results of image analyses of sintered samples.

Physical basis of grain growth

In this model the grains are approximated by polyhedra. The evolution of the microstructure is described by the motion of the polyhedra corners. The motion of a corner is calculated from the motion vectors of the grain boundaries close to the corner.

The driving force of the grain boundary motion is calculated using the sizes of the grains. Oel (1969) has set up a model for mass flux between two grains touching each other (Fig. 1). In his model the grains are approximated by spheres. The driving force between the two grains is proportional to the difference of their curvatures. He set up the equation (1) for the mass flux:

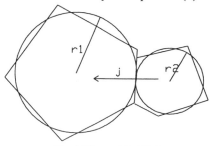

Fig. 1 The model of Oel

$$j = -u_k \Delta G = c_k \left(\frac{1}{r_2} - \frac{1}{r_1} \right) \qquad (1)$$

with $c_k = 2u_k \gamma V_m$

j	*mass flux*
u_k	*mobility of the atoms perpendicular to the grain boundary*
	(i.e. mobility of the grain boundary)
ΔG	*free enthalpy*
c_k	*rate constant for grain growth*
γ	*grain boundary energy*
V_m	*mol volume*

This equation is the basis of the kinetics of our model. However, the grains in our model are not described with universal radii but each grain shows different effective radii and different effective curvature, respectively in the directions of its different neighbors.

We have developed different methods to calculate the effective radii. One of them has shown to be as especially useful for the sintering simulation:

The effective radius is the distance between the point of gravity of the grain and the grain boundary. The magnitude of the motion vector is calculated using equation (1). The motion vector is perpendicular to the grain boundary and directed towards the smaller grain. (Fig. 2)

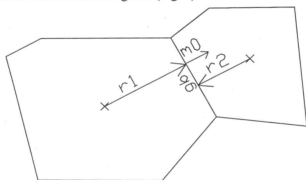

Fig. 2 Calculation of the grain boundary motion vector *m0* (2D version)

However the so calculated motion vector (*m0*) is not directly used. We assume that grains tend to become more spherical. It means grain boundaries have the tendency to align perpendicularly to the straight line joining the two point of gravity of the neighboring grains. So we calculate different motion vectors (in 2D two: *m1, m2*) instead of *m0* for the corners of each grain boundary. In this way the new grain boundary is perpendicular to the straight line joining the two centers of the neighboring grains, and the amount of recrystallizing mass is the same as calculated with the motion vector *m0*. Figure 3 shows a 2D example for this method. For the amount of transported mass holds in this case:

$$F = gb \; m0 \qquad\qquad (2)$$

gb   *length of the grain boundary*

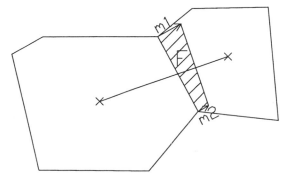

Fig. 3 Calculation of the motion vectors *m1* and *m2* (2D)

Physical basis of pore shrinkage

Pores may be found in different surroundings within the microstructure (Fig. 4):

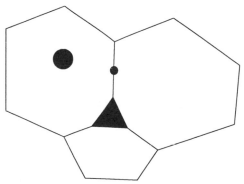

Fig. 4 Pores in the microstructure (2D)

1. Like grains pores may have more then two neighboring grains. In this case, the pores may be handled in an analogous manner as the grains with the distinction that another grain growth rate constant (c_k) must be used to calculate the grain boundary motion. Although the physical basis for both cases is not completely comparable, equation (1) gives a reasonably good description of the shrinkage behaviour of these kinds of pores.

2. Pores may be situated on grain boundaries. For these cases pores will be described like spheres on the grain boundaries. These pores can slow down the grain boundary motion. It means that the grain boundary motion vectors (see Fig. 3) are reduced in magnitude by a factor *f*. *f*

means the surface part of the grain boundary occupied by the pore (Equ. 3., inactive boundary surface).

$$f = \frac{2s}{gb} \text{ for 2D,} \qquad f = \frac{s^2\pi}{A} \text{ for 3D} \qquad (3)$$

s *radius of the pore*
A *surface of the boundary*

To derive the equations for pore shrinkage in this case, the following assumptions are made:

- The pores already have an approximate spherical shape
- The pores have no direct contact to each other
- The pores are empty: the influences of included gases can be neglected
- Driving force is the surface energy of the pores and the total grain boundary energy. The influence of the grain boundary energy on pore shrinkage can be neglected for smaller pores.
- Diffusion path is along the grain boundaries with sources of matter being the grain boundaries

Oel (1965) derived the following equation for pore shrinkage using the approximations mentioned above:

$$\frac{ds}{dt} = -\frac{c_p}{s^2} \qquad (4)$$

c_p *constant for pore shrinkage rate*

3.   A pore situated on a grain boundary can be detached from the grain boundary i.e. the pore will be encapsulated within the grain. In their papers Brook (1968, 1969) and Carpay (1979) report about the mechanism of the coupling of pore and grain boundary motion: a single pore is dragged along with a grain boundary, if its velocity is less than the maximum possible velocity of the pore. The following equation shows the condition for pore separation:

$$r = \left(\frac{\pi s}{r^2} + \frac{D_s \delta_s \Omega^{1/3}}{M_p s^3 kT} \right)^{-1} \qquad (5)$$

r *grain radius*
$D_s \delta_s$ *surface diffusion parameter*
Ω *atomic volume*
M_p *pore mobility*

With the aid of this equation it can be decided for each pore during the sintering simulation if there will appear pore drag or separation. The pores separated from the grain boundaries continue shrinking, but more slowly, and their influence on the further sintering process can be neglected.

The sintering model

In our model the sintering body is presented by four sets of geometrical objects. These are the grains (polyeders) the grain boundaries (polygons) the edges (lines) and the corners (points). The corners build the essential microstructure. The other geometrical elements represent the junctions between the corners.

To use the sintering model program we need a starting microstructure. It is very difficult to construct a 2D or 3D microstructure with a given distribution function of the building elements. With the aid of the stochastic geometry Voronoi mosaics can be built whose structures are very similar to that of the real sintering body. Since the original Voronoi mosaic is based on a statistically distributed set of points, the shape of the distribution functions of the building elements cannot be influenced. Using a point grid as a base for the construction of a Voronoi mosaic it is possible to control the shape of the distribution function by displacing the grid points to some extent.

The simulation program for grain growth and pore shrinkage then executes the following steps in each iteration loop:
1. calculation of the grain sizes
2. calculation of the motion vectors of the grain boundaries and the corners, respectively
3. calculation of the pore shrinkage
4. calculation of the new coordinates of the corners
5. handling of special cases: - disappearance of grains and, pores respectively
                                            - edge recombination

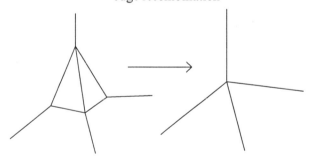

Fig. 5 Disappearance of grains and pores respectively

The advantage of this model compared with sintering simulations given by the literature is that this program uses a smaller amount of datas for modeling the microstructure because the shape of the grain boundary is not important. Moreover it is relatively simple to build further mechanisms (anisotropic crystal growth, sintering additives, chemical reactions on the grain boundary, ...) into the model because in this model the direct grain-grain and pore-grain interactions are considered.

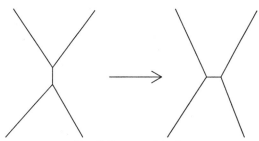

Fig. 6 Edge recombination

Figures 6-8 show the first results of a simulation process. Along the z-axis (p_0) there are plotted the number of grains and pores related to the initial number respectively.

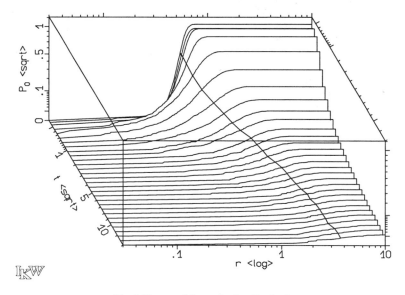

Fig. 6 Change of the grain size distribution

Practical works

To test our program and to make it practically usable we are going to determine material constants for different systems. With the aid of dense samples it is possible to calculate the material constant c_k by comparing theoretical and practical sintering results.

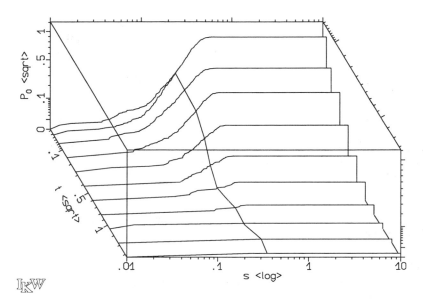

Fig. 7 Change of the pore size distribution of the pores dragged by the grain boundaries

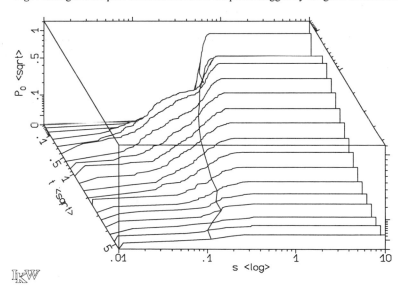

Fig. 8 Change of the pore size distribution of the pores having three or more neighboring grains

We must find different constants for the three different kinds of pores. For this reason, we have to distingish betwen the three kinds of pores during the image analysis of samples. Furthermore we have to take into account the fact, that the three kinds of pores are not independent from each other. That means, changes of one distribution function can also be caused by the change of the kind of the pore (pore separation).

Our aim is to build further different microscopical mechanisms into our model. There are many publications being engaged in examining microstructural processes during sintering. Most of them are supported by experimental results. These models can be used as stimulations for improving our model.

H. J. Oel, "Crystal Growth in Ceramic Powders" Materials Science Research, **4** 249-260 (1969)

R. J. Brook, "The Impurity-Drag Effect and Grain Growth Kinetics", Scripta Metallurgica, **2** 375-378 (1968)

R. J. Brook, "Pore-Grain Boundary Interactions and Grain Growth", Journal of the American Ceramic Society, **52** 56-57 (1969)

M. A. Carpay, "The Effect of Pore Drag on Ceramic Microstructures", Ceramic Microstructures '76, editors: R. M. Fulrath, J. A.. Pask 261-269 (1979)

C.H.Hsueh, "Microstructure Development During Final/Intermediate Stage Sintering - I. Pore/Grain Boundary Separation", Acta Metallurgica, **30** 1269-1279 (1982)

Computer Simulation of the Transient Grain Growth With Initial Uniform and Bimodal Size Distribution in Two Dimensional Model

M. S. Suh* , M. K. Kim**, and O. J. Kwon**

* Hyundai Motor Company
700 Yangjungdong, Ulsan 689-791 Korea

**Department of Metallurgical Engineering
Kyungpook National University, Taegu 702-701 Korea

Abstract

The grain growth from uniform and bimodal size distribution in the early period was simulated on a two-dimensional triangular lattice with Monte Carlo technique. Grains which were 3x3 and 6x6 were mapped onto 240 x 240 lattice for the uniform and bimodal size distributions. The vertex angles converged to a stable distribution with a peak at 120° by splitting from initial 4-grain vertices into 3-grain ones regardless of initial size distribution. A regular array of bimodal size distribution resulted in a narrower size distribution and subsequent retardation in grain growth after fast change from bimodal to monomodal at early stages. The initial microstructure can introduce a transient period, thus, steady-state grain size distribution should be confirmed to assure correct grain growth exponent.

1. Introduction

Many physical properties of polycrystalline materials are microstructure-dependent. Grain size is often the most important characteristic parameter in microstructure analysis. Mean grain size increases by growing of large grains at the expense of small grains.

From the earliest treatment of Beck, et al.(1948), it has been supposed that the grain growth obey exponential kinetics in a general form as

$$D^n - D_o^n = K t \qquad (1)$$

where D is the mean grain size at time t, D_o mean initial grain size t = 0, and K a constant. Assuming that the initial grain size is negligibly small compared to the grain size at time t, the equation (1) can be written as

$$D^n = Kt \qquad (2)$$

here n is termed grain growth exponent. For the parabolic law, n = 2.

Although many experimental data on the grain growth exponent have been set forth, the values vary from n = 2 to n = 4 even in the zone-refined ultrapure metals (Atkinson,1988). Vandermeer and Hu (1994) reanalyzed grain growth data on zone-refined iron previously made by Hu (1974) taking into account the initial grain size. They claimed that neglecting D_o can cause wrong value for n.

For the growth kinetics analysis to be accurate two prerequisite conditions should be satisfied: one is uniform appearance in grain shapes, and the other is statistical self-similarity during growth (Atkinson, 1988). These two conditions are very crucial for proper interpretation. Moreover, the proportionality constant for the relation of measured feature size on the metallographic surfaces and real spacial size is known to vary with grain shapes and grain size distributions (Han and Kim, 1995) (Louis and Gokhale, 1995). Thus, topological steady-state distribution must be proved to find out accurate value for grain growth exponent.

The aim of the present study is to simulate grain growth during transient period in a two-dimensional model by Monte Carlo technique and to study the effect of initial grain size distribution on grain growth.

2. Monte Carlo Simulation Procedure

Monte Carlo technique for the computer simulation of microstructural evolution was suggested by Srolovitz, et al. (1983). This model is based on probabilistic nature, and the migration of local grain boundaries is determined stochastically by applying a local rule. Simple description for its procedure is as follows.

The microstructure is mapped onto a triangular lattice. Each lattice site is assigned a number between 1 and Q corresponding to the crystallographic orientation. When neighboring sites have different orientations, a grain boundary is assumed to exist between two neighboring sites. The Metropolis algorithm (Metropolis, et al., 1953) is used to determine the orientation change of each site. A lattice site is selected at random and a new orientation is also chosen at random from the other (Q-1) possible orientations. The total energy change with the newly selected orientation at the lattice site is then found by counting the difference in the number of nearest neighbors having identical orientation. If the number increases, total energy, then, decreases and flips the orientation to new one.

For this study 240 x 240 triangular lattice was used. Initial grain size of 3 x 3 and 6 x 6 in square shape were mapped onto the lattice. For bimodal distributions, 3 x 3 and 6 x 6 grains were arranged regularly or at random with the area ratio of 1: 1 (grain number ratio, 4:1). To compare the effect of different initial grain shape on grain growth diamond-shaped grains were also used. Thus, six different initial microstructures were tested. Periodic boundary conditions were applied to reduce finite size effect. The total number of grain orientations, Q was set to 40. This value is sufficient to prevent from grain impingement (Srolovitz, et al., 1984). The data presented in this paper is an average of five simulation runs with the same initial microstructure.

The distributions of vertex angles, number of edges per grain, and normalized grain sizes were analyzed. Kolmogorov-Smirnov goodness-of-fit test (Press, et al., 1992) was used to determine if the grain size distributions had reached steady-state.

3. Results and Discussion
3.1 *Vertex morphology*

Fig. 1 shows the different grain shapes and size distributions in the starting microstructures. Each lattice shown is approximately 18 % in size of a complete lattice used. Fig.2 shows the simulated grain growth from the uniform initial grain size of 6 x 6.

The morphologies of the initial grains do not exist in the nature. However, these microstructures are energetically so unstable that local morphology changes fast especially in the vicinity of vertices. The results showed that at the very early stage of the growth (within 100 MCS) most 4-grain vertices transformed quickly into 3-grain ones, whatever the initial grain shape and size distribution had been . Four-grain vertices transform into two 3-grain vertices where each angle is 120° to decrease total grain boundary length (Isenberg, 1992) (Morgan, 1995). In the case of diamond-shaped grains, it reduces the boundary energy more drastically (Fradkov, 1993).

3.2 *Vertex angle distribution*

The vertex angle distributions became steady-state and identical within 1% error range of relative frequencies between 1000 MCS and 2000 MCS regardless of initial grain conditions with an exception. The vertex angles of regular array of bimodal square grains attained the steady-state around 4000 MCS. The relative frequencies of the three angles (60°, 120°, 180°) in the steady-state vertex angle distribution were 18%, 66%, and 16%, respectively. Thus, it is obvious that the average vertex angle is 120° and stable.

3.3 *Edge number distribution*

Topological distribution for the number of grain edges assumed steady-state between 400 MCS and 600 MCS in most cases. The steady-state distribution is in good agreement with other work (Grest, et al., 1986).

Fig. 3 shows the change of edge number distributions with time in the regular array of bimodal square grains. The initial bimodal distribution changed into a narrow one having a peak at the edge number of six. Subsequently, the narrow distribution curve was maintained for a long time after 600 MCS. Such a distribution having a peak at 6 is expected to be the main reason for the slowdown of grain growth after a fast initial growth in this particular array. According to Mullins-Von Neumann relationship (Von Neumann, 1952) (Mullins, 1956), the area growth rate of a certain grain is proportional to the number, N - 6, here N is the number of its edges. Thus, grains having six neighbors are very stable. Recently, Palmer, et al. (1994, 1995) observed experimentally that the relationship is valid in two-dimensional model system.

3.4 *Grain size distribution*

As shown in Fig. 4 , the initial uniform size distribution of 3 x 3 square grains became wider with time. The distribution attained steady-state around 1500 MCS. Since the grain size was so small that it reached steady-state first in comparison with other different initial microstructures. In contrast, size distribution of the regular array of bimodal square grains evolved as shown in Fig.5.   It remained narrow for a extended time. The distribution attained steady-state after 5000 MCS. In a regular array each large grain is believed to consume up small neighbor grains uniformly at the same rate and to result in approximately uniform size. Consequently, the temporary growth retardation was anticipated.

Kolmogorov-Smirnov goodness-of-fit   test showed that the size distributions of the six different initial conditions became identical after different transient periods which were dependent on the initial conditions. It is evident that any disturbance in size distribution at the initial stage of grain growth will disappear and approach to a unique steady-state distribution.

The grain size distribution in three dimension has been found to be approximately log-normal in several experimental studies (Okazaki and Conrad, 1972) (Rhines and Patterson, 1982) (Conrad, et al., 1985). Some researchers assumed log-normal distribution for their theoretical calculations (Takayama, et al., 1987) (Vaz and Fortes, 1988) (Han and Kim, 1995). But the result from this investigation showed that steady-state size distribution is in good agreement with the Rayleigh distribution as obtained by Srolovitz, et al. (1984).

3.5 *Grain growth kinetics*

As shown in Fig. 6, 3 x 3 square grain array had a linear relationship between average grain size (here, lattice points in a grain) and time (MCS) from the very early stage of growth.   Linear relationship notes that grain growth exponent, n is 2. From the present work it is clear that linear relationship does not always mean that the grain size distribution is steady-state. Even before the grain array attained steady-state size distribution around 1500 MCS, apparently this array showed linear growth kinetics.

On the other hand, in the case of the regular array of bimodal square grains, the steady-state was attained after 5000 MCS. As discussed previously, during the early stage of regular array of bimodal grains, large grains consumed smaller neighbor grains at the same rate and resulted in grains having six edges. Thus, growth rate became smaller. The grains in other bimodal arrays grew also faster than in uniform arrays during the early stage, and subsequently slowed down, too. But eventually all kinetics approached to the linear relationship with time, which meant grain growth exponent n became near 2.

4. Conclusion

Monte Carlo simulation of microstructural evolution during early stage of grain growth with several different initial microstructures showed that grains evolve to attain an identical steady-state size distribution regardless of the initial conditions. In this work the curvature of grain boundaries is assumed to be the unique driving force for grain growth. If grains attain narrow size distribution and have about six edges per grain, growth rate becomes smaller than steady-state. As linear growth kinetics is not consistent with steady-state size distribution at the early stage of growth, the steady-state size distribution must be confirmed first to determine accurate grain growth exponent experimentally.

Acknowledgments

One of the authors (OJK) would like to appreciate CALCE Electronic Packaging Research Center, University of Maryland at College Park, MD. This work was partly supported by the Center for Interfaces in Engineering Materials, KOSEF.

5. References

H.V. Atkinson, "Theories of Normal Grain Growth in Pure Single Phase Systems", *Acta metall.*, **36**, 3, 469-491 (1988)

P.A. Beck, J.C. Kremer, L.J. Demer and M.L. Holzworth, "Grain Growth in High Purity Aluminum and in an Aluminum-magnesium Alloy", *Trans. Am. Inst. Min. Engrs.*, **175**, 372 (1948)

H. Conrad, M. Swintowski, and S. Mannan, "Effect of Cold Work on Recrystallization Behavior and Grain Size Distribution in Titanium", *Metall. Trans.*, **16A**, 703 (1985)

V. E. Fradkov, M. D. Magnasco, D. Udler, and D. Weaire, "Determinism and stochasticity in ideal two-diemensional soap froths", *Phil. Mag. Lett*, **67**, 203 (1993)

G.S. Grest, M.P. Anderson, and D.J. Srolovitz, "Computer Simulation of Microstructural Dynamics", p.21-32, in *Computer Simulation of Microstructural Evolution*, edited by D.J. Srolovitz, TMS, Warrendale, PA (1986)

C. Isenberg, *The Science of Soap Films and Soap Bubbles*, p. 57, Dover Publications, Inc., New York, NY (1992)

J.H. Han and D.Y. Kim, "Analysis of the Proportionality Constant Correlating the Mean Intercept Length to the Average Grain Size", *Acta metall. mater.*, **43**, 8, 3185-3188 (1995)

H. Hu, *Canada Metall. Q.*, 13, 275 (1974)

P. Louis and A. Gokhale, "Can the Average Particle Section Size in a Metallographic Plane be Larger Than the True Average Particle Size in a Three-Dimensional Microstructure?", *Metall. Mater. Trans.*, **26A**, 1745 (1995)

F. Morgan, *Geometric Measure Theory, A Beginner's Guide*, 2nd ed., p. 111, Academic Press, San Diego, CA (1995)

N. Metropolis, A.W. Rosenbluth, M.N. Rosenbluth, A.H. Teller, and E. Teller, "Equation of State Calculations by Fast Computing Machines", *J. Chem. Phys.*, **21**, 1087 (1953)

W.W. Mullins, "Two-Dimensional Motion of Idealized Grain Boundaries", *J. Appl. Phys.*, **27**, 900-904 (1956)

K. Okazaki and H. Conrad, "Recrystallization and Grain Growth in Titanium: 1. Characterization of the Structure", *Metall. Trans.*, **3**, 2411 (1972)

M.A. Palmer, V.E. Fradkov, M.E. Glicksman, and K. Rajan, "Experimental Assessment of the Mullins-Von Neumann Grain Growth Law", *Scripta metall.*, **40**, 633 (1994)

M. Palmer, K. Rajan, M. Glicksman, V.Fradkov, and J. Nordberg, "Two-Dimensional Grain Growth in Rapidly Solidified Succinonitrile Films", *Metall. Mater. Trans.*, **26A**, 1061 (1995)

W.H. Press, S.A. Teukolsky, W.T. Vettering, and B.P. Flannery, *Numerical Recipes in C, the Art of Scientific Computing*, 2nd. Ed., p.623, Cambridge University Press, Cambridge, UK (1992)

D.J. Srolovitz, M.P.Anderson, G.S. Grest, and P.S. Sahni, "Grain Growth in Two Dimensions", *Scripta metall.*, *17*, 241 (1983)

D.J. Srolovitz, M.P. Anderson, P.S. Sahni, and G.S. Grest, "Computer Simulation of Grain Growth - II. Grain Size Distribution, Topology, and Local Dynamics", *Acta metall.*, **32**, 793 (1984)

Y. Takayama, T. Tozawa, and H. Kato, "Linear Intercept Length Distribution in a Grain Structure Model with Diameter Distribution of Log-Normal Form", *Trans. Japan Inst. Metals*, **28**, 631-643 (1987)

R.A. Vandermeer and H. Hu, "On the Grain Growth Exponent of Pure Iron", *Acta metall. mater.*, **42**, 9, 3071-3075 (1994)

M.F. Vaz and M.A. Fortes, "Grain Size Distribution: The Lognormal and the Gamma Distribution Functions", *Scripta metall.*, **22**, 35-40 (1988)

J. Von Neumann, *Metal Interfaces*, 108 (1952)

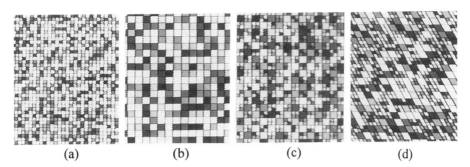

Fig. 1 Different initial microstructures used. (a) uniform 3 x 3 square grains,
(b) uniform 6 x 6 square grains, (c)regularly arrayed bimodal square grains,
and (d) randomly arrayed bimodal diamond-shaped grains.

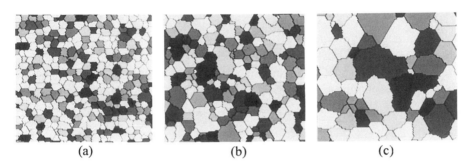

Fig. 2 Evolved microstructures from uniform 6 x 6 square grain arrays after
(a) 500 MCS, (b) 2000 MCS, and (c)6000 MCS.

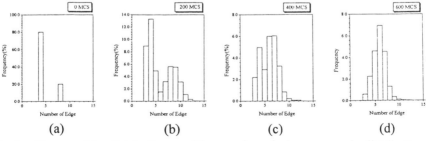

Fig. 3 Change of the distributions of the number of edges per grain distribution
along with MCS from regularly arrayed bimodal square grain
distribution. (a) 0 MCS, (b) 200 MCS,(c)400 MCS, and (d) 600 MCS.

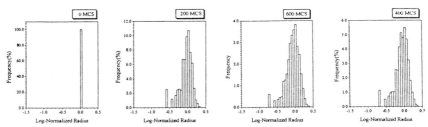

Fig. 4   The change of the log-normalized radius distribution of an initial uniform
          distribution of 3 x 3 square grains.

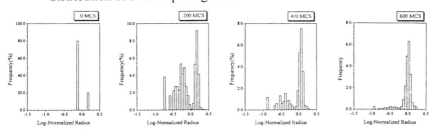

Fig. 5   The change of the log-normalized radius distribution of an initial uniform
         distribution of a regularly arrayed bimodal square grains.

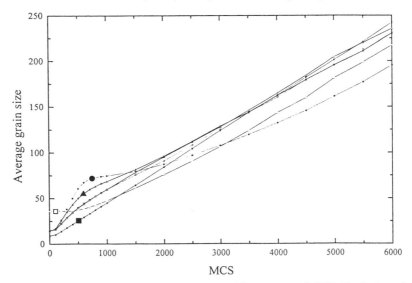

Fig. 6   The growth of average grain sizes with respect to MCS. ■ , 3x3 grains;
          □, 6x6 grains; ●, regularly arrayed bimodal square grains;   ▲,
          regularly arrayed diamond-shaped grains; ▵, irregular array   of
          bimodal diamond-shaped grains.

A Stochastic Simulation of Sintered Microstructures

G.B.Schaffer[1], T.B.Sercombe[1], J.A.Belward[2], I.Podolsky[2] and K.Burrage[2]

[1]Dept. of Mining and Metallurgical Engineering, [2]Dept. of Mathematics
The University of Queensland • Queensland 4072 • Australia

Abstract

Since solid state sintering is a diffusion controlled process, stochastic simulations using Monte Carlo techniques based on elements of lattice theory and cellular automata can be a powerful tool to investigate the operating mechanisms. Furthermore, such techniques also enable the direct simulation of the microstructures developed during sintering. Here we present results on the simulation of the sintering of copper wires. The initial microstructure is superimposed onto a lattice, where the energy of each lattice site is defined in terms of the interactions with neighbouring sites. Lattice sites are picked at random and re-oriented in terms of an atomistic model governing mass transport. This work indicates that the microstructural changes are strongly dependent on bulk and surface diffusion.

1. Introduction

The mechanical properties of crystalline materials are determined to a large extent by their microstructure, which is, therefore, of considerable importance in the materials sciences. Consequently, it would be extremely useful to be able to model microstructural development during materials processing and to simulate final microstructures. However, microstructural development is a function of a number of independent variables, some of which are poorly understood. While such models are as yet incapable of accurately predicting final microstructures, they can be used to test our understanding of microstructural development and the mechanisms of materials processing.

Monte Carlo techniques based on elements of lattice theory and cellular automata can be used to investigate the operating mechanisms of stochastic processes by direct simulation of microstructural development. This approach has been used previously to model soap froths (Weaire and Kermode, 1983, 1984), grain growth (Srolovitz *et al.* 1983; Anderson *et al.* 1984; Srolovitz *et al.* 1984; Kunaver and Kolar, 1993), recrystallisation (Srolovitz *et al.* 1986; Humphreys, 1992) and solidification (Brown and Spittle, 1989; Zhu and Smith, 1992; Rappaz and Gandin, 1993). Little has been done in the area of sintering, except in the final stages, when the principle of the conservation of mass was ignored (Hassold *et al.* 1990; Chen *et al.* 1990). Similar approaches have also been used recently to successfully simulate such disparate problems as grain flows in silos and problems within computational fluid dynamics (Somers and Rem, 1989). The philosophy behind this approach is similar to that used to interpret high resolution TEM micrographs (Spence, 1988) and to determine atomic co-ordinates using Rietveld techniques in X-ray analysis (Young, 1993), in that a simulated image or pattern is manipulated until it matches an experimental one. At this point, it is assumed that the model upon which the image was based must have been the correct one.

The model, in essence, considers a continuum structure, superimposed onto a lattice. Each lattice site corresponds to either an atom or a vacancy. The number of vacancies can vary with the simulation temperature, while a cluster of vacancies is a pore. The energy of each lattice site is defined in terms of the interactions with neighbouring sites. To simulate sintering, a lattice site is picked at random and re-oriented in terms of the atomistic model governing mass transport. The probability of the change being accepted is dependent on the change in energy of the site. The microstructure is updated and the process is then repeated. Progress is monitored by quantifying the microstructural changes.

Here, we report on initial results of the application of this technique to the sintering of copper wires.

2. The Model

The model is described in three parts. The first reviews the experimental analogue, the second considers the algorithm and in the third, we derive equations for measuring the microstructural parameters.

2.1 The Experimental Analogue

We simulated the sintering of copper wires on a mandrel, as examined experimentally by Alexander and Balluffi (1957). They wound 128μm diameter copper wire around a copper spool. This was sintered in hydrogen at 900°C for up to 600hrs. Cross sections of the wire windings were then examined metallographically (Figure 1). This classical experiment is essentially a two dimensional problem and can be simulated by a series of close packed circles.

2.2 The Algorithm

The simulation was performed, in the first instance, using *Matlab 4.0 for Windows*, a sophisticated numerical computation and visualisation software package, running on a 486DX40 PC. The model is initialised by mapping a close packed array of circles onto a square lattice. These circles represent a cross section through the wire windings, where each wire is a single crystal aligned along an <001> direction. Each lattice site is assigned an integer value between zero and three, representing free space, an atom, a vacancy or a triple point pore site, respectively. The simulation is based on the diffusion of vacancies, which is equivalent to the diffusion of atoms in the opposite direction. Bulk diffusion is permitted throughout the whole system while surface diffusion is only permitted around the internal pores. Bulk diffusion is assumed to occur in close packed directions only, whereas atoms on the internal pore surface are permitted to move in eight directions.

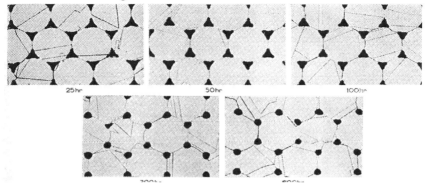

Fig. 1. An etched cross-section of 128μm diameter copper wire windings after sintering at 900°C (from Alexander and Balluffi, 1957).

After the simulation is initialised, a vacancy is chosen at random and a movement is attempted in a random direction. If there is an atom in the lattice site determined by the movement vector, then the atom and the vacancy change positions. The change in energy of the system, due to the change in the number of nearest neighbours, is then calculated. As the lattice is square and only nearest neighbours are assumed to influence the energy of a site, there can be no more than three neighbours to a moving atom. Additionally, since evaporation is not permitted, an atom must have at least one nearest neighbour. Hence the change in energy is the result of an atom gaining or losing up to two nearest neighbours.

If an atomic jump causes a decrease in energy (an increase in the number of nearest neighbours) then the atom is likely to remain in the new site. Conversely, if there is a decrease in the number of nearest neighbours, then the exchange is likely to be reversed. The probability, P, that the exchange is reversed is based on the change in nearest neighbours. The probability function for movement by bulk diffusion, P_b , is given by:

$$P_b = 1-0.5(1-\tanh(\Delta N)) \tag{1}$$

where ΔN is an integer function representing the change in nearest neighbours. We use an exponential function here because there is a exponential relationship between diffusion rate and the change in energy, here represented by a change in nearest neighbours. A third random number, R_n, is then generated ($R_n \in (0,1)$) and the exchange is reversed if $R_n < P_b$. For surface diffusion the exchange is reversed if $\Delta N < 0$. In this case, the probability function for reversion, P_s , satisfies $P_S = 1$, otherwise $P_S = 0$. The two probability functions are shown graphically in Figure 2.

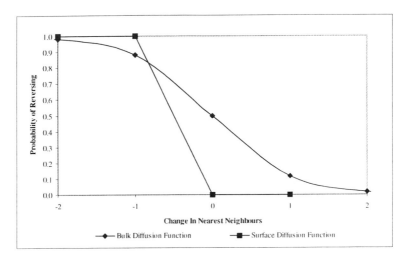

Fig. 2. The probability of an atomic jump being reversed for either bulk or surface diffusion.

The unit of time is the Monte Carlo Step (MCS) which is defined as N movement attempts, where N is the number of initial vacancies. The point at which a triple point pore site becomes an internal vacancy is defined as the moment where the vacancy ceases to have the pore as one of its nearest neighbours. Similarly, a vacancy becomes part of the free surface and the density of the array increases when one of it nearest neighbours is itself part of free space.

2.3 Quantifying the Microstructure

The microstructural features of interest are the pore size and the pore shape; the density, which is not strictly a microstructural feature, was not considered at this stage. The porosity is defined as the ratio of the pore area to the total area. In a discretised setting, this becomes:

Porosity = Number of triple point pore sites/Total number of sites   (2)

The pore morphology is quantified using the Rugosity Coefficient, R_c, which is defined as the perimeter of a pore divided by the circumference of a circle of equal area as the pore. The rugosity of a circle is therefore 1. In the model, rugosity is calculated by determining the length of the linear piece wise curve formed by the atoms on the pore surface, N_S, and the number of vacancies in the pore, N_p. Then R_c is given by:

$$R_c = N_S / \left(2\sqrt{\pi N_p} \right)$$ (3)

The porosity is normalised to the initial values. All results are the average of three simulation runs.

3. Results

Bulk diffusion results in pore shrinkage and limited pore rounding. Surface diffusion results in pore rounding, with no shrinkage. Combined, they cause both shrinkage and rounding. The microstructural development from a typical simulation run is shown in Figure 3, while the comparison of experimental results with calculated values of porosity and rugosity are shown in Figure 4. The variability in the calculated points is indicative of a stochastic process. There is an accurate fit between the two sets of data.

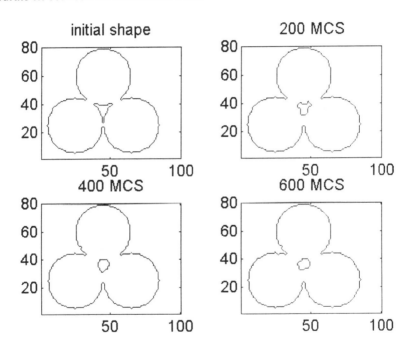

Fig. 3. The simulated evolution of pore size and shape during sintering, using both bulk and surface diffusion terms.

4. Discussion

It is evident that Monte Carlo methods can be used to accurately simulate microstructural evolution during simple solid state sintering events: the close fit between the predicted results and the experimental results validates the approach. However, this technique is not intended to provide a predictive tool but is simply a means to test current understanding. Before considering the metallurgical implications, we discuss the algorithm, the probability distributions and possible refinements to the model.

The algorithm involves selecting a vacancy and a movement direction at random, moving the vacancy, measuring the change in energy, determining the probability of the atom moving back to where it came from and then determining whether the move will be accepted or not. While this is not as computationally efficient as determining the new energy state and the likelihood of success before the atom is moved, it enables the determination of the mode of transport (surface or bulk) and allows the correct rule to be applied. An artefact of this method is that it more accurately mimics the random walk nature of diffusion, in that an atom is not allowed to "decide" whether it should move or not.

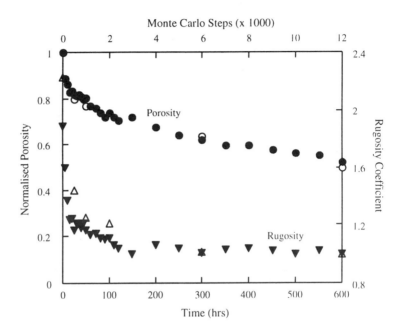

Fig. 4. A comparison between experimental data (hollow symbols, from Alexander and Balluffi, 1957) and calculated values (solid symbols) for the change in porosity and rugosity during the sintering of copper wires.

The correct choice of probability distribution is critical to the success of any Monte Carlo simulation. The distributions used here for both bulk and surface diffusion assume that the change in energy of a system due to an atom/vacancy exchange can be approximated by the change in number of nearest neighbours, because the excess energy of an atom adjacent to a vacancy is due to missing atomic bonds. Given that the energy of solid/vapour interfaces is predominantly due to missing nearest neighbours and that second nearest neighbours and entropy effects have a minor influence (Porter and Easterling,

1992), this seems a reasonable assumption. Surface diffusivity is higher than bulk diffusivity because the activation energy is lower for surface diffusion (the atoms are less tightly bound) (Kirkaldy and Young, 1987). It is therefore possible (at T $>0.8T_m$, where T_m is the melting point) for an atom to jump many times before it loses its kinetic energy and comes to rest (Shewmon, 1989). This is accounted for in the model by never reversing a jump if the change in energy was negative or zero. With the probability of reversal set to 0.5 when the change in energy was zero, the rate of pore rounding did not correlate with the rate of shrinkage. This suggests that the lower probability of reversal may be a reasonable means to simulate greater jump distances.

The simulations described here used three wires of 20 atoms in radius mapped onto a square lattice. The simulated wires are many orders of magnitude smaller than the real ones. This speeds computation but limits accuracy. While real sizes are not practical, an increase to a 100 atom wire would improve the accuracy of the initial rugosity measurement by ~10% and the initial porosity by ~5% (Podolsky et al, 1995). A change to a triangular lattice would better reflect the geometry of the model and would allow a similar description of atomic movements in the bulk and on the surface. These changes, however, require that the code be rewritten in a language that can be compiled, such as FORTRAN and migrated to a supercomputing platform. This will also be necessary for 3D modelling.

In the simplified model presented here, it appears that bulk diffusion can adequately account for pore shrinkage while surface diffusion is also required to explain pore rounding. This would not have been evident if pore morphology had not been monitored, although neck growth rates have been used in a similar way to come to similar conclusions using conventional modelling techniques (German and Munir, 1976).

Alexander and Balluffi (1957) noted that shrinkage stopped when grain boundaries disappeared and concluded that shrinkage in all stages was due to volume diffusion of atoms from grain boundary sources to the voids between the wires. Burke (1957) and Coleman and Beeré (1975) have also concluded that grain boundaries are necessary for pore shrinkage, although this view is not universal (Wilson and Shewmon, 1966). The current work indicates that pore shrinkage may indeed be independent of grain boundaries, although grain boundaries may be necessary for compact densification, which was not measured in the original experimental work and not considered in this simulation. A boundary needs to be included in the model to test this hypothesis. Grain boundaries enhance mass transport rates and provide vacancy sinks. This serves shrinkage of the whole system (densification), but not necessarily pore shrinkage. However, grain boundaries in the vicinity of a pore might prevent vacancies migrating back into the pore and/or reduce vacancy concentrations adjacent to the pore. Both factors would contribute to pore shrinkage. A grain boundary influence on pore shrinkage would indicate that the surface diffusion algorithm is inappropriate. This would require modification of the surface diffusion term in order for the three rates to correlate.

5. Conclusions

Stochastic simulations using Monte Carlo techniques can be used to model the microstructural changes which occur during sintering and to investigate sintering mechanisms. This work suggests that bulk diffusion may account for pore shrinkage and surface diffusion is responsible for pore rounding during sintering of close packed copper wires.

References

B.H.Alexander and R.W.Balluffi, (1957) *Acta Metall.* **5** 666.

M.P.Anderson, D.J.Srolovitz, G.S.Grest and P.S.Sahni, (1984) *Acta Metall.* **32** 783.

S.G.R.Brown and J.A.Spittle, (1989) *Mater. Sci. Tech.*, **5** 362.

J.E.Burke, (1957) *J. Am. Ceram. Soc.* **40** 80.

I-Wei Chen, G.N.Hassold and D.J.Srolovitz, (1990) *J.Am. Ceram. Soc.* **73** 2865.

S.C.Coleman and W.B.Beeré, (1975) *Phil. Mag.* **31** 1403.

R.M.German and Z.A.Munir, (1976) *J. Mater. Sci* **11** 71.

G.N.Hassold, I-Wei Chen and D.J.Srolovitz, (1990) *J.Am. Ceram. Soc.* **73** 2857.

F.J.Humphreys, (1992) *Mater. Sci. Tech.* **8** 135.

J.S.Kirkaldy and D.J.Young, (1987) "Diffusion in the Condensed State", p27, The Institute of Metals.

U.Kunaver and D.Kolar, (1993) *Acta Metall. Mater.* **41** 2255.

I.Podolsky, K.Burrage, G.B.Schaffer and J.A.Belward, in preparation.

D.A.Porter and K.E.Easterling, (1992)"Phase Transformations in Metals and Alloys", 2nd ed., p112-113 Chapman and Hall.

M.Rappaz and Ch.-A.Gandin, (1993) *Acta Metall. Mater.* **41** 345.

P.G.Shewmon, (1989) "Diffusion in Solids", 2nd ed., p218, TMS.

J.A.Somers and P.C.Rem, (1989)"A parallel cellular automata implementation on a transputer network for the simulation of small scale fluid flow experiments", Koninklijke (Shell-Laboratorium), Amsterdam.

J.C.H.Spence, (1988) "Experimental High-Resolution Electron Microscopy", OUP.

D.J.Srolovitz, M.P.Anderson, P.S.Sahni and G.S.Grest, (1984) *Acta Metall.* **32** 793.

D.J.Srolovitz, M.P.Anderson, G.S.Grest and P.S.Sahni (1983), *Scripta Metall.* **17** 241.

D.J.Srolovitz, G.S.Grest and M.P.Anderson, (1986) *Acta Metall.* **34** 1833.

D.Weaire and J.P.Kermode, (1983) *Phil. Mag. B* **48** 245.

D.Weaire and J.P.Kermode, (1984) *Phil. Mag. B* **50** 379.

T.L.Wilson and P.G.Shewmon, (1966) *Trans. Met. Soc. AIME* **236** 48.

R.A.Young ed., (1993) "The Rietveld Method", OUP.

Panping Zhu and R.W.Smith, (1992) *Acta Metall. Mater.* **40** 683.

Coarsening Behavior of An Alumina-Zirconia Composite (AZ50) Containing Liquid Phase

F. Jorge Alves, Helen M. Chan, and Martin P. Harmer

Materials Research Center
Lehigh University, Bethlehem, PA 18015

Abstract

Coarsening in a dual phase mixture (AZ50) of Al_2O_3 and c-ZrO_2 (cubic zirconia-yttria stabilised) is very limited. The addition of a liquid phase (anorthite-$CaO.Al_2O_3.2SiO_2$) to this mixture had the effect of promoting grain growth. The effect of varying the volume fraction (0.5-10 vol%) and composition (20 wt% Na_2O addition) of the liquid phase has been investigated. An interface reaction-controlled mechanism appears to govern the grain growth rate of the AZ50 composite.

1. Intoduction

Glass additions to dense single phase materials, such as Al_2O_3, MgO and c-ZrO_2, have been shown to decrease the grain growth rate constant (K) (Kaysser et al., 1987, Kim et al., 1993, and Lin, 1990). This is explained by the fact that the diffusion path length for atom transport, leading to grain growth, is increased through incorporation of a liquid phase. However, recent work by Dill et al. (1992), has shown that glass additions to a two-phase material had the opposite effect. Specifically, they observed that the addition of 1vol% anorthite glass ($CaO.Al_2O_3.2SiO_2$) promoted grain growth in a duplex mixture of 50 vol% Al_2O_3 + 50 vol% c-ZrO_2 (AZ50). The purpose of the present work is to clarify these seemingly contradictory results, and to determine the mechanism responsible for the enhanced coarsening in the AZ50 composite.

A series of grain growth studies have been conducted in which different percentages (0.5-10 vol%) and compositions (anorthite and anorthite with sodium oxide) of the liquid phase were added to AZ50 composite. From the results obtained, a mechanism for grain growth in this composite is proposed.

2. Experiments

Ultra high purity (99.995%) alpha alumina powder (AKP-HP, Sumitomo Chemical Company, Japan), cubic zirconia (8 mol% Y_2O_3, Tosoh Corporation, Japan) and anorthite ($CaO.Al_2O_3.2SiO_2$, Alcoa Industrial Chemicals, AR) glass were used as starting powders. Glass particles smaller than 1μm, were obtained by gravitational sedimentation. Powders were mixed in the correct proportions (AZ50; 50 vol% alumina and 50 vol% zirconia and AZ50-glass mixtures with 0.5 to 10 vol% glass) in 200-proof ethanol and ball milled (YTZ balls, Tosoh Corporation) for 24 hours. After milling, the slurry was magnetically stirred during drying and the dried cake was then crushed in a polyethylene bag with a teflon rolling pin.

Different powder compositions were uniaxially pressed in a cylindrical steel die (15 mm diameter), at 35 MPa, and isostatically pressed at 350 MPa. Pellets (2g weight) were cut in quarters, calcined in air at 950°C for 16 hours, then sintered in air at 1650°C for 1/2 hour. Densities were measured using the Archimedes principle (densities higher than 97% of the theoretical density were obtained for all samples). Pellets were then annealed in air at 1650°C for times ranging from 0 to 54 hours. All sintering and annealing treatments were conducted in alumina crucibles. Samples were surrounded with powder of the same composition, in order to minimize contamination and prevent volatilization.

Pellets were polished, using standard metallographic techniques, and thermally etched at 1350°C for times ranging from 2 to 3 hours. Scanning electron microscopy (SEM-JEOL 6300F) was used to characterize the microstructures. Grain size was measured using the method proposed by Wurst and Nelson (1972) that was adapted for measuring the individual grain sizes of the two phases. Because both phases show similar growth rates during all annealing times, the average grain size from alumina and zirconia phases $((GS_A+GS_Z)/2)$ was adopted.

3. Results and Discussion

Coarsening in AZ50 is severely limited. As shown in figures 1 and 2, after 54 hours of annealing at 1650°C, the average grain size only increased from approximately 1 to 2 μm. Grain growth inhibition, in this system, has been attributed to the limited mutual solubility, the limited long range interdiffusion, and the physical constraint provided by the interpenetration structure of the two phases (French et al., 1990).

Liquid phase ($CaO.Al_2O_3.2SiO_2$) additions to the AZ50 composite, had the effect of promoting grain growth. As the volume fraction of glass increased (0.5-10 vol%) larger grains were obtained (fig. 1). Figure 3 shows a typical microstructure of the AZ50+10 vol% glass. The fact that glass promotes grain

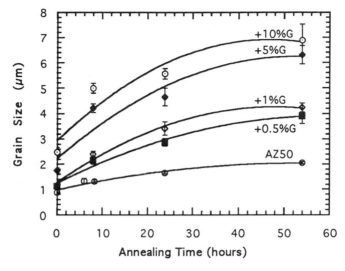

Figure 1 Effect of volume fraction of anorthite glass and annealing times, at 1650°C, on grain size of AZ50 composite.

growth in AZ50 is interesting, given that glass additions to dense single phase alumina (Kaysser et al., 1987) and single phase c-zirconia (Lin, 1990) inhibit grain growth. In single phase materials, the kinetics of grain growth are determined by the rate at which atoms can diffuse *across* the grain boundary. Liquid phase additions increase diffusion distances and slow down grain growth. On the other hand, in two-phase materials, grain growth occurs by atom diffusion *along* the grain boundaries (Ostwald ripening), which is increased as the thickness of the channels is enlarged by liquid phase additions.

The grain growth kinetics were analyzed according to the relation:
$$G^n - G_0^n = Kt$$
where G is the average grain size at time t, n is the grain growth kinetic exponent, G_0 is the grain size after sintering for 1/2 hour (defined as t=0), and K is the grain growth rate constant. Values for n between 2 and 3 gave good fits to the experimental data. Grain growth models predict n=2 for interface-controlled grain growth, and n=3 for diffusion-controlled grain growth. Figure 4 shows the experimental results, assuming an interface-controlled reaction (n=2). From the slopes of the lines, "K" values were obtained and plotted in figure 5, as a function of the volume fraction of glass.

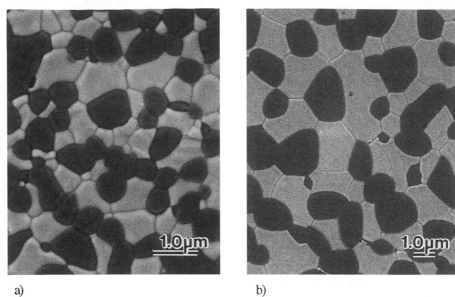

a)   b)

Figure 2 Effect of annealing time at 1650°C on the microstructure of AZ50 composite. a) 0 hours annealing; b) 54 hours annealing.

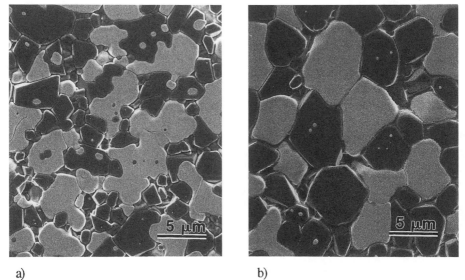

a)   b)

Figure 3 Effect of annealing time at 1650°C on grain size of AZ50+10 vol% glass composite. a) 0 hours annealing; b) 54 hours annealing.

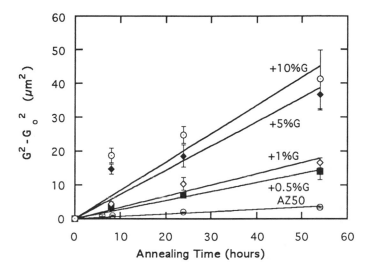

Figure 4 Grain growth kinetics of AZ50 and AZ50-glass mixtures.

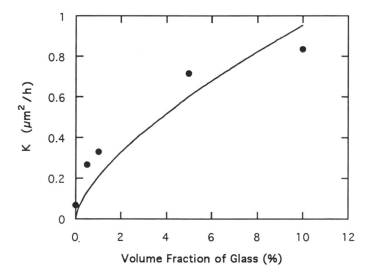

Figure 5 Grain growth constant (K) as a function of different volume fractions (V_f) of glass anorthite. Solid line indicates a two-thirds power relationship, $K \propto V_f^{2/3}$.

As shown in figure 5, $K \propto V_f^{2/3}$ provides a reasonable fit to the data. This result suggests that the kinetics of grain growth are consistent with an interface reaction-controlled mechanism. As the glass content increases (V_f), the fractional grain contact area decreases until the glass surrounds most of the grains. Further liquid additions cause the grain-boundary area, covered by bulk glass at three-and four grain junctions, to increase proportional to $V_f^{2/3}$ (Kwon, 1986). Furthermore, the viscosity of the glass, at the annealing temperature, is very low ($10^{1.22}$ Poises) (Hummel and Arndt, 1985) and diffusion along the liquid films and glass pockets should be very fast. Kwon and Messing (1990) found a $V_f^{2/3}$ dependence on the sintering rate during the densification of single phase alumina.

In order to achieve a better understanding of how the glass interacts with each phase, pellets of single phase alumina and c-zirconia were fabricated, following the same experimental procedures. Alumina had a much lower "K" value than c-zirconia, indicating control by the Al_2O_3 phase in the composite. Figure 6 shows the effect of 10 vol% glass on the microstructure of the single-phase materials.

The microstructures reveal that the zirconia grains are rounded, whereas the alumina grains are more faceted. Similar observations have been made by Lin

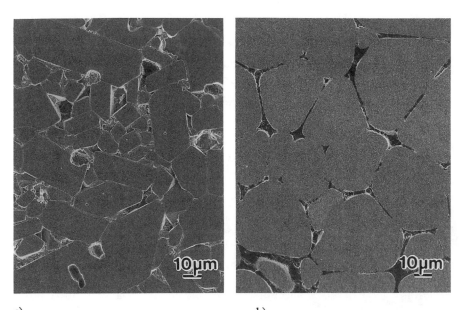

a)   b)

Figure 6 Effect of 10 vol% glass on the microstructures of single phase Al_2O_3 and ZrO_2, after 54h at 1650°C.a) Al_2O_3+10 vol% glass; b) ZrO_2+10 vol% glass.

(1990) for zirconia, and Kwon and Messing (1990) for alumina. Dong-Duk Lee et al. (1988), studying the mechanism of grain growth during liquid phase sintering of β'-Sialon, found that the growth of faceted grains was controlled by interface reaction while the growth of spherical grains was controlled by diffusion. Figure 7 shows the glass distribution in the AZ50 composite for 5 and 10 vol% glass. The microstructure of the composite reveals rounded grains for zirconia and faceted grains for alumina, as shown for single-phase materials. Also, as V_f increases, glass pockets are enlarged. Considering the above results, it is proposed that alumina grains are controlling the coarsening of the composite by an interface reaction-controlled mechanism.

Sodium addition (20 wt%) to the anorthite glass did not alter the grain growth kinetics, indicating that sodium had no significant effect on the interface reaction rate. Calculations based upon viscosity data from the literature (Babcock, 1977) also indicate that a 20 wt% addition of sodium to anorthite glass would not have a significant effect on the viscosity at 1650°C (Alves et al., 1994).

4. Conclusions

In contrast to single phase materials, liquid phase additions to AZ50 promoted faster grain growth. The fitting of the grain growth law; $G^2-G_0^2=Kt$, the appearance of faceted grains of alumina, and the observed relationship between the grain growth constant (K) and glass volume fraction (V_f): $K \propto V_f^{2/3}$, all suggested that an interface reaction-controlled mechanism is controlling the

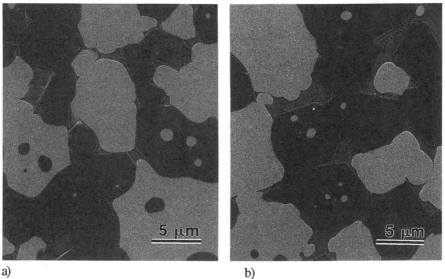

a)   b)

Figure 7   Glass distribution after 54h at 1650°C. Samples were not thermally etched. a) AZ50+5 vol% glass; b) AZ50+10 vol% glass.

coarsening behavior of the AZ50 composite. Sodium addition to the anorthite glass, did not change the grain growth rate.

Acknowledgments: Financial support provided by The National Science Foundation contract DMR 223663, the Air Force Office of Scientific Research contract AFOSR F49620-94-1-0284, JNICT (Junta Nacional de Investigação Cientifica e Tecnológica) and FLAD (Fundação Luso-Americana para o Desenvolvimento) are gratefully acknowledged.

References

W. A. Kaysser, M. Sprissler, C. A. Handwerker and J. E. Blendell, "Effect of Liquid Phase on the Morphology of Grain Growth in Alumina", J. Am. Cer. Soc. 70 [5] 339-43 (1987).

J. J. Kim, M. P. Harmer and T. M. Shaw, "Effect of Liquid Volume Fraction on Coarsening of MgO Grains in Molten CaMgSiO$_4$ Matrix", unpublished work (1993).

Yung-Jen Lin, "Silicate Grain Boundary Phases in Yttria-Zirconia", Ph. D. Thesis, University of Minnesota, Ann Arbor, MI, March (1990).

S. Dill, J. D. French, H. M. Chan and M. P. Harmer, "Coarsening of a Duplex Microstructure: Effect of a Liquid Phase", Ceramographic Poster and Winner of the Rowland B. Snow Award, Am. Cer. Soc. Annual Meeting, Minneapolis, April (1992).

J. C. Wurst and J. A. Nelson, " Lineal Intercept Technique for Measuring Grain Size in Two-Phase Polycrystalline Ceramics", J. Am. Cer. Soc. 55 [2] 109 (1972).

J. D. French, M. P. Harmer, H. M. Chan and G. A. Miller, "Coarsening-Resistant Dual-Phase Interpenetrating Microstructures", J. Am. Cer. Soc. 73 [8] 2508-10 (1990).

Oh-Hun Kwon, "Liquid Phase Sintering and Hot Isostatic Pressing of an Alumina-Glass Composite: Moddeling and Experiments", Ph. D. Thesis, The Pennsylvania State University, University Park, PA, August (1986).

W. Hummel and J. Arndt, "Variation of Viscosity With Temperature and Composition in the Plagioclase System", Contrib. Mineral Petrol. 90 83-92 (1985).

O.-Kwon and G. L. Messing, "Kinetic Analysis of Solution-Precipitation During Liquid-Phase Sintering of Alumina", J. Am. Cer. Soc. 73 [2] 275-81 (1990).

D.-D. Lee, S.-J. L. Kang and D. N. Yoon, "Mechanism of Grain Growth and $\alpha-\beta$' Transformation During Liquid-Phase Sintering of β'-Sialon", J. Am. Cer. Soc. 71 [9] 803-806 (1988).

Clarence L. Babcock, "Silicate Glass Technology Methods", edited by John Wiley & Sons, Inc., New York (1977).

F. J. Alves, H. M. Chan and M. P. Harmer, unpublished work (1994).

Computer Simulation of Microstructural Evolution in Two-Phase Polycrystals

Danan Fan and Long-Qing Chen
Department of Materials Science and Engineering,
The Pennsylvania State University,
University Park, PA 16802

Abstract

The microstructural evolution of a two-phase polycrystal was studied in two dimensions by computer simulations based on a continuum diffuse-interface model. The temporal and spatial evolution of orientation and concentration field variables were obtained through a numerical solution of the time-dependent Ginzburg-Landau (TDGL) equations. The microstructural evolution in a two-phase system with conserved volume fraction under different energetic conditions, i.e., different ratios of grain boundary energies and interfacial energies, was investigated. It was revealed that the grain topology and topological transformations in two-phase solids can be dramatically different from those in single-phase systems.

1. Introduction

The microstructural evolution and coarsening kinetics in advanced two-phase materials (French, 1990; Lange, 1987; Ankem, 1985), have been experimentally studied because material properties and performance are controlled by the size and distribution of the constituent phases. In these materials, different coarsening processes, such as grain growth and Ostwald ripening, occur simultaneously. Due to the complexity of microstructural evolution and coarsening behavior in two-phase solids, none of the current coarsening theories and models can realistically describe the microstructural evolution in two-phase materials since these theories cannot simultaneously consider the coupled grain growth, Ostwald ripening and diffusion processes. Computer simulation studies using the Q-state Potts model also assume that second-phase particles are immobile and cannot coarsen (Srolovitz, 1984, Hassold, 1990). The simulation of microstructural evolution in two-phase polycrystals by Holm *et al* (1993) did not take into account the diffusion and Ostwald ripening process and did not conserve the volume fractions of the two constituent phases.

Recently, Cahn (1991) proposed a theory for the stability of microstructures in a two-dimensional two-phase material, in which only the simplest case of a two-phase mixture, i.e., same concentration for the two phases (no diffusion) and no Ostwald ripening, was considered. In this theory, the microstructural stability and features were analyzed based on the energetic ratios R_α and R_β, as shown in Fig. 1. In this plot, $R_\alpha = \gamma_\alpha / \gamma_{\alpha\beta}$ and $R_\beta = \gamma_\beta / \gamma_{\alpha\beta}$, where the γ_α is the grain boundary energy of the α phase, γ_β is the grain boundary energy of the β phase and $\gamma_{\alpha\beta}$ is the interfacial energy between α and β phases. For $0 \le R_\alpha \le \sqrt{3}$, the trijunctions $\alpha\alpha\alpha$ are stable and when $R_\alpha > \sqrt{3}$ the $\alpha\alpha\alpha$ trijunctions are un-

stable with respect to the nucleation of β grains at these trijunctions (Cahn, 1991). Similarly, the trijunctions $\beta\beta\beta$ are stable for $0 \le R_\beta \le \sqrt{3}$ and are unstable with respect to the nucleation of α grains if $R_\beta > \sqrt{3}$. The trijunctions $\alpha\alpha\beta$ and $\alpha\beta\beta$ are stable under the conditions of $0 \le R_\alpha < 2$ and $0 \le R_\beta < 2$, respectively. The quadrijunctions $\alpha\beta\alpha\beta$ will become stable if the condition $R_\alpha^2 + R_\beta^2 \ge 4$ is satisfied (Cahn,

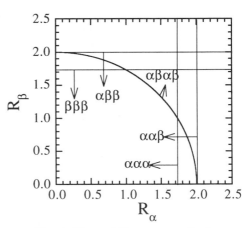

Fig. 1. The stability ranges of microstructural features in two phase systems. (Adapted from J. W. Cahn, Acta Metall. Mater., 39,2189 (1991).)

1991). Monte Carlo simulations (Holm, 1993) of this system, i.e. the phase volume nonconserved system and no diffusion involved, showed that microstructural features are indeed dependent on the energetic ratios (R_α and R_β) and quadrijunctions can be stable within a certain range of angles. While some common features may be found between conserved and nonconserved systems, many significant differences exist. Therefore, it is important to study systematically the microstructural evolution of the two-phase material with the volume conservation and diffusion.

The purposes of this paper are: (1) to propose a new computer simulation model for realistically simulating the kinetics of the coupled grain growth and Ostwald ripening; (2) to study the microstructural evolution of a two-phase material with the phase volume conservation and diffusion; and (3) to investigate the influence of energetic conditions on the microstructural features. This two-phase model is based on our recent computer simulation model for grain growth in single-phase systems (Chen, 1995; Fan, 1995; Chen,1994).

2. The kinetics model

2.1 Description of a two-phase microstructure

An arbitrary two-phase polycrystalline microstructure can be described by a set of continuous field variables (Chen, 1994, 1995; Fan, 1995),

$$\eta_1^{\alpha}(r), \eta_2^{\alpha}(r),..., \eta_p^{\alpha}(r), \eta_1^{\beta}(r), \eta_2^{\beta}(r),..., \eta_q^{\beta}(r), C(r) \qquad (1)$$

where η_i^{α} (i = 1, ..., p) and η_j^{β} (j = 1,..., q) are called orientation field variables with each orientation field representing grains of a given crystallographic orientation of a given phase. Those variables assume continuous values ranging from - 1.0 to 1.0. For example, a value of 1.0 for $\eta_i^{\alpha}(r)$ means that the material at position r belongs to an α-phase grain with the crystallographic orientation labeled as i; a value of -1.0, indicates that the α-phase grain at position r is 180°-rotation related to orientation i. At the grain boundary region, orientation variables will have absolute values intermediate between 0.0 and 1.0. $C(r)$ is the composition field which takes the value of C_{α} within an α grain and C_{β} within a β grain. Therefore, with the set of field variables, we can completely describe a microstructure with α and β-phase grains in different orientations, the α-α and β-β grain boundaries, and the α-β interphase boundaries.

Similar to the free energy description by Cahn and Hilliard for interphase interfaces (Cahn, 1958), the total free energy of a two-phase system, F, can be written as:

$$F = \int \left[f_o \left(C(r); \eta_1^{\alpha}(r), \eta_2^{\alpha}(r),..., \eta_p^{\alpha}(r); \eta_1^{\beta}(r), \eta_2^{\beta}(r),..., \eta_q^{\beta}(r) \right) \right.$$
$$\left. + \frac{\kappa_C}{2} \left(\nabla C(r) \right)^2 + \sum_{i=1}^{p} \frac{\kappa_i^{\alpha}}{2} \left(\nabla \eta_i^{\alpha}(r) \right)^2 + \sum_{i=1}^{q} \frac{\kappa_i^{\beta}}{2} \left(\nabla \eta_i^{\beta}(r) \right)^2 \right] d^3r \qquad (2)$$

,where κ_C, κ_i^{α} and κ_i^{β} are the gradient energy coefficients for the composition field and orientation fields, and p and q represent the number of orientation field variables for the α and β phases. All coupled terms between those gradient terms have been neglected in this model.

The energy of a planar grain boundary, σ_{gb}^{α}, between an α-grain of orientation 1 and another α-grain of orientation 2 can be calculated as

$$\sigma_{gb}^{\alpha} = \int\limits_{-\infty}^{+\infty} \left[\Delta f\left(\eta_1^{\alpha}, \eta_2^{\alpha}, C\right) + \frac{\kappa_C}{2}\left(\frac{dC}{dx}\right)^2 + \frac{\kappa_1^{\alpha}}{2}\left(\frac{d\eta_1^{\alpha}}{dx}\right)^2 + \frac{\kappa_2}{2}\left(\frac{d\eta_2^{\alpha}}{dx}\right)^2 \right] dx \quad (3)$$

in which

$$\Delta f\left(\eta_1^{\alpha}, \eta_2^{\alpha}, C\right) = f_o\left(\eta_1^{\alpha}, \eta_2^{\alpha}, C\right) - f_o\left(\eta_{1,e}^{\alpha}, \eta_{2,e}^{\alpha}, C_{\alpha}\right)$$

$$-\left(C - C_{\alpha}\right)\left(\frac{\partial f_o}{\partial C}\right)_{\eta_{1,e}^{\alpha}, \eta_{2,e}^{\alpha}, C_{\alpha}} \quad (4)$$

where $f_o\left(\eta_{1,e}^{\alpha}, \eta_{2,e}^{\alpha}, C_{\alpha}\right)$ represents the free energy density minimized with re-

spect to η_1^{α} and η_2^{α} at composition C_{α}. Similarly, the interphase boundary en-

ergy between an α-grain and a β-grain can be defined.

2.2 The kinetic equations

The kinetics of microstructural evolution of a two-phase system can be
described by the spatial and temporal evolution of orientation and composition
field variables. The evolution of these field variables are described by the time-
dependent Ginzburg-Landau (TDGL) (Allen, 1979) and Cahn-Hilliard (1958) equations,

$$\frac{d\eta_i^{\alpha}(r,t)}{dt} = -L_i^{\alpha}\frac{\delta F}{\delta\eta_i^{\alpha}(r,t)}, \qquad i = 1, 2, ..., p \quad (5)$$

$$\frac{d\eta_i^{\beta}(r,t)}{dt} = -L_i^{\beta}\frac{\delta F}{\delta\eta_i^{\beta}(r,t)}, \qquad i = 1, 2, ..., q \quad (6)$$

$$\frac{dC(r,t)}{dt} = \nabla\left\{L_C\nabla\left[\frac{\delta F}{\delta C(r,t)}\right]\right\} \quad (7)$$

where L_i^{α}, L_i^{β} and L_C are kinetic coefficients related to grain boundary mobilities
and atomic diffusion coefficients, which may be functions of local orientation and
composition field variables, t is time, and F is the total free energy given in equation (2).

Once the free energy functional, f_o, is constructed, the gradient en-
ergy coefficients can be fitted to the grain boundary energies of the α and β phases
as well as the α/β interfacial energy by numerically solving equations (3) to (7).

The kinetic coefficients, L_i^α, L_i^β and L_C, can be determined from grain boundary mobility and atomic diffusion data. To obtain the kinetics of the microstructural evolution, the kinetic equations (equation 5-7) are numerically solved by discretizing them with respect to space and time.

3. Simulations of microstructural evolution

The computer simulations were conducted in two-dimensions and a system with 256 x 256 grid points and periodic boundary conditions were employed in all simulations. $L_\eta^\alpha = L_\eta^\beta = 1.0$ (isotropic grain boundary mobility), $L_C = 0.5$, $C_\alpha = 0.05$, $C_\beta = 0.95$ and $C_m = 0.5$ were chosen as the basic parameters. All other parameters were determined through calculating the grain boundary energies and the interfacial energy as well as the energy ratios R_α and R_β.

Systems with 50% volume fraction of each phase and $R_\alpha = R_\beta$ were studied. The simplest case is $R_\alpha = R_\beta = 1$, in which the interfacial energy between α grains and β grains is equal to the grain boundary energies of α grains and β grains. Hence, $\alpha\alpha\alpha$, $\alpha\alpha\beta$, $\alpha\beta\beta$ and $\beta\beta\beta$ trijunctions are equally favored during the microstructural evolution. Here, $\alpha\alpha\alpha$ represents the trijunctions formed by three α grains and $\alpha\alpha\beta$ are the trijunctions formed by two α grains and a β grain, and so on. The microstructure with $R_\alpha = R_\beta = 1$ and 50% of each phase is shown in Fig. 2(a). It can be seen that two-phase grains are randomly distributed and trijunctions, $\alpha\alpha\alpha$, $\beta\beta\beta$, $\alpha\alpha\beta$ and $\alpha\beta\beta$, are stable in the microstructure.

Fig. 2(b) is a microstructure in the system with $R_\alpha = R_\beta = 0.6$. In this system, all trijunctions are energetically stable and the interfacial energy $\gamma_{\alpha\beta}$ is larger than the grain boundary energies γ_α and γ_β. Therefore, the system tends to form as many grain boundaries as possible in order to minimize the total free energy of the system during coarsening. It can be seen that grain growth occurs in both phases and grains of the same phase evolve to form clusters to eliminate the α - β interfaces. However, this process is controlled by the long distance diffusion in the volume conserved system, and a result, the formation of clusters of grains of the same phase is difficult in an initially separated two-phase microstructure. Instead, grains of the same phase try to link together to maximize the grain boundary area and the chain structures of grains of the same phase are formed.

The microstructure in a system with $R_\alpha = R_\beta = 1.6$ is shown in the Fig. 2(c). In this system, $R_\alpha = R_\beta < \sqrt{3}$ and $R_\alpha^2 + R_\beta^2 = 5.12$, i.e., the condition $R_\alpha^2 + R_\beta^2 \geq 4$ is satisfied. Hence, all trijunctions and the quadrijunctions $\alpha\beta\alpha\beta$ are energetically stable in this system. It can be seen that the microstructure consists of trijunctions of $\alpha\alpha\beta$ or $\alpha\beta\beta$ and quadrijunctions $\alpha\beta\alpha\beta$, and the two phases are alternatively distributed to minimize the high energy grain boundaries and very few grain boundaries are observed in this microstructure. Even though the

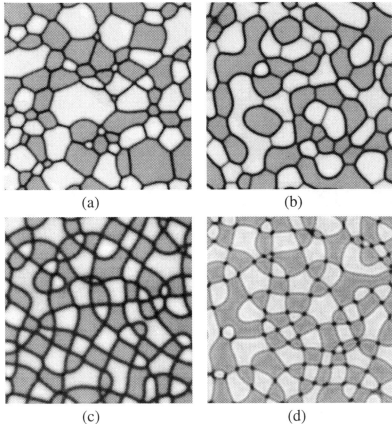

(a)   (b)

(c)   (d)

Fig. 2. The effect of energetic ratio on the microstructures of
two-phase materials. (a) $R_\alpha = R_\beta = 1.0$; (b) $R_\alpha = R_\beta = 0.6$; (c)
$R_\alpha = R_\beta = 1.6$; (d) $R_\alpha = R_\beta = 2.05$.

trijunctions $\alpha\alpha\alpha$ and $\beta\beta\beta$ are still thermodynamically stable in the condition, they
are seldom observed in the microstructures because of the high grain boundary
energies.

    For the extreme condition, $R_\alpha > 2.0$ and $R_\beta > 2.0$, the so called "double"
wetting will occur. Under this condition, no trijunctions are stable and the only
stable surfaces are the α-β interfaces because the two phases will wet each other
due to the extremely high grain boundary energies. The microstructure in the
system with $R_\alpha = R_\beta = 2.1$ is shown in Fig. 2(d). It can be seen that the microstructures are comprised only of quadrijunctions $\alpha\beta\alpha\beta$ and α-β interfaces during
coarsening. No trijunctions are observed in the microstructure and the angles of
the quadrijunctions vary within a certain range, i.e., there is no thermodynamically
fixed angle for quadrijunctions. While the above facts are consistent with the

thermodynamic predictions by Cahn (1991), many important information can be extracted from the kinetic simulations. First of all, the boundaries crossing the quadrijunctions do not have to be straight lines, which is assumed by the thermodynamic analysis (1991). Actually, most of the interfaces are curved to fit into the quadrijunctions and the balance of surface tensions. The only requirement for the energetic balances at quadrijunctions is that the tangents of those boundaries crossing a quadrijunction balance each other. The topological events in all quadrijunction microstructures are very different from those in single phase grain growth and in two-phase systems with only trijunctions. It is observed that grains with two quadrijunctions and grains with three quadrijunctions can vanish directly during coarsening in all quadrijunction system. The vanishing of a grain with two quadrijunctions, which is surrounded by two interfaces with two grains of the other phase and two quadrijunctions, brings the four grains (two of each phase) together. As a result, two quadrijunctions disappear and a new quadrijunction is formed while the two grains of the other phase lose one quadrijunction each. The vanishing of three quadrijunction grains results in the six grains (three of each phase) coming together, which is highly unstable and quickly splits into two new quadrijunctions. The result is that each of the two adjacent grains of the other phase loses one interface and a quadrijunction respectively and one of the same phase grains gains one interface and a quadrijunction whereas other grains remain unchanged in term of the topology.

4. Conclusions

A continuum diffuse-interface model is proposed to simulate the coupled grain growth and Ostwald ripening processes in two-phase solids of conserved volume fractions. It is shown that the ratios of grain boundary energies and the interfacial energy play a important role in the microstructural topology of a two-phase solid. It is demonstrated that quadrijunctions may become stable in the two-phase materials if the energetic condition $R_\alpha^2 + R_\beta^2 \geq 4$ is satisfied, confirming previous thermodynamic analysis. New topological transformations are observed in the quadrijunction systems, which are uniquely different from those in single phase grain growth and systems with only trijunctions.

Acknowledgments: The work is supported by the National Science Foundation under the grant number DMR 93-1898 and the simulations were performed at the Pittsburgh Supercomputing Center.

References

J. D. French, M. P. Harmer, H. M. Chan, and G. A. Miller, "Coarsening-Resistant Dual-Phase Interpenetrating Microstructures", J. Am. Ceram. Soc., 73, 2508 (1990).

F. F. Lange and M. M. Hirlinger, "Grain Growth in Two-Phase Ceramics: Al_2O_3 Inclusions in ZrO_2", J. Am. Ceram. Soc., **70**, 827 (1987).

S. Ankem and H. Margolin, "Grain Growth Relationships in Two-Phase Titanium Alloys", Proc. Fifth Int. Conf. on Titanium, Munich, 1984. Published by Deutsche Gesellsschaft Fur Metallkunde E. V., F.R.G., Vol. 3, 1705 (1985).

D. J. Srolovitz, M. P. Anderson, G. S. Grest, and P. S. Sahni, "Computer Simulation of Grain Growth-III. Influence of a Particle Dispersion", Acta metall., **32**, 1429 (1984).

G. N. Hassold, E. A. Holm, and D. J. Srolovitz, "Effects of Particle Size on Inhibited Grain Growth", Scripta metall., **24**, 101 (1990).

E. A. Holm, D. J. Srolovits, J. W. Cahn, "Microstructural Evolution in Two-Dimensional Two-Phase Polycrystals", Acta metall. mater. **41**, 1119 (1993)

J. W. Cahn, "Stability, Microstructural Evolution, grain Growth, and Coarsening in a Two-Dimensional Two-Phase Microstructure," Acta Metall. Mater., 39, 2189-99 (1991).

L. Q. Chen, "A Novel Computer Simulation Technique for Modeling Grain Growth", Scr. metall.et Mater., **32**, 115 (1995).

Danan Fan and L. Q. Chen, "Computer Simulation of Grain Growth Using a Continuum Field Model," submitted to Acta Metall. Mater., 1995

L. Q. Chen and W. Yang, "Computer Simulation of the Domain Dynamics of a Quenched System with a Large Number of Nonconserved Order Parameters: Grain Growth Kinetics", Phys. Rev. B, **50**, 15 752 (1994).

J. W. Cahn and J. E. Hilliard, "Free Energy of a Nonuniform System. I. Interfacial Free Energy", J. Chem. Phys. **28**, 258 (1958).

S. M. Allen and J. W. Cahn, "A Microscopic Theory for Antiphase Domain Boundary Motion and Its Application to Antiphase Domain Coarsening", Acta metall. **27**, 1085 (1979).

Grain Growth Control via In-Situ Particle Formation and Zener Drag (Pinning) During the Sintering of ZnO

R. C. Bradt, S. I. Nunes[*], T. Senda[+], H. Suzuki[#] and S. L. Burkett

College of Engineering
The University of Alabama
Tuscaloosa, AL 35487-0202 U.S.A.

Abstract

Sintering of ZnO ceramics for most commercial varistor applications is accomplished in the presence of a Bi_2O_3 liquid phase. These electrical ceramics also contain numerous oxide additions, some are to promote appropriate varistor action and others function to reduce the ZnO grain growth during the firing process. One group, or series of the latter oxides reacts with the ZnO during firing to form a stable spinel crystalline phase. These fine spinel crystals are located at the ZnO grain boundaries and serve to limit the grain growth of the ZnO during the firing process, very likely through a Zener pinning or drag type of mechanism. Examination of the grain growth kinetics in these ZnO-based liquid phase sintered systems reveals that the different spinel forming oxide additions result in different ZnO grain growth characteristics, i.e. different kinetic activation parameters for the ZnO grain growth. These parameters are examined and discussed with regard to the ZnO grain growth process. It is concluded that the ZnO grain growth is probably controlled by the cation diffusion in the spinel particles and not by diffusion within the ZnO grains themselves.

Introduction

Metal oxide varistors are highly non-linear electronic devices that are referred to as MOV's, or metal oxide varistors. These ceramic devices are usually categorized by their major component, either ZnO, or $SrTiO_3$. The ZnO varieties of varistors are commonly utilized in surge protection devices, taking advantage of their highly non-linear current voltage characteristics. Most electric power distribution systems employ these MOV's as protection against lightning strikes. Okinaka and Hata (1) have recently reviewed these ceramics with particular emphasis on their manufacturing in Japan. Gupta (2) has previously reviewed them from a more detailed technical perspective. Levinson and Phillip (3) emphasized the importance of the microstructure of the varistor ceramics to the electrical properties of the devices, noting that the current-voltage characteristics are directly related to the number of ZnO grains per unit thickness in the device. This is because the electrical performance of these devices is directly related to the electrical activity of the ZnO-ZnO grain boundaries. It is thus evident that grain growth control of the ZnO crystals during the firing of these ceramics is a critical aspect to their successful manufacture and electrical performance.

The actual commercial compositions of ZnO based MOV's are rather complicated, having been developed on an empirical basis rather than through any specific chemical design critieria. In addition to the basic component material ZnO, a number of other additional oxides are always present, often in only a fraction of a mole percent. While the independent roles of these numerous oxide additions remains somewhat speculative in many instances, particularly within the complicated commercially produced varistors compositions, it does appear that the additive oxides can be generally classified into several groups. One of these additive groups is that of the varistor-action oxides: BaO, Bi_2O_3, Pr_6O_{11} and La_2O_3. The cations of these oxides are all much larger than the Zn^{2+} cation and thus are not expected to exhibit very much solubility within the ZnO grains. It is the grain boundary segregation of these oxides that is primarily responsible for the fundamental varistor action of ZnO. This has been discussed by Cordaro et al. (4) and more recently by Alim (5). A recent phase equilibrium study by Hwang et al. (6) has confirmed the minimal solubility of Bi_2O_3 in ZnO, even at 1000° C. It thus is not surprising that the electrical properties of these ceramics are grain boundary controlled, for these varistor action oxides do not exhibit significant solubility in the ZnO structure.

Other groups of additive oxides are often contained in commercial ZnO varistor compositions for specific electrical effects. One of these groups consists of some of the 3-d transition metal oxides. The 3-d transition metal oxides are believed to directly influence the leakage current characteristics and are also considered to affect the current-voltage characteristics. Another group is the trivalent extrinsic donor group of oxides such as Al_2O_3, Ga_2O_3 and In_2O_3. These trivalent cations are soluble in the ZnO lattice and increase the native free electron concentration in the semiconducting ZnO, thus increasing its conductivity. Monovalent cations, Na_2O, K_2O, etc may also be present (7). Unfortunately, the equilibrium defect chemistry of ZnO is not fully understood, nor is it even certain that equilibrium is achieved under commercial firing conditions, therefore the precise details of the above oxide additions remains less than completely defined.

The final group of oxide additions, which are the specific focus of this paper, are those which react in-situ with the host ZnO during firing to form spinel phases at the grain boundaries and inhibit the grain growth of the ZnO during firing, thus controlling the grain size of the ZnO in the fired ceramic. These oxides may be primarily considered as ZnO grain growth control additives. As an auxiliary benefit, these oxide additions also appear to result in a more uniform ZnO microstructure during sintering, although the specific ZnO grain size distributions do not appear to have been determined and reported. The general reaction for the formation of these spinels may be written as:

$$ZnO + M_2O_3 = ZnM_2O_4 \qquad (1)$$

This is the exact reaction when the additive is an M_2O_3 oxide such as Al_2O_3, but it becomes somewhat more complicated for the zinc-antimonate, zinc-titanate and the zinc-niobate spinels, as the cations do not have the +3 valence. These latter spinels may be considered as defect spinels in that the usual number of tetrahedral and octahedral interstitial sites in the structure are not necessarily filled. This is a consequence of the cation valence states and the necessity for charge balance within the structure. In addition to the above four oxide additives, one can easily envision the formation of zinc-spinels with numerous other oxides, including Cr_2O_3 and Ga_2O_3 among the group.

Although it is possible to react the component oxides to form the spinels completely independent of the commercial varistor formulations, it is illustrative to consider the reaction process for the formation of the spinel phases within the scope of the varistor composition and the ceramic firing process. In fact, since the batch formulations consist of the individual oxide components, these oxides must react during firing to form the spinel phases. The reaction sequence has only been documented in detail for the formation of the Zn_2TiO_4 spinel, however, the other spinels may be expected to follow similar, if not exactly analogous routes (8). Within the ZnO-Bi_2O_3 varistor compositions, the Bi_2O_3 is the lowest melting constituent, $T_{mp}=$ 825°C. Kirkpatrick and Mason (9) confirm that there is a binary eutectic at 740°C and with the presence of additional oxide components, it is reasonable to expect the formation of a Bi_2O_3-rich liquid phase at a temperature of about that magnitude, well below the temperature range for extensive solid state reaction of the ZnO and the additive oxides for spinel formation to proceed (10). As molten Bi_2O_3 is an extremely reactive liquid, it would be expected to react with and dissolve the ZnO and the other additive oxides present. Whether the ZnO and the oxide additions react wholly within the liquid phase to form the spinel, or if the dissolved oxide reacts with the surface of the remanent ZnO grains to form the spinel is not specifically defined. However, the distinctive geometric shapes of the spinel grains at the ZnO grain boundaries, as illustrated in Figure 1, strongly suggest that the reaction to form the spinel occurs within the Bi_2O_3-rich liquid. The research of Peigney, et al (11) addressing the titania doped varistor system strongly substantiates the above liquid phase reaction sequence for spinel formation.

Grain growth can be addressed on two different levels. One is the migration of individual grain boundaries and the other is the change of the average grain size within an assemblage of grains. The latter is more readily applied to the grain growth of a fine grain size system such as the ZnO varistor compositions. Within that condition, grain growth is described by the kinetic equation:

$$G^n - G_o^n = K t \exp\text{-}\{Q/RT\}. \tag{2}$$

Here G is the grain size at the time t, while G_o is the original grain size. The K is a geometrical preexponential constant and Q is the activation energy for grain growth. R and T have their usual meanings. In order to determine the activation parameters for the grain growth process, the above equation is rearranged, neglecting the G_o^n term, for the case where the initial grain size is much smaller than the grain size at time, t. The grain growth exponent, n, can then be readily determined from the slope of a graph of the equation:

$$\log G = (1/n) \log t + \log K \exp\text{-}\{Q/RT\}, \tag{3}$$

where a plot of (log G) versus (log t) has a slope of (1/n). Once the grain growth exponent, or n-value has been determined, then the activation energy, Q, can be determined from the standard Arrhenius plot of the above equation rearranged as:

$$\log (G^n/t) = \log (K) - Q/RT, \tag{4}$$

where the activation energy, Q, is determined from the slope and the preexponential geometrical constant, K, can be determined from the intercept (12).

Discussion

Results describing the effect of spinel particle Zener drag on the grain growth of ZnO exist for four systems. One system has been developed wholly within the solid state, that of the zinc-antimonate spinel ($Zn_7Sb_2O_{12}$) (13). Three have been developed in the presence of a Bi_2O_3-rich liquid phase ($ZnAl_2O_4$, Zn_2TiO_4 and $Zn_3Nb_2O_8$) (8,14,15). The spinel formation was of the in-situ variety in all four of these systems as the constituents were added as the individual oxides prior to the firing of the ceramic. These results can then be directly compared with the results for the grain growth of "pure" ZnO sintered in the solid state (12) and for ZnO sintered in the presence of a Bi_2O_3-rich liquid, but without any spinel forming additives present for grain growth control (16). Table I summarizes the grain growth activation parameters for ZnO in the six systems and a simulated commercial varistor composition(17).

It is a logical sequence to first compare the grain growth of the "pure" ZnO during solid state sintering with the same system containing an in-situ developed spinel phase, the zinc-antimonate, which inhibits grain growth during firing through a Zener drag or pinning mechanism. As noted in Table I, solid state sintered ZnO exhibits a grain growth exponent of 3, an activation energy of about 220 kJ/mol and a K value of about 10^{11}. When Sb_2O_3 is added to form a binary system, it develops the $Zn_7Sb_2O_{12}$ spinel during firing. The spinel particles affect a Zener pinning or drag type of mechanism which reduces the rate of ZnO grain growth. The grain growth exponent is experimentally observed to increase to 6, the activation energy for ZnO grain growth increases to about 600 kJ/mol and the preexponential geometric factor, K, increases to 10^{27}. The ZnO grain size after firing decreases with increasing spinel phase content in agreement with the general concepts of Zener pinning as recently reviewed by Liu and Patterson (18). There is considerable evidence that the reduced ZnO grain size in the presence of these zinc-antimonate spinel particles is the result of a Zener pinning or drag mechanism, although the equilibrium condition of a final grain size is probably not achieved within the time frame of the actual experiments (13).

Zener pinning or drag is not usually considered to persist in the presence of liquid phase sintering. The phenomenon is not even referenced in the text by German (19) addressing liquid phase sintering. Nevertheless, the incorporation of spinel forming oxide additions within liquid phase sintered ZnO-Bi_2O_3 ceramics does inhibit grain growth. Directly related results on actual varistor compositions by Chen et al. (17) also suggests that the process reduces ZnO grain growth in the complicated commercial varistor compositions. The success of Zener pinning or drag for the liquid phase sintered compositions of ZnO varistors may, in fact, be a consequence of the thickness of the grain boundary liquid phase layer. Unlike some of the liquid phase sintered systems, the thickness of the Bi_2O_3-rich grain boundary liquid layer in the ZnO varistor compositions is only a few 10's of nm in width (20,21), thus the micron size spinel particles are readily able to actually bridge the liquid film present at the grain boundary.

The grain growth of ZnO in binary ZnO-Bi_2O_3 ceramics has been studied for Bi_2O_3 levels to 12 wt %, suggesting that there is a rate controlling mechanism transition from interface reaction control to that of diffusion through the liquid boundary layer at about the 3-4wt% level. At the lower Bi_2O_3 levels, the activation energy for ZnO grain growth is only about 150 kJ/mol, whereas the higher contents have an activation energy of about 270 kJ/mol. These are both significantly different from the value of 224 kJ/mol for the solid state grain growth of pure ZnO. The ZnO grain growth exponent in the presence of the Bi_2O_3-rich liquid phase is 5 and the K values are about 10^{13} . Typical ZnO varistor compositions correspond to the lower liquid phase contents, thus suggesting the first set of activation parameters may be most appropriate for comparison. By contrast the studies of the ternary systems with individual Al_2O_3, Nb_2O_5 and TiO_2 additions had slightly higher liquid phase contents.

Experimental results for the grain growth activation parameters in the three ternary systems containing spinel forming additions are summarized in Table I. All three have grain growth exponents greater than 3 and activation energies for ZnO grain growth significantly larger than the binary $ZnO-Bi_2O_3$ systems. The activation energies are about 400, 366 and 335 kJ/mol, respectively, while the preexponential geometric constants, or K values, are between 10^{14} - 10^{19}. These values, as well as a decreasing ZnO grain size with increasing spinel-former content confirm an inhibition of the ZnO grain growth process in these compositions. As the relationships are exactly analogous to the effect on ZnO grain growth exhibited in the solid state Zener drag (pinning) situation, it is only logical to assume that the same process is operative in these liquid phase containing compositions.

It is further illustrative to examine the results of a similar grain growth analysis for much more complex commercial-like compositions. Chen, et al (17) have completed such an analysis for a multicomponent ZnO-based system, specifically focusing on the effects of Mn and Co oxide additions. They observed a grain growth exponent of 6 and an activation energy for ZnO grain growth of about 300 kJ/mol, a trend exactly parallel to that for the solid state Zener drag (pinning) effect relative to the solid state sintering of "pure" ZnO. They did not report a K value within the above format. In their paper they concluded that the ZnO grain growth was controlled by the diffusion of Zn vacancies created by the substitution of the Co^{3+} within the ZnO lattice, yet noted a finer ZnO grain size. Inexplainably, they did not mention the spinel Zener pinning or drag even though the compositions contained considerable amounts of both Sb_2O_3 and Cr_2O_3, both well established spinel forming oxides. It appears likely that the ZnO grain growth in their system was inhibited by the presence of the $Zn_7Sb_2O_{12}$ and $ZnCr_2O_4$ spinel phases creating a Zener drag (pinning) effect.

Nunes and Bradt (14) in their study of the effect of the formation of the $ZnAl_2O_4$ spinel on the grain growth of ZnO in a Bi_2O_3-rich liquid phase sintered system, examined the activation parameters and compared the activation energies with virtually all of the possibilities in the system. While the activation energies for diffusion are well known for ZnO (22) and the previous studies of the $ZnO-Bi_2O_3$ system have determined the values for the interface reaction and diffusion in the grain boundary liquid phase, the diffusion coefficients in the various spinel structures have not generally been measured. However, the general trends of the diffusion coefficients in complex oxides, including spinels as have been summarized by Freer (23) suggest that the diffusion activation energies for these systems are similar in magnitude to the ones listed for ZnO grain growth in the spinel containing systems. A logical extension of these concepts is that the different activation energies for ZnO grain growth in the different spinel particle containing systems suggests that it is the diffusion within the pinning spinel particles which determines the activation energy for the grain growth of ZnO in these Zener drag (pinning) grain growth inhibited systems.

Table I. Summary of ZnO Grain Growth Activation Parameters

ZnO System	n	Q (kJ/mol)	K	Reference
"Pure" ZnO	3	224 ± 16	10^{11}	12
ZnO - Sb$_2$O$_3$	6	600 ± 65	10^{27}	13
ZnO - Bi$_2$O$_3$ (<4wt%)	5	150 ± 30	10^{12}	12
ZnO - Bi$_2$O$_3$ (>4wt%)	5	270 ± 33	10^{14}	16
ZnO - Bi$_2$O$_3$ - Al$_2$O$_3$	4	400 ± 45	10^{19}	14
ZnO - Bi$_2$O$_3$ - Nb$_2$O$_5$	5	366 ± 58	10^{19}	15
ZnO - Bi$_2$O$_3$ - TiO$_2$	6	335 ± 27	10^{14}	8
"Commercial" Varistor	6	301 ± 35	none	17

Figure 1. SEM micrograph of sintered ZnO - Sb$_2$O$_3$ illustrating the several spinel particles at the grain boundaries of the ZnO grains. Note the voids left by the spinel particles which broke away with the matching surface.

Conclusions

The effects of Zener drag (pinning) was examined for the instance of in-situ formed spinel particles in solid state sintered and also for liquid phase sintered ZnO through activation analysis of the grain growth parameters. For both the solid state sintering of ZnO and also during liquid phase sintering in the $ZnO-Bi_2O_3$ binary system the presence of zinc-spinels created during firing reduced or inhibited the rate of ZnO grain growth. This was reflected by changes in the activation parameters for grain growth, usually an increase in the grain growth exponent, or n value, an increase in the activation energy for the ZnO grain growth and also an increase in the K value, the preexponential geometric constant. It is suggested that the activation energy for ZnO grain growth during liquid phase sintering in the presence of spinel particles is related to diffusion within the spinel particles. It is evident that the Zener drag (pinning) process is a suitable technique for the control of grain growth during liquid phase sintering, especially when the grain boundary liquid phase is of a thickness comparable to the pinning particle diameter, or less.

Acknowledgements

The authors acknowledge discussions with and assistance from a number of individuals including; D. Dey, T. K. Gupta, Y. Han, L.M. Levinson, B. R. Patterson and T. Yamaguchi, several of whom have been kind enough to provide preprints prior to publication.

References

(1) H. Okinaka and T. Hata, "Varistor and Thermistor Manufacturing in Japan", Amer. Cer. Soc. Bull. 74 (2) 62 - 66 (1995).

(2) T. K. Gupta, "Application of Zinc Oxide Varistors", J. Amer. Soc. Soc. 73 (3) 1817 - 1840 (1990).

(3) L. M. Levinson and H. R. Philipp, "Zinc Oxide Varistors - A Review", Amer. Cer. Soc. Bull. 65 (4) 639- 646 (1986).

(4) J. F. Cordaro, Y. Shim and J.E. May, "Bulk Electron Traps in Zinc Oxide Varistors", J. Appl. Phys. 60, 4186 - 4192 (1986).

(5) M. A. Alim, "Influence of Intrinsic Trapping on the Performance Characteristics of $ZnO-Bi_2O_3$ Based Varistors", Active and Passive Elec. Comp. 17, 99 - 118 (1994).

(6) J.H. Hwang, T. O. Mason and V. P. Dravid, "Microanalytical Determination of ZnO Solidus and Liquidus Boundaries in the $ZnO-Bi_2O_3$ System", J. Amer. Cer. Soc. 77 (6) 1499 - 1504(1994).

(7) T. Watari and R. C. Bradt, "Grain Growth of Sintered ZnO with Alkali Oxide Additions", J. Cer. Soc. Japan 101 (10) 1085 - 1089 (1993).

(8) H. Suzuki and R. C. Bradt, "Grain Growth of ZnO in $ZnO-Bi_2O_3$ Ceramics with TiO_2 Additions", J. Amer. Cer. Soc. 78 (5) 1354 - 1360 (1995).

(9) K. S. Kirkpatrick and T. O. Mason, "Impedance Spectroscopy Study of Sintering in Bi-Doped ZnO", J. Amer.Cer. Soc. 77 (6) 1493 - 1498 (1994).

(10) T. Tsuchida, "Formation of $ZnAl_2O_4$ in the presence of Cl_2", J. Amer. Cer. Soc. 71 (9) C404 - C405 (1988).

(11) A.Peigney, H. Anderianjatuvo, R. Legros and A. Rousset, "Phase Evolution during the Sintering of Bi-Ti-Doped Zinc Oxide", J. Eur. Cer. Soc. 11, 533 - 543 (1993).

(12) T. Senda and R. C. Bradt, "Grain Growth in Sintered ZnO and $ZnO-Bi_2O_3$ Ceramics", J. Amer. Cer. Soc. 73 (1) 106 - 114 (1990).

(13) T. Senda and R. C. Bradt, "Grain Growth of Zinc Oxide During the Sintering of Zinc Oxide - Antimony Oxide Ceramics", J. Amer. Cer. Soc. 74 (6) 1296 - 1302 (1991).

(14) S. I. Nunes and R. C. Bradt, Grain Growth of ZnO in $ZnO-Bi_2O_3$ Ceramics with Al_2O_3 Additions", J. Amer. Cer. Soc. 78 (9) 2469 - 2475 (1995).

(15) S. I. Nunes and R. C. Bradt, "Grain Growth of ZnO in $ZnO-Bi_2O_3$ Ceramics with Nb_2O_5 Additions", (submitted to Amer. Cer. Soc.).

(16) D. Dey and R. C. Bradt, "Grain Growth of ZnO during Bi_2O_3 Liquid-Phase Sintering", J. Amer. Cer. Soc. 75 (9) 2529 - 2534 (1992).

(17) Y. C. Chen, C. Y. Shen and L. Wu, "Grain Growth Processes in ZnO Varistors with Various Valence States of Manganese and Cobalt", J. App. Phys. 69 (12) 8363 - 8367 (1991).

(18) Y. Liu and B. R. Patterson, "Stereological Analysis of Zener Pinning", (accepted by Acta Metallurgica).

(19) R. M. German, "Liquid Phase Sintering", Plenum Press, N.Y., N.Y. (1985).

(20) D. R. Clarke, "The Microstructural Location of the Intergranular Metal Oxide Phase in a Zinc Oxide Varistor", J. App. Phys. 49 (4) 2407 - 2411 (1978).

(21) E. Olsson, "Interfacial Microstructure in ZnO Varistor Materials", Ph. D. thesis in Department of Physics, Chalmers University of Technology, Goteborg, Sweden (1988).

(22) J. H. Hoffman and I. Lauder, "Diffusion of Oxygen in Single Crystal Zinc Oxide", Trans. Faraday Soc. 66 (10) 2346 - 2353 (1970).

(23) R. Freer, "Bibliography of Self-Diffusion and Impurity Diffusion in Oxides", J. Mat Sc. 15, 803 - 824 (1980).

---

* S.I. Nunes is presently with the Univ.Fed.Sao Carlos; Sao Carlos, Brazil.

+T. Senda is presently with the Ship Research Institute, Tokyo, Japan.

H.Suzuki is presently with Toshiba Corp., Kawasaki, Japan.

S. L. Burkett is with the Dept. of Electrical Engineering and R.C. Bradt is with the Dept. of Metallurgical and Materials Engineering at the University of Alabama.

Sintering, Grain Growth and Homogeneity of Soft Lead Zirconate Titanate Prepared by Different Routes

M.Demartin[1], M.Kosec[2], *C.Carry[3], S.Rimlinger[4], B.Malič[2], G.Drazic[2]

1 Swiss Federal Institute of Technology, Materials Department, Ceramics Laboratory, CH-1015 Lausanne, Switzerland

2 "Josef Stefan" Institute, University of Ljubljana, Ljubljana, Slovenia

3 ISMA, Université Paris Sud, F 91405 Orsay ·Cedex, France

4 Quartz and Silice, B.P.102, F 77792 Nemours Cedex, France

Abstract

Powders of Nb doped $Pb(Zr,Ti)O_3$ of composition close to the morphotropic phase boundary have been prepared by two different solid state routes and by a sol-gel method. The densification and grain growth have been studied during natural sintering,and annealing treatments in order to get dense fine grained samples. Ceramics with grain sizes ranging from 1 to 10 μm and relative densities above 98% have been obtained. The results are discussed in terms of densification and grain growth , with a focus on local fluctuations of Zr/Ti ratio and their homogenisation versus sintering conditions.

1. Introduction

Lead zirconate titanate ceramics (PZT) of composition close to the morphotropic phase boundary (MPB) are widely used for various piezoelectric applications (Jaffe 1971). A trend toward miniaturisation leads to a need for fine grain size materials. However, several studies have shown a decrease of the dielectric and the piezoelectric properties of fine grain materials, sintered at low temperature (Martirena 1974, Okasaki 1973, Yamamoto 1992). For compositions close to the MPB, the phase coexistence range has been shown to depend upon powder preparation, calcination and sintering conditions (Lucuta 1985). The wideness of this phase coexistence region has been related to the presence of local compositional fluctuations (namely the distribution of Zr and Ti ions) (Kakegawa 1982, Thomann 1985, Lucuta 1985) resulting from the powder elaboration conditions and from the effect of internal stresses of mechanical or electrical origin (Kala 1983, Amin 1986, Isupov 1980, Kakegawa 1982). As the dielectric and piezoelectric properties show a sharp maximum around the MPB, they are very sensitive to any compositional change.

In the case of a conventional reaction of the initial oxides, the reaction sequence has been showed (Hiremath 1983, Hankey 1981) to start with the formation of lead titanate followed by the formation of different phases of PZT that finally interdiffuse to give an homogeneous composition. But the process is very sensitive to the calcining and sintering parameters and this homogenisation might not be completed for low temperature processed materials. Improved powder preparation methods (B-site precursor, chemical methods) have been used to enhance the properties of fine grain PZT ceramics (Yamamoto 1992, Lal 1989, Tashiro 1990). The B-site precursor method (Shrout 1990, Kim 1991) consists of a reaction between the pre-reacted $(Zr_x,Ti_{1-x})O_2$ and PbO leading to the formation of perovskite without intermediate phases. Sol-gel routes may lead to very fine powders and a fine dispersion of the Ti and Zr ions. In both cases, the homogeneity of the material should be improved.

However, the extend of the inhomogeneities and their influence on the properties is far from being fully understood. Moreover, as doped compositions are of great technological importance, care should be taken to the effect of the dopant. The aim of this study is to investigate the effect of the compositional inhomogeneities of different powders of morphotropic Nb doped PZT and how homogenisation proceed throughout sintering and grain growth.

2. Experimental procedure and powder characteristics

A soft PZT (53/47) doped with 2 at% Nb has been prepared by three different routes : A conventional mixed oxide route (MO) in which the oxides PbO, TiO_2, ZrO_2, Nb_2O_5 are simultaneously calcined after thorough milling. A B-site precursor route (DC) in which TiO_2, ZrO_2 and Nb_2O_5 are pre-reacted. The resulting compound is then reacted with PbO. These two powders contain 1.2 at% excess lead. A sol-gel powder (SG) for which anhydrous lead acetate is reacted with zirconium n-butoxide, titanium n-butoxide and niobium ethoxide in butanol solution under reflux conditions. After distillation the solution is hydrolyzed with water butanol solution at room temperature using ten mols water per mols PZT. After drying, the powder is thermally treated at 600°C for 5 h in O_2 and milled for 2 h in planetary mill in butanol.

Powder characteristics : The MO powder is composed of grains between 0.6 and 0.8 μm with larger grains around 1 μm, the DC powder shows grains between 0.4 and 0.7 μm with a few grains around 1 μm but with a narrower size distribution, while the sol-gel prepared powder shows grains between 100 and 500 nm. The X-Rays diffraction patterns of the MO and DC powders show the coexistence of a tetragonal (splitting of the 200 / 002 peak) and a rhombohedral phase (unique 200 peak). No second phase can be detected. The spectra of the sol-gel prepared powder shows a single pseudocubic phase. The presence of a pyrochlore phase can be detected.

The dry prepared powders are uniaxialy pressed under 320 MPa. The green density obtained is 68% (8.0 g/cm^3 being taken as the theoretical density) . The sol-gel prepared dry powder is isostaticaly pressed with 500 MPa. The green density obtained is 63 %. The microstructures of compacts are shown on fig.1. Dilatometry has been performed on a NETZSCH402E.

The dry prepared samples are sintered in sealed crucibles using a compensating atmosphere of $PbZrO_3 + 10\%$ ZrO_2 to ensure a constant lead oxide activity (Kingon 1983). The sintering temperatures range between 1050 and 1280°C and the soaking times between 10 min and 12 h. The heating rate is 2°C/min. The sol-gel samples are sintered in O_2 with a heating rate of 10°C/min at various temperatures for 2 h. Densities are measured by Archimede's method. Grain sizes

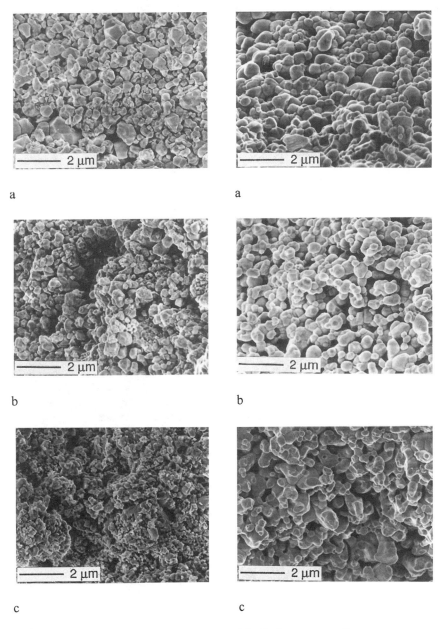

Fig.1 : Green compacts
a) MO powder
b) DC powder
c) SG powder

Fig.3 : Fracture surfaces.
a) MO 990°C - 5 min.
b) DC 990°C - 5 min.
c) SG 950°C - 60 min.

are measured on polished and chemically (HCl +HF) etched surfaces by measuring the mean grain area and using the formula d=1.38√s. EDX(Link AN10000) analyses have been performed on a TEM Jeol 200FX. X-Ray diffraction has been performed on crushed samples with a SIEMENS Kristalloflex 805 using a 0,02° step and 6 s count time.

3. Results and discussion

Fig.2 shows the densification behaviour of the three powders. For the three different powders, the main densification starts at 1000°C and shows a maximum rate around 1100°C, whatever the initial particle size. In the case of the conventional (MO) powder, a first densification stage occurs already between 740°C and 1000°C. Fig.3 shows the microstructural evolution during sintering. For the two powders MO and DC, no significant grain growth takes place under 1000°C. Densification and grain coarsening become significant above 1100°C. At 1050°C the relative densities are

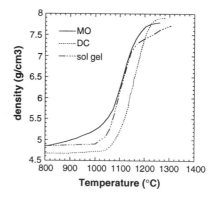

Fig.2 : Densification behaviour of the three powders

>97% after 90 min and at 1150°C after 20 min. The weight losses are below 0.5wt% in any case. Grain sizes range between 1.2 and 10 μm. In the case of the B-site precursor route, the densities are >97°C after 20 min at 1150°C. The grain sizes obtained are similar. For the sol-gel prepared samples, in spite of the smaller grain size of the powder, 100 to 500 nm (fig.1.c), the densification temperature and the resulting grain sizes are comparable to the ones of the dry processed materials. During sintering, for the three powders, there is a change of the phases in presence, as shown in fig.4 for the MO powder. Sintering at higher temperature leads to an increase of the tetragonal phase.

3.1 MO powder :
This first densification stage is related to the presence of a small amount of Pb rich liquid phase (fig.3.a). TEM analyses showed that samples sintered at 1050°C still show Pb rich phase in some triple points. However samples sintered at higher temperature show clean triple points. This phase appears to be transient as it is the case in PLZT (Snow 1973; Wolfram 1978). Rossner (1985) showed that such a phase

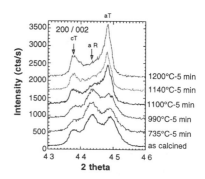

Fig.4 : Structure change during sintering of the MO powder

leads to a start of the densification around 730°C for Nd doped samples, however he couldn't observe it microscopically. The amount of liquid phase depends on the PbO excess of the powder. In our case, it is probable that a small difference in the PbO content of the MO and the DC powders due to the different calcination step, leads to the formation of a liquid phase in the case of the MO powder. Under 1000°C, the presence of liquid phase induces a rearrangement of the grains by capillarity inducing little densification and no grain growth. The main densification and grain growth takes place through normal solid state sintering above 1000°C.

The powder is shown to consist of two different phases. Up to 1000°C no change occur in the X-Rays diffraction pattern. Between 1000°C and 1150°C the ratio of tetragonal phase to rhombohedral phase increases and above 1200°C, the sample shows almost only tetragonal phase (fig.4). In terms of Zr/Ti fluctuation, the comparison between a sample of 4.5 µm (1250°C-120 min) and a sample of 1.8 µm (1150°C-20 min) has been done in TEM by EDX. The large grain sample shows very little fluctuations between the different grains (standard deviation of 5% on the Zr/Ti ratio). Moreover the average composition corresponds to the nominal composition. The fine grain sample shows much larger fluctuations, many grains being clearly Ti rich (standard deviation of 13% on the Zr/Ti ratio).

Below 1000°C, a rearrangement of the small grains occurred but without any structure change. It shows that, in the beginning of sintering, the presence of liquid phase only leads to rearrangement by capillarity, without mechanisms of dissolution-reprecipitation that would lead to homogenisation of composition between grains. Above 1100°C, a rapid densification occurs together with grain growth and a change in the structure. During grain growth, the Ti ions redistribute in surrounding grains, leading to a more tetragonal structure. To see whether homogenisation was possible without changing the grain size, long time annealing treatments, from 1 to 15 days at 800°C and at 1000°C were performed on dense samples sintered at 1050°C, 90 min with a grain size of 1.2 µm.

At 800°C, after 15 days, neither the grain size, nor the ratio of the phases in presence were changed. At 1000°C, slow grain growth occurs, following a law $G \propto t^{1/n}$ with n between 2 and 3 (Fig.5). The ratio of the phases in presence also changes, showing an increase of the tetragonal phase for increasing time and grain size.

Comparing the X-Rays diffraction pattern of samples of same grain sizes obtained by normal one step sintering or by sintering and annealing at 1000°C shows that the same ratio of the two phases are present. It confirms that the homogenisation proceed

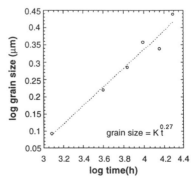

Fig.5: Grain growth at 1000°C for the MO powder

throughout grain growth and is not uniquely temperature dependent. As homogenisation proceed trough grain growth, it seems necessary to improve the homogeneity of the initial powder to get homogeneous fine grain samples.

3.2 B-site precursor route :
This route has been used to get more homogenous dispersion of the B-site cations as the reaction doesn't lead to the formation of intermediate phases. The second calcination step has been studied for pure $(Zr_{0.5},Ti_{0.5})O_2 + PbO$. Calcination between 800°C and 900°C leads to two perovskite phases, tetragonal and rhombohedral. After calcination at 1000°C for 1h, the rhombohedral phase has almost entirely disappeared and the tetragonal structure of the composition $Pb(Zr_{0.5}Ti_{0.5})O_3$ is obtained. The intermediate presence of a two phases material indicates that this route may also lead to local inhomogeneities.

The presence of local inhomogeneities in the DC powder is then possible and on sintering, the change of structure can be understood in a similar way as for the MO powder. The homogenisation of composition also goes along with grain growth, being significant above 1000°C, despite the fact that the scale of the inhomogeneities should be finer, and more likely within grains. The microstructure at 990°C shows that grain rearrangement and rounding occurred (fig.1.b). As surface mechanisms alone don't lead to homogenisation (no occurrence below 1000°C), the homogenisation seems to proceed by grain boundary motion and grain boundary diffusion. Redistribution by bulk diffusion within grains is unlikely. The phase coexistence in such a powder is then similar to a normally prepared powder. But as the scale of the fluctuations is probably finer the effect on the properties of fine grain samples is lesser.

3.3 Sol-gel powder
For the sol-gel prepared samples, in spite of the smaller grain size of the powder, the densification temperature is comparable to the ones of the dry processed materials. Below 1000°C there is no sintering of the fine aggregates. The presence of a small amount of pyrochlore phase in the powder, difficult to avoid for Nb doped sol-gel powders could explain the delay. SEM observation shows that this pyrochlore phase seems to form a very fine second phase at the surface of the grains. X-Rays diffraction shows that this phase decomposes between 1000 and 1050°C (fig.6). The densification is probably inhibited as long as the second phase is present. Above 1000°C, the densification is then extremely fast (fig.7). It is possible that the lead contained in the pyrochlore phase, enhances the sintering by forming a liquid phase after the pyrochlore decomposition. Such observations have been reported for Sb and Nb doped PZT (Kosec 1990). As the temperature of densification is not lowered, the final grain sizes of dense samples start at 1 μm. The role of the pyrochlore phases is underlined by the fact that an undoped powder of similar characteristics sinters 200°C lower (Malic 1995).

The powder shows a dominant rhombohedral structure and a small tetragonal component. Sintering below 1150°C leads to a single rhombohedral phase material. The presence of the pyrochlore phases can account for this difference.

Part of the Nb ions are concentrated in the parasite phase. During sintering, this phase decompose, leading to a distribution of the Nb in the PZT. The Nb has been shown to shift the MPB towards lower Zr contents (Kosec 1990).. As the sintered PZT globally contains more Nb, the structure tends to be more rhombohedral. Sintering at higher temperatures leads to a reapparition of a tetragonal phase. A possible explanation is that a small lead loss occurred, leading to the precipitation of some ZrO_2 and a matrix enriched in Ti. Moreover a precipitation of the Nb in linear or planar defects of the structure can not be excluded and further work is needed to clarify this point.

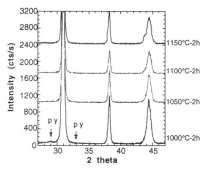

Fig.6 : Structure change during sintering of the SG powder

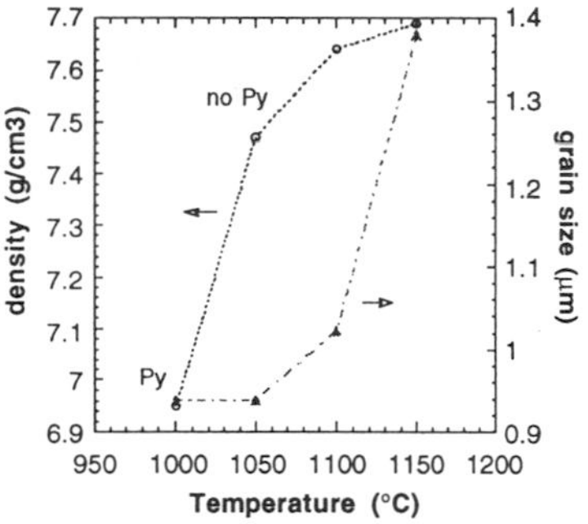

Fig.7 : Densification and grain growth of the SG powder

4. Concluding remarks and prospects

For a conventionally prepared powder, a first sintering step with liquid phase contribution can be recognised but only leading to a rearrangement of the grains without any grain growth and homogenisation. The main densification takes place together with grain growth, by classical solid state sintering.

The homogenisation of composition is connected with grain growth making difficult the elaboration of homogeneous fine grain samples.

The sintering of the fine grain sol-gel prepared sample, takes place at the same temperature as the dry prepared powders in spite of a fine starting powder. A parasite pyrochlore phase at the surface of the grains inhibits an earlier densification and governs the sintering behaviour.

5. References

• Jaffe, Cook and Jaffe, "Piezoelectric Ceramics". New York, Academic Press. (1971).
• H.T. Martirena and J.C. Burfoot, "Grain Size Effects on Properties of Some Ferroelectric Ceramics.", *J.Phys.C.: Solid State Phys.* 7 3182-3192H (1974).
• K. Okasaki and K. Nagata, "Effects of Grain Size and Porosity on Electrical and Optical Properties of PLZT Ceramics", *J.Am.Ceram.Soc.* 56 [9] 82-86 (1973).
• T. Yamamoto, "Optimum Preparation Methods for Piezoelectric Ceramics and their Evaluation", *Ceramic Bulletin* 71 [6] 978-985 (1992).

• P.G. Lucuta, F. Constantinescu and D. Barb, "Structural Dependence on Sintering Temperature of Lead Zirconate-Titanate Solid Solutions", *J.Am.Ceram.Soc.* 68 [10] 533-537 (1985).

• K. Kakegawa, J. Mohri, S. Shirasaki and K. Takahashi, "Sluggish Transition Between Tetragonal and Rhombohedral Phases of Pb(Zr,Ti)O3 Prepared by Application of Electric Field", *J.Am.Ceram.Soc.* 65 515-519 (1982).

• H. Thomann and W. Wersing, "Principles of Piezoelectric Ceramics for Mechanical Filters"., Gordon and Breach Science Publishers. (1985).

• T. Kala, "Contribution to the Study of Tetragonal and Rhombohedral Phase Coexistence in the PbZrO$_3$-PbTiO$_3$ System", *Phys.Stat.Sol.* 78 [a] 277-282 (1983).

• A. Amin, R.E. Newnham and L.E. Cross, "Effect of elastic boundary conditions on morphotropic Pb(Zr,Ti)O3 piezoelectrics", *Physical Review B* 34 [3] 1595-1597 (1986).

• V.A. Isupov, "Reasons for discrepancies relating to the range of coexistence of phases in lead zirconate-titanate solid solutions", *Sov. Phys. Solid State* 22 [1] 98-101 (1980).

• B.V. Hiremath, A.I. Kingon and J.V. Biggers, "Reaction Sequence in the Formation of Lead Zirconate -Lead Titanate Solid Solution: Role of Raw Materials", *J.Am.Ceram.Soc.* 66 [11] 790-793 (1983).

• D.L. Hankey and J.V. Biggers, "Solid-State Reactions in the System PbO-TiO2-ZrO2", *Comm. Am. Ceram. Soc.* 64 C172-173 (1981).

• R. Lal, N.M. Gokhale, R. Krishnan and P. Ramakrishnan, "Effect of sintering parameters on the microstructure and properties of strontium modified PZT ceramics prepared using spray-dried powders", *J.Mat.Science* 24 2911-2916 (1989).

• S. Tashiro, Y. Kotani, H. Igarashi, K. Hidaka and K. Fukai, "Effect of ZrO$_2$ powder on sintering of PZT ceramics" 7th ISAF. (1990).

• T.R. Shrout, P. Papet, S. Kim and G.-S. Lee, "Conventionally Prepared Submicrometer Lead-Based Perovskite Powders by Reactive Calcination", *J.Am.Ceram.Soc.* 73 [7] 1862-1867 (1990).

• S. Kim, G.-S. Lee, T.R. Shrout and S. Venkataramani, "Fabrication of fine grain piezoelectric ceramics using reactive calcination", *J.Mat.Science* 26 4411-4415 (1991).

• A.I. Kingon and J.B. Clark, "Sintering of PZT Ceramics : I Atmosphere Control.", *J.Am.Ceram.Soc.* 66 [4] 253-256 (1983).

• G.S. Snow, "Fabrication of Transparent Electrooptic PLZT Ceramics by Atmosphere Sintering", *J.Am.Ceram.Soc.* 56 [2] 91-96 (1973).

• G. Wolfram, "Grain Growth in PLZT Ceramics", *Ber.Dt.Keram.Ges.* 55 [7] 365-367 (1978).

• W. Rossner, "Sinterverhalten und Elektrische Eigenschaften von Neodym Dotierter Bleizirconat-Titanat-Keramik, Hergestellt nach dem Mixed-Oxide-Verfahren" Thesis, Universität Erlangen, (1985)

• M. Kosec, M. Dvorsek, M. Pristavec, V. Krasevec and D. Kolar, "Effect of antimony oxide and niobium oxide on the sintering of PZT ceramics" Poster presented at 6th CIMTEC, Montecatini, Italy. (1990).

• B. Malic, D. Kolar and M. Kosec, "Anomalous densification of complex ceramics in the initial sintering stage" Conference on Sintering, Penn State. (1995).

ENHANCED SINTERING OF IRON COMPACTS BY THE ADDITION OF TiO$_2$

Y. C. Lu, H. C. Chen, and K. S. Hwang

Institute of Materials Science and Engineering
National Taiwan University, Taipei, Taiwan, R.O.C.

ABSTRACT

TiO$_2$ was introduced into Fe compacts by adding primary TiO$_2$ powders and by forming oxides from Ti containing chemicals. The processing methods included dry pressing and powder injection molding. It was found that sintered densities varied with the process and the characteristics of TiO$_2$. An increase of 5% in sintered density, contrary to previous studies, was obtained by adding 0.025wt% TiO$_2$ and using the powder injection molding technique. Dilatometry and micro-structures showed that TiO$_2$ enhanced densification by impeding early stage sintering, delaying phase transformation, and preventing exaggerated grain growth.

1 INTRODUCTION

Carbonyl iron powder compacts which are made by either powder injection molding (PIM) or dry pressing usually have 90% density when sintered at 1200°C for 1 hr. The main reasons for obtaining such a low density include the occurrence of exaggerated grain growth during alpha-gamma phase transformation, the low volume diffusion rate in the gamma phase, and some trapped gases in the isolated pores (Hayashi and Lim, 1991, and Lenel, 1980).

To improve densification, one approach is to inhibit the grain growth by dispersing fine oxides in the iron matrix. In an early study carried out by Singh and Houseman (1971), 0.005-0.040 μm alumina, titania, and zirconia in amounts of 0.5-5.0% were added, and inhibited grain growth was observed. However, the densities of compacts sintered between 1300°C and 1490°C were not increased except when compacts contained 0.5-1.0% TiO$_2$ and were sintered at 1350°C.

They postulated that the enhanced densification was
due to the formation of Fe_2TiO_5 between iron and TiO_2
particles, but no detailed mechanisms were given.

The purpose of the present study was to show the
influence of TiO_2 on the sintered densities of carbonyl iron powders using different processes and different TiO_2 powders and to understand the role of TiO_2 in
the sintering of iron. The densification results
obtained are discussed in relation to the Fe/TiO_2
reaction, microstructure, the particle size of the
TiO_2, and the uniformity of particle distribution.

2 EXPERIMENTAL PROCEDURE

The base powder used in this study was a $5\mu m$
carbonyl iron powder, CIPS-1641 grade, supplied by the
ISP Corp. In the PIM process, iron powder was kneaded
with a multicomponent binder which consisted of high
or low density polyethylene (HDPE and LDPE, USIFE Co.
Taipei), Acrawax (Glyco Inc., Norwalk, CT), and stear-
ic acid (Nakarai Chem., Kyoto). The iron powder content was 92% (58.5 vol%), and the remaining 8% consisted of the aforementioned polymers. The feedstock
was then injection molded into tensile bar specimens.
Thermal debinding was carried out in H_2 with a step-
wise heating schedule up to 450°C and then isothermally heated at 700°C for 3 hours. Sintering was carried
out at 1200°C for 1 hr in H_2. For the dry pressing
process, specimens were also compacted to a green
density of 58.5%.

Three methods were used to produce TiO_2 dispersed parts. First, Ti containing HDPE was used as
one of the binder components to form TiO_2 during
debinding and sintering. Other methods were to add
into the LDPE binder system 0.4 μm primary TiO_2 powders (R-KB-6, Bayer AG) or 0.2 μm TiO_2 ashes which
were produced by burning off pure HDPE using the
previously described thermal debinding schedule.

In order to see whether TiO_2 reacts with Fe and
changes the composition of the iron matrix, Induction
Coupled Plasma (ICP) analysis was employed. Samples
were immersed in heated 50% hydrochloric acid, and the
solution was analyzed for Ti in the solid solution.

The undissolved particles were collected and were analyzed for the amount of TiO_2 using the fusion technique.

3 RESULTS

3.1 <u>BINDER</u> <u>RESIDUES</u> The thermogravimetric analyses on binders showed that a small amount of residue was left after the HDPE was heated to 700°C. The particle size of the residue was 0.2 μm and the X-ray diffraction patterns indicated that these ashes were TiO_2 rutiles. These oxides are very likely converted from the titanium chloride which is frequently used as one of the coordination initiators, or Ziegler-Natta initiators, in producing HDPE (Odian, 1981). As HDPE decomposes during debinding, Ti reacts with the moisture in the atmosphere (-50°C dew point) and/or the oxygen in the iron powder (0.9%) and forms TiO_2 residues.

3.2 <u>SINTERED</u> <u>DENSITIES</u> Table 1 lists the sintered densities of PIM specimens. An increase in density from 89.5% to 95.1% was obtained by simply changing the binder content from 3.55% LDPE to 3.55% HDPE. To examine whether the TiO_2 ash left from HDPE would influence the sintering of iron powder, the ashes were ball milled with iron powder in alcohol, and the powder mixture was then dried and pressed into discs. The sintered densities of these discs (Table 2), were also improved but were lower than those of PIM HDPE specimens. Another experiment using the PIM process and the LDPE binder system with various amounts of primary TiO_2 powders also showed an increase in sintered density of 3% when 0.135% TiO_2 was added.

TABLE 1: Sintered densities of HDPE and LDPE PIM specimens

Binder	PE% : Acrawax% :	SA%	TiO_2 contained	Density
LDPE	3.55LDPE 3.55	0.90	0%	89.5%
HDPE	1.92HDPE 5.18	0.90	0.013	89.5
	2.80 4.30	0.90	0.019	92.0
	3.55 3.55	0.90	0.025	95.1
	4.36 2.74	0.90	0.030	93.5
	5.12 1.98	0.90	0.036	91.6

TABLE 2: Sintered densities of pressed compacts
containing TiO$_2$ binder ashes

TiO$_2$ ashes, %	Sintered Density, %
0.00 (pure Fe)	90.3
0.05	93.7
0.10	93.7
0.15	93.7
0.50	92.4
1.00	92.0

The above results indicate that using the PIM
process and HDPE gives the best results. This is
because iron powders in the feedstock are coated with
HDPE and the TiO$_2$ particles thus formed during debinding are finer in size and are more uniformly distributed on powder surfaces than those specimens which
contain primary TiO$_2$ powders or ashes.

3.3 <u>DILATOMETRY</u> <u>ANALYSIS</u> The dilatometric curves in
Figure 1a show that the LDPE sample shrank much faster
in the alpha range than did HDPE, but its densification rate slowed down dramatically above 912°C. This
behavior is similar to that of pure iron compacts, in
which exaggerated grain growth occurs during phase
transformation (Lenel, 1980). In contrast, HDPE gave
more continuous shrinkage throughout the phase
changes. It was also noticed that the phase transformation temperature shifted upward on HDPE specimens.

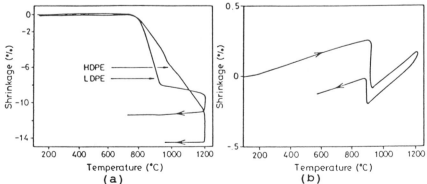

Figure 1. Dilatometric curves of (a) Fe compacts made
from HDPE and LDPE binder systems, and (b) HDPE samples which had undergone the heat excursion in (a).

However, when these specimens were reheated in the dilatometer, the phase transformation temperature returned to 912°C, as shown in Figure 1b.

3.4 CHEMICAL ANALYSIS To determine whether the shift in the transformation temperature of HDPE specimens was caused by the composition change in the iron matrix, ICP analysis was carried out. The results showed that Ti was present in the sintered parts. Similar results were also obtained in samples which contained primary TiO_2 particles, as shown in Table 3.

Since it has been suggested that TiO_2 is not stable at high temperatures under a hydrogen atmosphere (Matzke, 1981), a test was carried out by heating primary TiO_2 powders at 1200°C for 1 hour in hydrogen. The color of the powder changed from white to dark blue after heating. The X-ray diffraction patterns showed that the intensities of TiO_2 peaks were reduced and a wide range of new weak peaks was present. This indicated that some TiO_2 were reduced and became TiO_{2-x}, but no Fe_2TiO_5 which was found by Singh and Houseman was observed.

To examine how TiO_2 might affect the sintering behavior and the microstructure, iron compacts were covered with TiO_2 and then sintered at 1200°C for 1 hour in hydrogen. Figure 2a shows two different types of grains; columnar grains near the Fe/TiO_2 interface and equiaxed grains in the center. This compact also showed decreasing Ti concentration away from the Fe/TiO_2 interface, as illustrated in Figure 2b. These results suggest that as parts were cooled to room temperature the gamma→alpha phase transformation

TABLE 3: The ICP analysis of TiO_2 containing specimens showing that elemental Ti is present in the iron

Source of Ti	Dissolved Ti	Ti in the form of TiO_2
Pure Fe Powder	3 ppm	7 ppm
3.55% HDPE Binder	87 ppm	161 ppm
0.135% Primary TiO_2	341 ppm	443 ppm

a 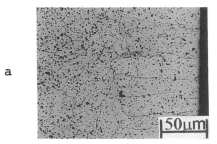 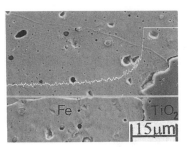 b

Figure 2. (a) The metallograph of iron compacts which were covered by TiO_2 powders during sintering. (b) the Ti concentration profile.

started at the interface where Ti content was higher. This is because the alpha-gamma transformation temperature of iron increases sharply with increasing Ti content. As temperature passed 912°C, the center portion which had little Ti also transformed into alpha phase and formed equiaxed grains.

3.5 <u>METALLOGRAPHY</u> Figures 3 and 4 compare the microstructure of samples which were removed from the furnace at 935°C and 1200°C during heating. The grains in specimens which were made from LDPE had already grown extensively at 935°C and that most pores were trapped inside the grains. This breakaway of pores from grain boundaries caused the sluggish densification of LDPE compacts above 912°C. In contrast, the HDPE samples showed little sintering at 935°C and the grain size of the parts cooled from 1200°C is smaller.

4 DISCUSSION

As TiO_2 was reduced to TiO_{2-x}, some Ti became dissolved in iron particles and formed a Ti-rich outer layer. This layer had a higher phase transformation temperature than did pure iron as was verified in Figure 2a by the presence of columnar grains near the Fe/TiO_2 interface. Thus, when the center portion of iron particles changed from alpha to gamma at 912°C during heating, the exaggerated grain growth was inhibited by the Ti-rich layer because this layer was still very stable in the alpha phase. Thus, phase transformation did not occur at one temperature;

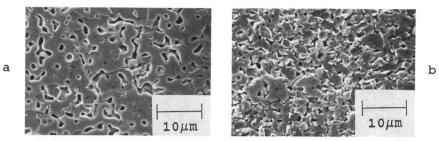

Figure 3. The metallographs of PIM specimens cooled from 935°C. (a) LDPE system, and (b) HDPE system.

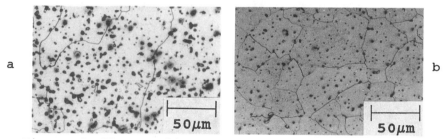

Figure 4. The metallographs of PIM specimens cooled from 1200°C. (a) LDPE system, and (b) HDPE system.

instead, it occurred in a range. As sintering continued, specimens became homogenized. Thus, when sintered parts were reheated, the sharp phase transformation at 912°C was observed again, as shown in Figure 1b.

The above dilatometric curves (Figure 1) and metallographs (Figure 3) also suggest that TiO_2 inhibits the initial stage sintering. Thus, most driving forces, i.e., surface area, are not wasted by the non-densification surface diffusion mechanisms but are retained for densification mechanisms such as the grain boundary diffusion mechanism at high temperatures. Moreover, it has been demonstrated by Hänsel et al. (1981) that the grain boundary diffusion rate is increased in pure iron of which the impurities are scavenged with Ti. Since some elemental Ti is present in TiO_2 containing iron as shown in Table 3 and Figure 2b, it is likely that enhanced grain boundary diffusion also occurred in this study.

5 CONCLUSION

By using a binder which can leave fine TiO_2 ash after debinding, the sintered densities of powder injection molded iron parts could be increased from 89.5% to 95.1%. When primary TiO_2 particles or TiO_2 ashes were added into iron, density increase of 3% was also obtained. These density increases are accompanied by retarded grain growth and inhibited initial stage sintering. ICP analysis indicated that some Ti was dissolved in iron during sintering, possibly due to the reduction of TiO_2 to TiO_{2-x}. This reaction seemed to give a non-uniform Ti concentration in the matrix and cause the phase transformation to occur in a temperature range. As a result, inhibited grain growth and enhanced densification were obtained.

6 Acknowledgment

This work was supported by the National Science Council of R.O.C. under contract NSC82-0405-E-002-406.

7 References

1. K. Hayashi and T.W. Lim, "Role of Equilibrium Pressure of Gas in Sintering Densification of Carbonyl Iron Powder for Metal Injection Molding", *Mat. Trans. JIM* **32**[4] 383-388 (1991).
2. F.V. Lenel, "Sintering of a Single Phase Powder-Experimental Observations", pp.220-221, *Powder Metallurgy Principles and Applications*, MPIF, Princeton, NJ, (1980).
3. B.N. Singh and D.H. Houseman, "The Influence of Oxide Particles on the Sintering Characteristics of Carbonyl Iron", *Powder Met. Int.*, **3**[1] 26-29 (1971).
4. G. Odian, "Ziegler-Natta Polymerization of Nonpolar Vinyl Monomers", pp.591-592, *Principles of Polymerization*, John Wiley & Sons, 2nd ed., NY, (1981).
5. Hj. Matzke, "Diffusion in the Rutile Structure: TiO_2", pp.198-210, *Nonstoichiometric Oxides*, edited by O.T. Sorensen, Academic Press, NY, (1981).
6. H. Hänsel, L. Startmann, H. Keller, and H. J. Grabke, "Effects of the Grain Boundary Segregants P, S, C and N on the Grain Boundary Self-Diffusivity in α-Iron", *Acta Metall.*, **33**[4] 659-665 (1985).

Computer Simulation of Grain Growth during Liquid Phase Sintering

Hideaki Matsubara and Richard J. Brook*

Japan Fine Ceramics Center, 2-4-1 Mutsuno Atsuta-ku Nagoya, 456, Japan
* University of Oxford, Department of Materials, Parks Road Oxford OX1 3PH,
United Kingdom

Abstract

A computer simulation study was applied to the modeling of grain growth during liquid phase sintering. The two mass transfer routes through solid boundaries and a liquid phase were designed in the grain growth simulation based on the Monte Carlo method in two dimensional triangle lattices. The parameters of the fraction of a liquid phase (f_L) and the energy ratio of the solid/liquid interface and solid/solid boundaries (γ_{SL}/γ_{SS}) were changed in the simulation. The characteristic features by the change of these parameters were clearly shown in the simulated results.

I. Introduction

Many systems in ceramics and cermets are fabricated by liquid phase sintering [1, 2]. The grain growth of solid particles in a liquid phase is called "Ostwald ripening" or "solution-reprecipitation process" and has been often studied theoretically and experimentally in relation to microstructural control of these materials [1-9]. In most materials of ceramics and cermets, the volume fraction of a liquid phase is so small that the pure theories of Ostwald ripening can not be directly applied to the microstructural design. In other words, the grain growth of solid particles with a small amount of liquid can be considered to occur through the mixed or complicated mechanism of grain growth in solid state and Ostwald ripening in liquid. In this study, the computer simulation of the grain growth under the existence of a liquid phase is studied mainly by the technique based on the Monte Carlo method where the grain growth is regarded as the process of decreasing a total energy on microstructures such as boundaries and interfaces in materials.

II. Simulation Method

The simulation algorithm in this study is principally based on the Monte Carlo method of grain growth developed by Srolovitz et al. [10, 11]. The temperature term in the Boltzman equation is set at zero; the probability of an increasing energy trial is zero. The two dimensional 200 x 200 triangular lattices including solid particles and a liquid phase were used as the array for the simulation. The number of crystal orientations, Q, in solid grains was fixed at 64.

Two kinds of mass transfer routes were designed in the simulation program for the grain growth under the existence of a liquid phase. If a solid lattice is selected in the array, it tries the mass transfer through grain boundaries in solid owing to the same way by Srolovitz's algorithm [10]. If a liquid lattice is selected, the mass transfer through a liquid phase occurs due to the following process.

This process is followed by the change from the selected liquid lattice to the solid lattice with the Q to give the minimum energy and then the energy change, ΔG_1, is calculated. Liquid lattices neighboring solid particles are advantageous to success such a trial. The next stage of the process is designed to be the random walk from the selected lattice point without back steps in a liquid phase. The random walk is continued till it hits a solid lattice. The hit solid lattice is changed to a liquid lattice and then its energy change, ΔG_2, is calculated. If the total energy change is less than zero, $\Delta G_1 + \Delta G_2 \leq 0$, this try is in practice; otherwise it is canceled. The important parameters in this study are the energy of the interface between solid particles and a liquid phase, γ_{SS}, the energy of grain boundaries of solid particles, γ_{SL}, and the fraction of a liquid phase, f_L.

III. Results and Discussions

Figure 1 shows the simulation results of the case of a high liquid content, $f_L = 80\%$, and a low energy ratio, $\gamma_{SL}/\gamma_{SS} = 0.5$, as a function of Monte Carlo steps (MCS). In these simulated microstructures, solid particles and a liquid phase are white and dark-gray phases, respectively. The growth of particles, which are surrounded by liquid, is mainly due to the mass transfer of solution-reprecipitation mechanism through liquid.

Figure 2 shows the results of a low amount liquid, $f_L = 20\%$. The main process of mass transfer for grain growth in this case is similarly the solution-reprecipitation mechanism. However, some particles in Fig. 2 have grain boundaries indicated by black lines and the mass transfer of grain boundary movement in solid state contributes to the grain growth in such a material.

Figure 3 is a series of the simulated microstructures of $\gamma_{SL}/\gamma_{SS} = 0.5$ of 500MCS as a function of f_L. The case of $f_L = 0\%$ is the grain growth free from liquid. It is clear in Fig.3 that the microstructures are strongly influenced by the liquid content; especially, the growth enhancement by the liquid phase existence is marked around $f_L = 20\%$. It is also noted that a liquid phase forms thin films in the structures of lower liquid content. The value of the energy ratio in this simulation means the relation that the so-called dihedral angle is zero. There is the tendency for liquid to penetrates grain boundaries in solid to form solid/liquid interfaces. Therefore, the grain growth in lower liquid content occurs by the

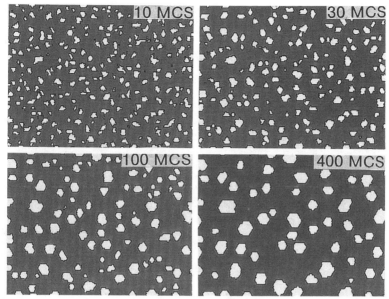

Fig. 1 Simulated microstructures of growth behavior of solid particles in a liquid phase as a function of Monte Carlo steps (MCS). The fraction of liquid phase (f_L) is 80%. The ratio of solid/liquid interface energy to solid-solid grain boundary energy $(\gamma_{SL}/\gamma_{SS})$ is 0.5.

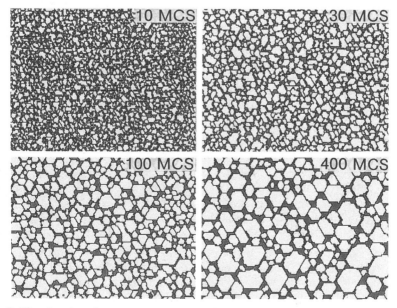

Fig. 2 Simulated grain growth for $f_L = 20\%$ and $\gamma_{SL}/\gamma_{SS} = 0.5$.

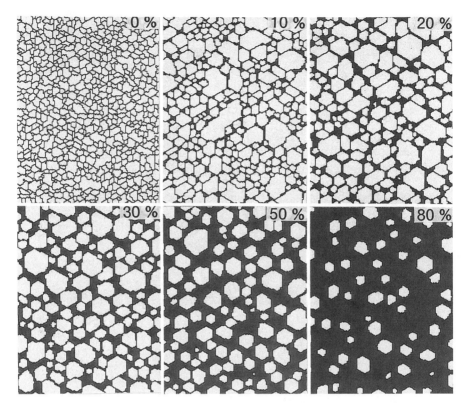

Fig. 3 Simulated microstructures at 500MCS for $\gamma_{SL}/\gamma_{SS} = 0.5$ as a function of f_L.

combined process of the two mass transfer mechanisms through solid boundaries and solid/liquid interfaces under the morphological interaction.

Figure 4 illustrates the simulated microstructures at 500MCS for a lower energy ration of $\gamma_{SL}/\gamma_{SS} = 0.3$ as a function of f_L. The microstructures of $\gamma_{SL}/\gamma_{SS} = 0.3$ at a high f_L of 50~80% are similar to those of $\gamma_{SL}/\gamma_{SS} = 0.5$. The microstructures of $\gamma_{SL}/\gamma_{SS} = 0.3$ at a low f_L of 10~20% have the characteristic features comparing to those of $\gamma_{SL}/\gamma_{SS} = 0.5$; thin films of a liquid phase more easily form in the microstructures of $\gamma_{SL}/\gamma_{SS} = 0.3$ and $f_L = 10~20\%$.

The simulation results of higher energy ratio, $\gamma_{SL}/\gamma_{SS} = 0.55$ and 1.0 as a function of f_L are shown in Fig. 5 and Fig. 6, respectively. In high liquid content of 80%, the growth behavior and the structure are similar to those of $\gamma_{SL}/\gamma_{SS} = 0.5$. However, the features in low or medium liquid content are characteristic in the structure comparing with the cases of $\gamma_{SL}/\gamma_{SS} = 0.5$. The existence of liquid is not film-like but rounded and it is isolated from solid. The grain growth enhancement by liquid is not strong at low liquid content in high energy ratios of $\gamma_{SL}/\gamma_{SS} = 0.55$

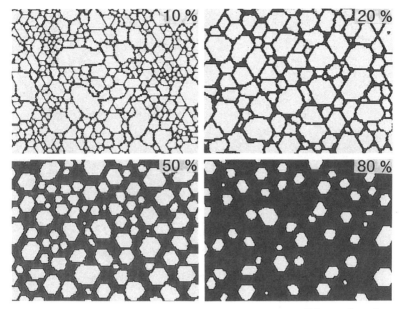

Fig. 4 Simulated microstructures at 500MCS for $\gamma_{SL}/\gamma_{SS} = 0.3$ as a function of f_L.

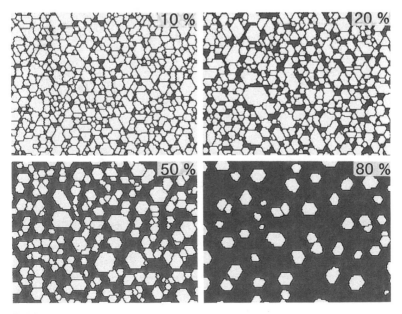

Fig. 5 Simulated microstructures at 500MCS for $\gamma_{SL}/\gamma_{SS} = 0.55$ as a function of f_L.

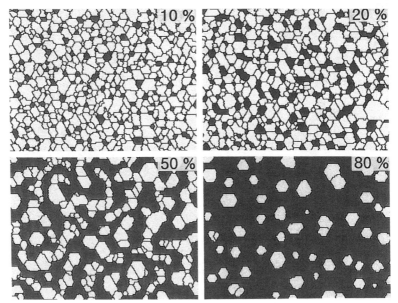

Fig. 6 Simulated microstructures at 500MCS for $\gamma_{SL}/\gamma_{SS} = 1.0$ as a function of f_L.

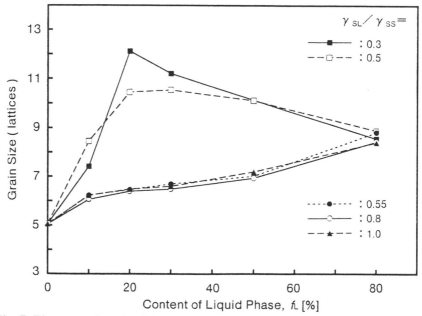

Fig. 7 The mean size of solid grains in simulated microstructures as a function of f_L.

and 0.8. It is also noted that particle coalescence or contiguous solid boundaries are observed in at the middle content of $f_L = 50\%$ of the high energy ratios.

Figures 7 and 8 show the mean size of solid grains in the simulated structure at 500MCS as a function of liquid phase content, f_L, and energy ratio, γ_{SL}/γ_{SS}, respectively. The grain size of low energy ratio of 0.3 and 0.5 has a peak around $f_L = 20\%$, but the grain sizes of high energy ratios of 0.55~1.0 have a similar tendency of a slight increase with f_L. The influence of energy ratio is that grain size is remarkable decreasing from $\gamma_{SL}/\gamma_{SS} = 0.5$ to 0.55 at low f_L of 20~50%, but that grain size change by γ_{SL}/γ_{SS} is very small at high f_L of 80%.

It has been founded that these simulated microstructures and results clearly illustrate the effect of the parameters such as liquid phase content and the energy balance of interfaces and grain boundaries. However, these simulations are under the limited parameters or conditions in two dimensional lattice. Some problems concerning lattices dimension, growth kinetics, simulation algorism mass transfer mechanism, etc. should be further investigated.

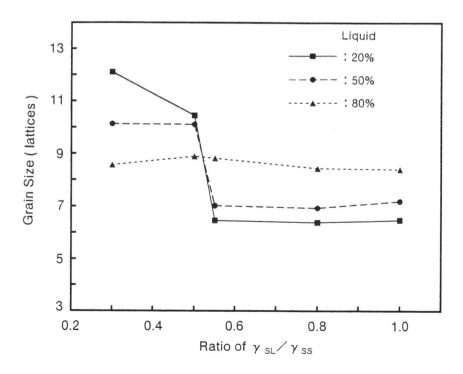

Fig. 8 The mean size of solid grains in simulated microstructures as a function the enegy ratio of γ_{SL}/γ_{SS}.

IV. Conclusion

The computer simulation of the grain growth during liquid phase sintering was performed mainly by the technique basing on the Monte Carlo method in two dimensional triangle lattices. The two mass transfer routes through solid boundaries and a liquid phase were designed in the simulation. The important parameters such as the fraction of a liquid phase (f_L) and the energy ratio of the solid/liquid interface and solid/solid boundaries (γ_{SL}/γ_{SS}) were changed in the simulation. In the simulated microstructures of low energy ratio, the liquid phase was easy to form a thin film in a low liquid content and the growth enhancement by the existence of liquid had the maximum at a low liquid content. In the high energy ratio, the liquid was rounded and isolated from solid and the particle size had a slight tendency to increase with liquid content.

This work has been entrusted by NEDO as part of the Synergy Ceramics Project under the Industrial Science and Technology Frontier (ISTF) Program promoted by AIST, MITI, Japan. H.M. is the member of the Joint Research Consortium of Synergy Ceramics.

References
1). G. Petzow and W.A. Kaysser, "Basic Mechanisms of Liquid Phase Sintering", Sintered Metal-Ceramic Composites, Ed.by G.S. Upadhyaya, Elsevier Science Publishers, 51-70 (1984).
2). R.M. German, *Liquid Phase Sintering*, Plenum Press, 1985.
3). G.W. Greenwood, "The Growth of Dispersed Precipitates in Solutions", *Acta Metall.*, **4** [5] 243-248 (1956).
4). I.M. Lifshitz and V.V. Slyozov, "The Kinetics of Precipitation from Supersaturated Solid Solutions", *J. Phys. Chem. Solids*, **19**[Nos.1/2]35-50 (1961).
5). C. Wagner, "Theorie der Alterung von Niederschlagen durch Umlosen", *Z. Elektrochem.*, **65** [Nr7/8] 581-591 (1961).
6). P.W. Voorhees and M.E. Glicksman, "Ostwald Ripening during Liquid Phase Sintering-Effect of Volume Fraction on Coarsening Kinetics", *Metall. Trans.*,**15A** [6] 1081-1088 (1984).
7). S. Takajo, W.A. Kaysser and G. Petzow, "Analysis of Particle Growth by Coalescence during Liquid Phase Sintering", *Acta Metall.*, **32** [1] 107-113 (1984).
8). R. Warren and M.B. Waldron, "Carbide Grain Growth", *Powder Metall.*, **15** [30] 180-201(1972).
9). H. Matsubara, S-G. Shin and T. Sakuma, "Growth of Carbide Particles in TiC-Ni and TiC-Mo$_2$C-Ni Cermets during Liquid Phase Sintering", *Metall. Trans.*, **32** [10] 951-956 (1991).
10). M.P. Anderson, D.J. Srolovitz, G.S. Grest and P.S. Sahni, "Computer Simulation of Grain Growth - I. Kinetics", *Acta Metall.*, **32** [5] 783-791 (1984).
11). G.S. Grest, D.J. Srolovitz and M.P. Anderson, "Computer Simulation of Grain Growth - IV. Anisotropic Grain Boundary Energies", *Acta Metall.*, **33** [3] 509-520 (1985).

A NEW COMPUTER MODEL OF OSTWALD RIPENING

Zoran S. Nikolic[1], Richard M. Spriggs[2], and Momcilo M. Ristic[3]

[1]Faculty of Electronic Engineering
University of Nish, 18000 Nish, Yugoslavia
[2]CACT Center, Alfred University, Alfred, NY 14802, USA
[3]Serbian Academy of Sciences and Arts
Joint Laboratory for Advanced Materials, 11000 Belgrade, Yugoslavia.

Abstract

From experiments with mixtures of small and large single particles annealed in the presence of a liquid phase, it was concluded that during liquid phase sintering small particles partially dissolve and the solid phase precipitates on the large solid particles. Due to the reprecipitation process large particles form polyhedral shapes. Simultaneously, the number of small particles decreases due to the dissolution and reprecipitation processes. The dissolution of small particles leads to further densification due to rearrangement of small and large particles. This rearrangement process was simulated in the model system. The purpose of this paper is two-dimensional computer modeling of the effect of Ostwald Ripening or microstructural evolution of liquid phase sintered material.

1. Introduction

For different sintered materials a number of common basic mechanisms of liquid phase sintering exist which determine their microstructural development and final properties. Investigation of liquid phase sintering has shown that differences in chemical potential exist in solid-liquid systems [1-6], which causes atoms to dissolve from the solid phase, to diffuse through the liquid phase, and to reprecipitate onto the grains, resulting in sintering. Since the chemical gradients must be continuous, it is implied the corresponding gradients in the concentration of the solid phase dissolved in the liquid. The explanation of these phenomena is largely based on empirically established laws [7-11]. If it is assumed that the liquid phase is in equilibrium with the solid phase, with which it is in contact, the concentration gradient between solid phase particles which dissolve and the solid phase formed as a reprecipitation product leads to material transport through the thin layer of the liquid.

To realize the computer simulation of diffusion during liquid phase sintering, we will use several results already available in the literature, plus several new results not previously reported. Initially we must first define the new initial state of the liquid-solid system [12]. It is also necessary to define the model-system, keeping in mind the convenience for the application of the simulation method and the possibility of the simulation of all characteristic phenomena during the sintering process.

2. Numerical Solution of the Diffusion Equation

If $C = C(x, y, t)$ is the concentration of the liquid, and D_L is the concentration independent diffusion coefficient in the liquid, then the diffusion through the liquid phase is defined by the partial differential equation of parabolic type (two-dimensional case)

$$\frac{\partial C}{\partial t} = D_L \left(\frac{\partial^2 C}{\partial x^2} + \frac{\partial^2 C}{\partial y^2} \right).$$

(1)

Let there be a network of grid points which is established throughout the rectangular region (experimental domain). Suppose, we have two distance coordinates x and y, and time t as independent variables, and that the respective grid spacings are Δx, Δy, and Δt. Subscripts i, j, and k may then be used to denote the space point having coordinates $i\Delta x, j\Delta y$, and $k\Delta t$, for the grid-point (i, j, k).

For an approximation solution of the equation (1) we will use the classical five points approximation [13]

$$C_{i,j,k+1} = (1 - 2\lambda_1 - 2\lambda_2)C_{i,j,k} + \lambda_1(C_{i+1,j,k} + C_{i-1,j,k})$$
$$+ \lambda_2(C_{i,j+1,k} + C_{i,j-1,k})$$

(2)

$$(i = 2, n-1; \; j = 2, m-1; \; k = 0,1,...)$$

where

$$\lambda_1 = D_L \Delta t / (\Delta x)^2 \text{ and } \lambda_2 = D_L \Delta t / (\Delta y)^2.$$

If all the $C_{i,j,k}$ are known of any time level t_k, equation (2) enables $C_{i,j,k+1}$ to be calculated directly at the time level t_{k+1} for all i and j. For reasons of computational stability, values of distance and time interval (Δx, Δy, and Δt) must also be taken so that $\lambda_1 + \lambda_2$ does not exceed 0.5.

In problems having simple model geometry, certain grid points can lie on the boundaries. In cases in which the boundary does not fall on regular grid points, the boundary is considered to be irregular. The initial and boundary conditions as in [14] were used in this paper.

3. Process Model Development

From many experiments with mixtures of small and large single particles annealed in the presence of liquid phase, it was concluded that shrinkage is directly linked to grain growth. During liquid phase sintering, small particles partially dissolved and the solid phase precipitated on the large solid particles. Due to the reprecipitation process that large particles form polyhedral shapes. Simultaneously, the number of small particles decreases also due to coarsening. The dissolution of small particles leads to further densification by rearrangement of small and large particles. The initial system geometry may change either by large particles growing during the Ostwald ripening process or by shape accommodation.

Let there be a model system of N contours (closed boundaries, two-dimensional particles representation) of solid phase in the liquid and let

$$D_s = \left\{ \left(x_s(\ell), y_s(\ell), \Theta_s(\ell) \right), \ \ell = 1, 2, ..., n_s \right\},$$

where $\{x_s(\cdot)\}$ and $\{y_s(\cdot)\}$ represent coordinates, $\{\Theta_s(\cdot)\}$ their angle positions, and n_s boundary points number for s-th contour, denote the domains of boundary points for each contour. Then the initial concentration distribution could be defined by the concentration of a solid in contact with the liquid in all boundary points of solid phase and the equilibrium concentration of liquid phase in liquid phase region.

4. Process Simulation

Suppose that N solid particles (contours) exist in a liquid, the difference in chemical potential causes solid phase dissolution and which is then transported through a liquid phase and is precipitated at the place of alloy formation. When taking the above mentioned fact as the starting assumption and using the simulation method and by modification of algorithms [15] a general algorithm will be as follows:

• Computation of a new concentration $C_{i,j,k+1}$ from the concentration $C_{i,j,k}$ in all points of liquid phase.

• Computation of the flux through the finite element $(\Delta x \times \Delta y)$ of solid phase boundary point (x_i, y_j)

$$J_{i,j} = -D_L \left(\frac{C_{i,j} - C_{i-1,j}}{\Delta x} + \frac{C_{i,j} - C_{i+1,j}}{\Delta x} + \frac{C_{i,j} - C_{i,j-1}}{\Delta y} + \frac{C_{i,j} - C_{i,j+1}}{\Delta y} \right),$$

and the mass flow, dM/dt, in all boundary points of solid phase, $\forall (x_i, y_j) \in D_s$.

• Determination of a new geometry of the model system

$$\left\{ \left(x_s(\cdot), y_s(\cdot), \Theta_s(\cdot) \right) \; s = 1, 2, \ldots, N \right\}$$

including current results of dissolution $\left(dM/dt < 0 \right)$ and precipitation $\left(dM/dt > 0 \right)$ processes.

This procedure ends on reaching the required simulation time. Having established Δx, and Δy by rectangular experimental domain, and by mesh size $n \times m$, Δt can be determined from the stability relationship.

5. Results and Discussion

We will use a finite difference model for studying the sintering of a W-Ni system. The alloy parameters used in the model are: the composition of the precipitated alloys: 99.55 at.% W [16]; the composition of the liquid in contact with these alloys: 35 at.% W; the composition of liquid in contact with pure is W: 35.16 at.% W; the diffusion coefficient in the liquid, $D_L = 10^{-5} \; cm^2/s$ [17]. All calculations were performed on 100×100 mesh points.

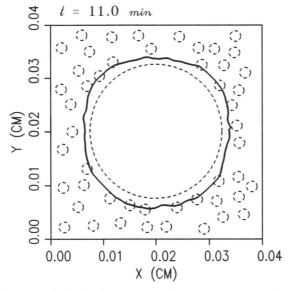

Fig. 1 The morphological evolution of model system with mixtures of the large and small W particles after 11 *min.* sintering time (solid line); dashed line denotes starting model geometry.

The starting model system geometry with mixtures of the large W particle of radius 125 μm and 49 small W particles of radius 10 μm located at randomly in the experimental region is shown in Fig. 1. It can be seen that the particles of different sizes in the liquid show a tendency for growth into their neighbors. Hence the smaller ones dissolve in liquid and the largest one grows. We used the assumption that at any given moment the particles smaller than the critical size $r^* = 9\bar{r}/8$ [7], where $\bar{r}$ is the arithmetic mean particle size, will dissolve surrounding themselves with a zone of excess solute which will migrate to the particles larger than r^*, and these ones therefore grow. Because of a high concentration gradient in this area, dissolution and reprecipitation processes are very fast. The concentration of liquid phase in contact with the small particles is greater than that in contact with the large particle. Therefore the small ones dissolve in the liquid matrix and dissolved atoms precipitate as W(Ni) solid solution on the large particle. After 11 *min.* simulation time all small particles completely dissolved into the liquid.

The next model system geometry with mixtures of two large W particles both of radius 125 μm and 48 small W particles of radii 10-20 μm all located at randomly in the experimental region is shown in Fig. 2. It can be seen that after 61 *min.* all small particles disappear and the two large ones grow. Shown in Fig. 3 is the time dependent morphological evolution of an enlarged part of model system.

The very good model system for simulation of Ostwald ripening is shown on Fig. 4. Four large particles of radii 125 μm show a tendency for growth into their neighbors, 46 small particles of radii 10-20 μm. If we assume that the growth kinetics of particles is diffusion controlled, then the growth rates [18] for the larger particles are functions of simulation time, according to an amount of dissolved solid phase into liquid or liquid phase concentration profile between larger particles. Shown in Fig. 5 are the computed average growth rates for all boundary points of large particles numbered 1, 2, 3 and 4. The growth results depend on number and distribution of the surrounding small particles which dissolve. Because of decreasing amount of solid phase due to dissolution, after 10 *min.* the growth rate decreases.

This phenomenon was previously described by Yoon and Huppmann [19]. On the basis of experiments with mixtures of large spherical and small W particles which were sintered in liquid Ni they concluded that the main material flux during liquid phase sintering is from small to large grains and that this grain growth leads to grain shape accommodation. Our simulation results obtained are very similar to the characteristic microstructural changes observed in their experiments.

There has been much work on the modeling and simulation of Ostwald ripening. So, for example, in paper [20] it was shown by simple calculations that shape change towards an improved shape accommodation may be caused by geometrical conditions in the neck areas of large growing particles. In paper [21] each particle in model system was defined by a set of surface points. The Ostwald ripening problem is a typical multibody free boundary problem in which the domains alter their morphologies in response to the diffusion field. In that sense,

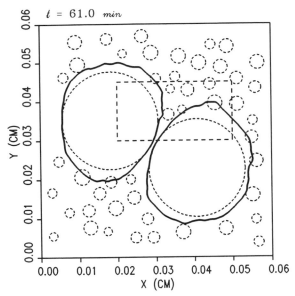

Fig. 2 The morphological evolution of model system with mixtures of two large and small W particles after 61 *min.* sintering time (solid line); dashed line denotes starting model geometry.

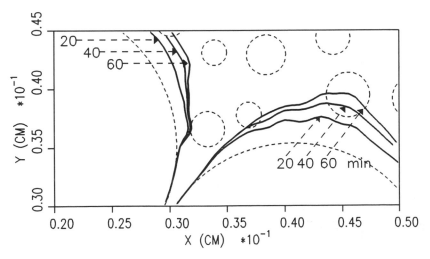

Fig. 3 The enlarged part of model system from Fig. 2 (dashed box) at various sintering times.

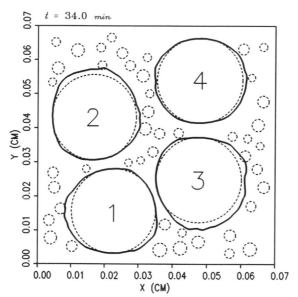

Fig. 4. The morphological evolution of model system with mixtures of four large W particles and 46 small W particles after 34 *min.* sintering time; dashed line denotes starting model geometry.

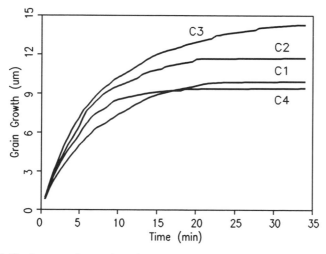

Fig. 5 Grain growth vs. sintering time for large particles (P1, P2, P3 and P4) of model system from Fig. 4.

our model systems study the morphological evolution of particles in liquid phase sintering conditions by solution of diffusion equation in the interparticle medium. Because of the large number of algebraic equations needed for solution in this method, our approach is limited to two dimensional models. Theoretical basis of this analysis is general and applicable to any binary system which satisfy three general requirements: there is a liquid phase at the sintering temperature, the solid phase is soluble in the liquid and the liquid wet the solid. The time dependent model system geometry is determined by the initial particle distribution, the size and shape distribution of the precipitates, their growth kinetics and the transport properties of the liquid phase. The accuracy and efficiency of the simulation method depend only on the accuracy of compositions of the solid phase and the diffusion coefficient in the liquid.

References

1. D.N. Yoon and W.J. Huppmann, *Practical Metallography* **15** 399 (1978).
2. D.N. Yoon and W.J. Huppmann, *Acta Metall.* **27** 973 (1979).
3. R.B. Heady and J.W Cahn, *Metall. Trans.* **1** 185 (1970).
4. J. W. Cahn and R.B. Heady, *J. Am. Ceram. Soc.* **53** [7] 406 (1970).
5. R. Raj and C.K. Chyung, *Acta Metall.* **29** 159 (1980).
6. G. M. Pharr and M.F. Ashby, *Acta Metall.* **31** [1] 129 (1983).
7. H. Fischmeister and G. Grimvall, in: *Sintering and Related Phenomena*, G. C. Kuczynski (ed.), Plenum Press, New York, 119 (1973).
8. N.C. Kothari, *J. Less-Common Metals* **13** 457 (1967).
9. T.K. Kang and D.N. Yoon, *Metall. Trans.* **9A** 433 (1978).
10. R. Warren and M.B. Waldron, *Powder Met.* **15** 180 (1972).
11. W.J. Huppmann and G. Petzow, *Ber. Bunsenges. Phys. Chem.* **82** 308 (1978).
12. Z.S. Nikolic and M.M. Ristic, *Science of Sintering* **13** 91 (1981).
13. G.E. Forsythe and W.R. Wasow, "Finite Difference Methods for Partial Differential Equations", John Wiley, New York (1960).
14. Z.S. Nikolic and R.M. Spriggs, *Science of Sintering* **26** (1) 1 (1994).
15. Z.S. Nikolic, M.M. Ristic and W.J. Huppmann, in: *Modern Developments in Powder Metallurgy*, H.H. Hausner, H.W. Antes and G.D. Smith (eds.), **12** 497 (1981).
16. W.J. Muster, D.N. Yoon and W.J. Huppmann, *J. Less-Common Metals* **65** 211 (1979).
17. Y. Ono and T-Shigematsu, *J. Japan Inst. Met.* **41** 62 (1977).
18. Z.S. Nikolic, R.M. Spriggs and M.M. Ristic, *Science of Sintering* **24** 1 49 (1992).
19. D.N. Yoon and W.J. Huppmann, *Acta Metall.* **27** 693 (1979).
20. W.A. Kaysser, M. Zivkovic and G. Petzow, *J. Mat. Sci.* **20** 578 (1985).
21. J.M. Chaix, M. Guyan, J. Rodriguez and C.H. Allibert, *Scripta Metall.* **22** 71 (1988).

Plasma Activated Sintering (PAS) of $\beta-$ Fe(Mn)Si(Al)$_2$ Prepared by Mechanical Alloying (MA)

Shinya Shiga [1], Keiich Masuyama [2], Minoru Umemoto [2], and Kazuo Yamazaki [3]

1) Dept. of Materials Engineering, Niihama National College of Technology, 7−1 Yagumo−cho, Niihama, Ehime 792, Japan.

2) Dept. of Production systems Engineering, Faculty of Engineering, Toyohashi University of Technology, Toyohashi, Aichi 441, Japan.

3) Dept. of Mechanical, Aeronautical, and Materials Engineering, University of California at Davis, Davis, CA 95616−5294.

Abstract

Consolidation of Fe$_{28}$Mn$_2$Si$_{67}$Al$_3$ powder mixture by plasma activated sintering (PAS) was carried out. The powders used for PAS were prepared by mechanical alloying (MA) starting from pure Fe, Mn, Si and Al powders. $\beta-$ FeSi$_2$, which is one of the promising thermoelectric generator materials, was obtained by PAS the MA powder at 1123K for 300s under 58.8MPa. The production process of the bulk specimen of $\beta-$ FeSi$_2$ phase becomes simpler by using PAS than that by hot− pressing (HP) which requires long time sintering and homogenization. The seebeck potential of the PAS specimen followed by β annealing is almost equal to that of the HP specimen. However, the specific electrical resistance of the PAS specimen is about 10 times larger than that of the HP specimen. As a result the effective maximum power of the PAS specimen is smaller than that of the HP specimen. The high specific electrical resistance of the PAS specimen is caused by the low density. Relative densities of the PAS and HP specimens were 83.4% and 95.2%, respectively.

1. Introduction

Plasma activated sintering (PAS) is given attention because the PAS technique perhaps makes the sintering temperature lower or the sintering time shorter. The PAS process generates plasma among powder particles to be sintered and utilizes the activation of particle surfaces caused by the plasma. The plasma is generated by instantaneous electric pulsed—power application control with high current and adequate voltage settings. We tried to produce thermoelectric materials by using PAS in order to simplify the production process and improve the thermoelectric properties. The increase in Seebeck coefficient and the reduction of specific electrical resistance and thermal conductivity improve the thermoelectric properties. Because of its large thermoelectric power and phase stability at high temperature[1,2], β—FeSi$_2$ is considered to be one of the promising thermoelectric generators. Fig.1 shows equilibrium phase diagram in Fe—Si system[3]. The eutectic (L → α+ϵ) and peritectoid (α+ϵ → β) reactions from liquid to β phase make it difficult to form a bulk material with homogeneous β phase. We have reported earlier that the simplification of production process and the improvement of thermoelectric properties of β—FeSi$_2$ can be obtained by using mechanical alloying (MA) followed by hot—pressing (HP), because of the fine grain size which reduces the lattice thermal conductivity[4]. MA is a solid state alloying process[5], by which alloying proceeds with flattening, laminating and homogenization. Nonequilibrium phases such as supersaturated solid solution, metastable intermetallic compound, amorphous phase can be synthesized, as well as

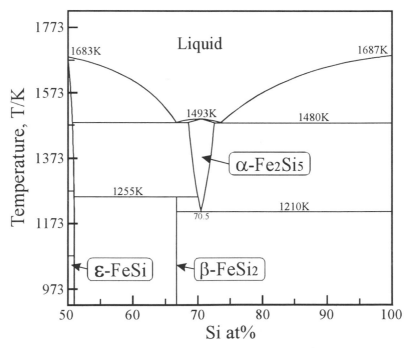

Fig.1   Equilibrium phase diagram in Fe—Si system.

fine crystal size, by MA. The purpose of this paper is to make clear the effect of the PAS on production process and thermoelectric properties of β-FeSi$_2$. PAS of β-FeSi$_2$ powder, which was prepared by MA starting from pure Fe and Si doped with Mn and Al as acceptors, was carried out. The thermoelectric properties of PAS specimen was compared with those of HP specimen.

Table 1 Conditions of plasma activated sintering (PAS).

PAS on time	99sec	Sintering temp.	1023~1223K
PAS puls time	30msec	Sintering time	300sec
PAS voltage	40V	Sintering pressure	58.8MPa
PAS current	750A	Vacuum degree	9.3Pa

Puls time
30msec

PAS on time
99sec

2. Experimental Procedure

Elemental powders of Fe(over 99.9% purity, <150μm), Si(over 99.9% purity, <10μm), Mn(over 99.9% purity, <75μm), and Al(over 99.9% purity, <180μm), were subjected to mechanical alloying (MA) by horizontal ball mills for 720ks under an atmosphere of Ar gas. Mixtures of these powders were prepared in atomic proportions of Fe$_{28}$Mn$_2$Si$_{67}$Al$_3$. The vials were 1700ml in volume and balls were 9.6mm in diameter. The mill was rotated at 95rpm. The powder charge was 36g, with a ball to powder weight ratio of 100:1. Milling was performed with the addition of 2wt% methyl alcohol. After MA processing powders were consolidated by hot-pressing (HP) or plasma activated sintering (PAS). Part of the MA powder was HP sintered at 1373K for 1.8ks under 28MPa into a disc with 60mm diameter which is followed by annealing at 1033K for 360ks to transform from (α-Fe$_2$Si$_5$ + ε-FeSi) to β-FeSi$_2$ phase. The other part of MA powder was PAS sintered at 1023~1223K for 300s under 58.8MPa. During the PAS process, the chamber was evacuated to about 9Pa. Details of the PAS conditions are listed in Table 1. Phases in the consolidated products were characterized by means of X-ray diffraction (XRD). Hardness of the consolidated products was measured by Vicker's hardness tester, and the relative density was estimated from the weight and volume. Seebeck potential (E_0) and specific electrical resistance (ρ) were measured to get the effective maximum power ($P_{eff}=E_0^{2}/(4\rho)$) using specimens with about $5 \times 5 \times 20\text{--}40$mm cut from HP and PAS samples by applying temperature difference (ΔT) between specimen ends (Fig.2).

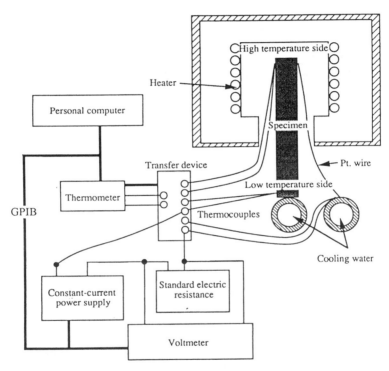

Fig.2   Schematic illustration of the equipment of
thermoelectric properties measurement.

3. Results and discussion

3.1 Phase, hardness and relative density

MA powders with starting composition of $Fe_{28}Mn_2Si_{67}Al_3$ were PAS sintered at 1023~1223K, which is below the peritectoid point, for 300s under 58.8MPa. Fig.3 shows the XRD patterns taken from the PAS sintered specimens prepared from 720ks milled powder. The XRD pattern of the PAS specimen sintered at 1223K is identified to be $\alpha-Fe_2Si_5$ phase. According to the equilibrium phase diagram (Fig.1), $\beta-FeSi_2$ is expected to form at this temperature and composition. The sintering temperature

(1223K) may be less than actual temperature of the sample, because the sintering temperature was measured by thermocouple located inside of carbon mold. The XRD patterns of the PAS specimens sintered at 1123 and 1023K are identified to be almost β phase with a small fraction of $\alpha-Fe_2Si_5$. If we use the conventional casting and powder metallurgy technique to get bulk specimen with β phase, higher sintering temperature and longer heat treatment time is necessary. This result indicates that production process of the bulk specimen with almost β phase can be simplified by using MA and PAS technique. Phase, hardness and relative density observed in PAS sintered specimens are listed in Table 2. To compare with these values, the MA powder was sintered by hot–pressing (HP) as well as PAS. Phase, hardness and relative density of the HP specimen are listed in Table 3. Here, the MA powder was HP sintered at 1373K for 1.8ks under 28MPa, which is followed by β annealing at 1033K for 360ks. Both the hardness and relative density of the PAS specimens are smaller than those of the HP specimen. Although, high density can make the specific electrical resistance lower, the relative density of the PAS specimen is more than 10% lower than that of HP specimen.

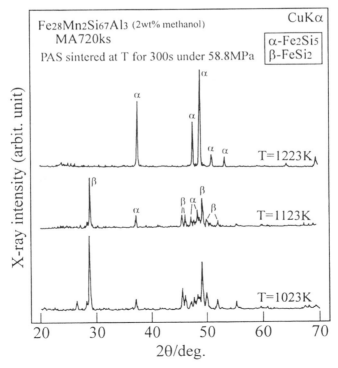

Fig.3 X–ray diffraction patterns of the PAS specimens sintered at T.

Table 2   Phase, hardness and relative density of PAS sintered specimen. The powder sintered is $Fe_{28}Mn_2Si_{67}Al_3$ milled for 720ks.

Sintering temperature (K)	Phase	Hardness (Hv)	Relative density (%)
1223	α	687 584~780	-
1123	β + (α)	566 413~743	83.4
1023	β + (α)	432 351~508	77.1

α : 4.992 g/cm³, β : 4.929 g/cm³

Table 3   Phase, hardness and relative density of HP sintered specimen. The powder sintered is $Fe_{28}Mn_2Si_{67}Al_3$ milled for 720ks.

Sintering temperature (K)	Phase	Hardness (Hv)	Relative density (%)
1373	β	1129	95.2

3.2   Thermoelectric properties

Thermoelectric performance, which depends on the seebeck potential (E_0), specific electrical resistance (ρ) and thermal conductivity (K), is often evaluated by effective maximum power ($P_{eff}=E_0^2/(4\rho)$).

E_0, ρ and P_{eff} of the PAS and HP sintered specimens are shown in Fig.4 (a), (b) and (c), respectively. In those figures, the values indicated "1st run" and "2nd run" are the data corresponding respectively to the first and second measurement after PAS process. The values indicated "β annealed" are the data of PAS specimen which is followed by β annealing at 1033K for 360ks. E_0 and ρ of 2nd run are larger than those of 1st run. After β annealing, E_0 and ρ become larger. These results show that during 1st measurement and β annealing, transformation into $\beta-FeSi_2$ of a small amount of metallic $\alpha-Fe_2Si_5$ which exists with $\beta-FeSi_2$ in PAS specimen may be taking place. $\alpha-Fe_2Si_5$ is not a semiconductor but a metallic phase.

E_0 of the β annealed PAS specimen is almost equal to that of HP

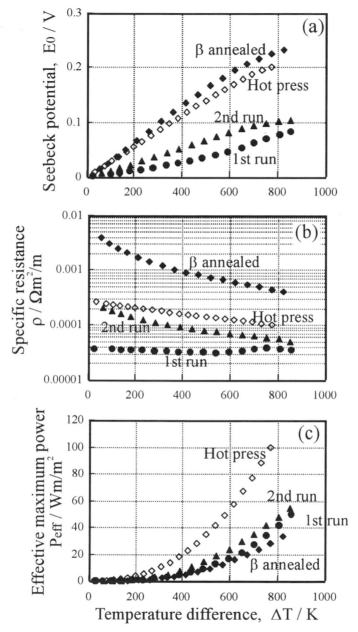

Fig. 4   Seebeck potential (a) specific resistance, (b) and effective maximum power of (c) the PAS and HP sintered specimens.

specimen. Since ρ of the β annealed PAS specimen, which is about 10 times larger than that of HP specimen, the observed P_{eff} of PAS specimen is less than 50% of that of HP specimen. It is considered that this large ρ of PAS specimen is caused by the low density. To get improved thermoelectric properties, the PAS specimen must be densified more.

4. Conclusions

Consolidation of mechanically alloyed $Fe_{28}Mn_2Si_{67}Al_3$ powder by plasma activated sintering (PAS) was carried out to get $\beta-FeSi_2$ thermoelectric generator. Phase, hardness and relative density and thermoelectric properties of the PAS specimens were measured. The main results obtained are as follows :

(1) The production process of bulk specimen with almost $\beta-FeSi_2$ phase becomes simpler by using PAS than that by hot−pressing (HP) which requires long time sintering and homogenization.

(2) Small amount of metallic phase ($\alpha-Fe_2Si_5$) existing with $\beta-FeSi_2$ in PAS specimen transforms into $\beta-FeSi_2$ during the measurement of the thermoelectric properties when the specimen gets heated to about 1000K.

(3) The effective maximum power of the PAS specimen followed by β annealing is smaller than that of the HP specimen. This is because specific electrical resistance of the PAS specimen is about 10 times larger than that of the HP specimen. This high specific electrical resistance of the PAS specimen can be attributed to its lower density.

Acknowledgments

The authors would like to thank the IMRA MATERIAL R&D CO., LTD for providing materials used in this study.

References

[1] R.M.Ware and D.J.McNeill, "Iron disilicide as a thermoelectric generator Material", *Proc. IEE*, III , 178− 182 (1964).

[2] U.Birkholz and J.Schelm, "Mechanism of Electrical Conduction in $\beta-FeSi_2$", *Phys. Stat. Sol.* **27** 413− 425 (1968).

[3] O.Kubaschewski, "Iron− Binary Phase Diagram", Spninger− Verlag, New York, 136 (1982).

[4] S.Shiga, K.Fujimoto and M.Umemoto, "Thermoelectric Properties of $FeSi_2$ Compound Prepared by Mechanical Alloying", *Proc. of the 12th Int. Conf. on Thermoelectrics*, 311− 315 (1993).

[5] J.S.Benjamin, "Dispersion Strengthened Superalloys by Mechanical Alloying ", *Met. Trans.* **1** 2943− 2951 (1970).

MICROWAVE SINTERING OF SEMICONDUCTING PEROVSKITE AND THEIR ELECTRICAL PROPERTIES

I-Nan Lin*, Horng-Yi Chang
Materials Science Center, National Tsing-Hua University,
Hsinchu, Taiwan, R.O.C., 30043

Abstract

Densification behavior of semiconducting perovskite materials processed by microwave sintering technique is investigated. The temperature required to densify the materials in this process is generally around 200°C lower than that used in the conventional furnace sintering process. The soaking time deemed necessary is also substantially shorter. The activation energy for diffusion of the species is estimated to be one order of magnitude smaller in the microwave-sintering process. Moreover, the electrical properties of the materials were substantially improved by the post-processes after microwave sintering, which is not possible in the conventional process. The defect chemistry involved in this process is also discussed.

I. Introduction

Since the positive temperature coefficient of resistivity (PTCR) of semiconducting $BaTiO_3$ materials above the ferroelectric Curie temperature (T_c) was first observed in 1955 [1,2], more thoroughly understanding the conduction mechanisms [3,4] and developing various applications of these materials have received much attention [5,6]. Conventionally, these materials are prepared by mixed oxide method and sintered by a resistance-furnace. A high sintering temperature (1260°C~1350°C) and long soaking period (0.5h~1h) are required to densify the materials and produce a suitable microstructure [7,8]. Close control of the sintering atmosphere is deemed necessary to prevent the loss of Pb-species in Pb-containing materials, even when the liquid phase sintering aids are applied [7,8].

On the other hand, the microwave sintering process has been observed to densify the ceramic materials in a very rapid rate and at a substantially lower temperature [9,10]. Moreover, this process is capable of rapidly cooling after sintering, thereby making possible the study of the defect chemistry having occurred at grain boundary. This technique is therefore adopted in this study in an attempt to rapidly sinter the donor-doped PTCR materials. Moreover, the effects of processing parameters on the PTCR properties of these materials, i.e., the cooling rate and the post-heating schemes, are systematically examined. The densification behaviors of these materials undergoing microwave heating process are also presented.

II. Experimental

PTCR powders containing semiconductive dopants and sintering aids were prepared by the conventional mixed oxide method. The nominal compositions of the samples were $(Sr_{0.2}Ba_{0.8})TiO_3$ or $(Sr_{0.4}Pb_{0.6})TiO_3$ and were abbreviated as SBT or SPT, respectively. The SBT materials contained 0.5mol% Sb_2O_3 as dopants and 5mol% AST (i.e., Al_2O_3:SiO_2:TiO_2=5:9:3) as sintering aids, while SPT materials contained 1mol% Y_2O_3 as dopants and 8mol% ST (SiO_2:TiO_2=5:3) as sintering aids. The mixtures consisting of high purity oxides of proper proportion were ball-milled in a plastic jar, using plastic-coated iron balls, with deionized water, for 8 h. They were then calcined at $1000°C$ in air for 4 h, followed by pulverization in a ball-mill for 8 h to around 1 μm size.

The green compacts made of the SBT or SPT powders were microwave sintered at $1130°C$ or $1100°C$, respectively, for 10 min in an applicator made of WR284 waveguide. The 2.45 GHz microwave generated from the commercial source (Gerling GL107 magnetron) was used to heat up the samples. The temperature profile was measured using a Pt-13%Rh thermocouple, and then placed near the sample surface. To facilitate the comparison, the green compacts of the same materials were also densified by conventional sintering process. In this process, the samples were sintered in a resistive heating furnace either at $1300°C$ (2h) or at $1250°C$ (0.5h) for SBT materials or SPT materials, respectively. The heating and cooling rates were controlled at 5°C/min in both the microwave and conventional sintering processes.

The crystal structure and microstructure of the sintered samples were examined using Rigaku D/max-IIB X-ray diffractometer and Jeol JSM-840A scanning electron microscope (SEM), respectively. The resistivity-temperature (ρ-T) properties of these samples were measured using a HP3457A multimeter after the In-Ga alloy was rubbed onto the sample surface to function as electrodes. The frequency dependence of the resistance (R) and reactance (ImZ) of the samples were measured using a HP4194A impedance analyzer.

III. Results and Discussion

(a) Material characteristics

The explanations on the densification behavior of the ceramics processed by microwave sintering technique are contradictory. Measurement uncertainty of the sintering temperature is one of the difficulties in accounting for this behavior. In this work, a preliminary experiment was performed to resolve the ambiguity in temperature measurement. For this purpose, TiO_2 pellets of anatase structure were microwave sintered at 920°C or 940°C for 30 min. Figure 1 shows that the samples sintered at 940°C have been fully transformed into rutile phase, while those sintered at 920°C still contain a large amount of anatase phase. The samples placed at the top of the stack contain a larger proportion of anatase phase than those placed at the bottom (Fig.1b). This finding reveals that the former have experienced a lower sintering temperature than the latter and is probably caused by a larger heat dissipation rate to the ambient of the top pellets. These results indicate that the actual sintering temperature experienced by the samples is very close to the anatase-to-rutile transformation temperature of TiO_2 materials, i.e., 920°C, which is not far from that measured by the thermocouple. Therefore, the accuracy of temperature measurement is estimated to be around ±10°C.

The success of applying microwave sintering technique to densify the ceramic materials relies on the sample's capability to self-generate a sufficient

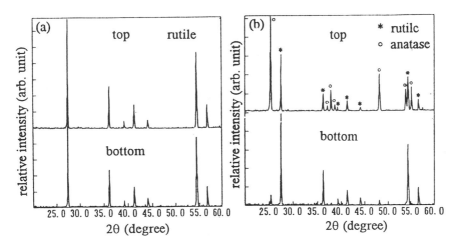

Fig.1 X-ray diffraction patterns of TiO_2 samples placed at top and bottom of the stacks, microwave sintered at (a) 940°C and (b) 920°C.

amount of heat by effectively absorbing the microwave power. This behavior was examined by measuring the time evolution of the temperature (T) as soon as the microwave was applied. Figure 2a shows the T-t curves of the empty susceptors and that loaded with SBT samples, heated by microwave of the same power profile. These curves indicate that the samples were indirectly heated by the SiC susceptors below 800°C and started to absorb the microwave power and, subsequently, self-generate the heat. Consequently, the temperature of the samples markedly deviated from that of the susceptors above this temperature. This behavior suggests that the dielectric loss factor of SBT materials is significantly larger than that of the SiC materials for a temperature higher than 800°C such that the SBT materials completely take over the role of absorbing microwave power. This temperature is defined as ignition temperature (T_i) for microwave sintering process.

Figure 2a shows that the microwave power should be rapidly reduced as soon as the samples started to self-generate the heat. Sintering temperature can be increased by maintaining the steady state microwave power at a higher level. This temperature, however, is limited by a thermal runway phenomenon, which occurs when the dielectric loss factor increases abruptly. It is defined as

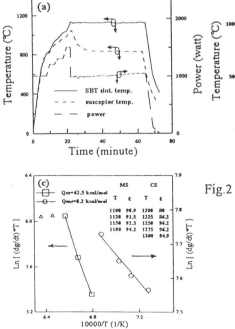

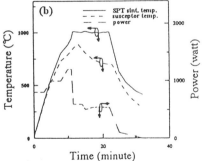

Fig.2 Temperature-time profiles of either empty or sample loaded susceptors: (a) SBT materials, (b) SPT materials, with the corresponding profile of microwave power and (c) Arrhenius plots of densification rate of SBT materials.

the thermal runaway temperature (T_r). A material is expected to behave satisfactorily in the microwave sintering process whenever the ignition temperature (T_i) is lower than that of the thermal runaway temperature (T_r). Similar analysis indicates that SPT materials also are well behaved materials, as shown in Fig.2b. In contrast, materials such as $Ba(Mg_{1/3}Nb_{2/3})O_3$ are not microwave sinterable because of the phenomenon that their dielectric loss factor is always low compared to that of the susceptors up to thermal runaway temperature.

Microwave sintering process still requires careful control, even for a well behaved materials such as SBT. Too low of a sintering temperature ($<$ 1100°C) results in insufficient densification, while too high of a sintering temperature (> 1200 °C) induces thermal runaway. Densification rate is, however, significantly enhanced in the microwave sintering process. The activation energy (Q) can be derived from the temperature dependence of densification rate, following the model;

$$dg/dt=KD_o\exp(-q/kT)\text{--------------}(1)$$

where K is a geometric factor, D_o is the pre-exponential term of diffusivity, k and T are the Boltzmann constant and temperature, respectively, and dg/dt is the densification rate of the samples. Figure 2c indicates that $(Q)_{ms}$ is 8.2 kcal/mole. This value is substantially lower than the activation energy calculated from densification rate of the same materials sintered by the conventional sintering process $((Q=62.5)_{cs}$ kcal/mole).

(b) Electrical properties

The electrical properties of the PTCR materials are normally characterized by the temperature dependence of resistivity. Figure 3a reveals that the SBT materials sintered at a low temperature (1130°C) and short interval (40 min) by microwave sintering (ms) process possess the same marvelous ρ-T properties as those sintered at a high temperature (1300°C) and a long soaking time (2h) by conventional sintering (cs) process, although the grain size of the former is substantially smaller than that of the latter. They are around 5μm and 15μm, respectively, for ms- and cs-SBT samples. On the other hand, the 1000°C(10 min) microwave sintered SPT materials possess significantly different ρ-T properties from those of the 1220°C(0.5h) conventionally sintered samples (Fig.3b), indicating that rigorous reaction has occurred in the grain boundary region of these samples during microwave sintering process. The grain size is around 5μm and 10μm, respectively, for ms- and cs-SPT samples.

To investigate the nature of the reaction, the materials were rapidly cooled after microwave sintering and then post-annealed. The cooling rate and post-annealing conditions are listed in the corresponding ρ-T curves. Figure 3a reveals that the maximum resistivity (ρ_{max}) of the SBT sample is markedly lowered due to rapid cooling (~154°C/min) process (ms1) and is resumed by the post-annealing in air, 1250°C-2h (ms2). The ρ_{min}-value is maintained at the same level and the PTCR property, i.e. ρ_{max}/ρ_{min} ratio, has been increased to 10^7. These results suggest that the intrinsic electrical property, i.e., the bulk resistivity of the materials is not significantly changed, while the extrinsic one, i.e., the grain boundary resistivity, is markedly altered by the post-sintering processes. Restated, defect chemistry only occurred along grain boundary region that substantially modifies the PTCR property of the SBT materials.

    In contrast, rapid cooling (~126°C/min) after microwave sintering substantially lower the minimum resistivity, but insignificantly changes the

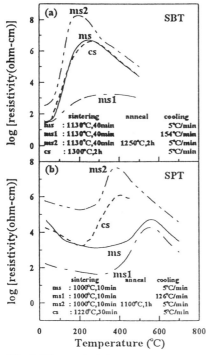

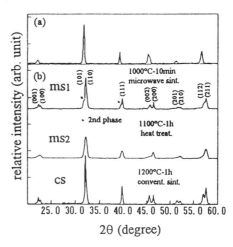

Fig.4 X-ray diffraction patterns of microwave sintered, annealed and conventionally sintered (a) SBT and (b) SPT materials.

Fig.3 The resistivity-temperature (ρ-T) properties of microwave sintered (a) SBT and (b) SPT materials, with sintering temperature and post-sintering conditions indicated (dotted curves are the ρ-T properties of conventionally sintered materials).

maximum resistivity of the SPT samples (ms1, Fig.3b). The resistivity jump (ρ_{max}/ρ_{min}) of the samples having rapidly cooled after microwave sintering is the same as that of the samples conventionally sintered. Interestingly, the ms1-samples posses a markedly higher critical temperature (T_c) of resistivity jump than the cs-sample, viz. $(T_c)_{ms1}\sim380°C$ and $(T_c)_{cs}\sim210°C$. Post-annealing at $1100°C(30min)$ resumes the low-T_c value of SPT materials (ms2, Fig.3b). The over-all resistivity of the samples, however, has been pronouncedly increased.

The possible reaction that occurs during processing was further investigated by examining the phase constituents of the samples. Both the microwave and conventionally sintered SBT samples are of pure perovskite structure (Fig.4a). In contrast, rapid cooling and post-annealing processes imposed a marked effect on the phase constituent of Pb-containing samples. Extra peaks corresponding to perovskite structure of a slightly larger lattice constant were observed in XRD pattern of quenched (ms1) SPT samples. They were eliminated after post-annealing. A single phase perovskite structure similar to that of the conventionally sintered samples was resumed, as indicated by $(XRD)_{ms2}$ and $(XRD)_{cs}$ patterns in Fig.4b. Restated, the samples are of "core-shell" microstructure, in which a core deficit in Pb is surrounded by a shell enriched in Pb. Both phases are of perovskite structure. Either slow cooling or post-annealing process induces the interdiffusion between the "core" and the "shell", that eventually results in a sample with a uniform composition. Evolution of microstructure during slow cooling or post-annealing process is schematically shown in Figs. 5a and 5b, accompanied with the expected ρ-T behavior.

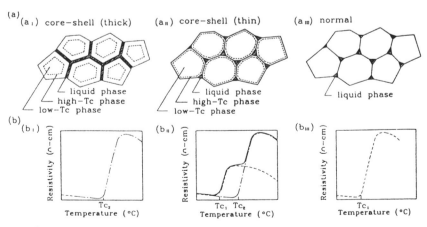

Fig.5 Schematic diagram showing (a) the core—shell microstructure and (b) the corresponding ρ—T properties.

How the core-shell structure forms must be fully addressed. Recall that only the Pb-containing materials possess a core-shell structure. Meanwhile, it is known that Pb-species interacts with SiO_2 more rigorously than the Sr- or Ba- species. The $PbTiO_3$-SiO_2 eutectic point is $1000°C$, while the $BaTiO_3$-SiO_2 and $SrTiO_3$-SiO_2 eutectic points are, $1320°C$ and $1260°C$, respectively. We can, therefore, assume that, at sintering temperature, the liquid phase preferentially reacts with the Pb-species. The dissolved Pb-species segregates out during cooling, reacts with the matrix and forms a Pb-rich phase enveloping the grains. A core-shell microstructure is subsequently produced.

IV. Conclusion

Microwave sintering behavior and the corresponding electrical properties of semiconducting SBT and SPT perovskites were investigated. The temperature required is significantly lower and soaking time required is markedly shorter in the microwave sintering process than that in the conventional sintering process. Post-sintering processes such as rapid cooling and annealing substantially modify the resistivity- temperature (ρ -T) properties of the samples. For SBT materials, these processes only modify the defect chemistry along grain boundary region and insignificantly change the phase constituent of the samples. The resistivity ratio, i.e., ρ_{max}/ρ_{min}, has been increased to 10^7. On the other hand, single high-T_c characteristics were obtained for rapidly cooled SPT samples. Post-annealing resumed the T_c value to a temperature similar to that commonly obtained by the conventional sintering process. The phenomenon is accounted for by the formation of a Pb-rich shell due to preferential reaction of Pb-species contained in SPT materials with the sintering aid, SiO_2.

V. References

[1] O. Saburi, J. Phys. Soc. Japan, **14[9]**, 1159(1959).

[2] O. Saburi, J. Am. Ceram. Soc., **44[2]**, 55(1961).

[3] W. Heywang, Solid-State Electron., **3**, 51(1961).

[4] G.H. Jonker, Solid-State Electron., **7**, 895(1964).

[5] E. Andrich, Electr. Appl., **26[3]**, 123(1965/66).

[6] K. Wakino, N. Fujikawa, Electron. Ceram., **2[5]**, 73(1971).

[7] C.K. Lee, I.N. Lin and C.T. Hu, J. Am. Ceram. Soc., **77[5]**, 1340(1994).

[8] H.F. Cheng, T.F. Lin. C.T. Hu and I.N. Lin, J. Am. Ceram Soc., **76[4]** 827(1993)

[9] W.H. Sutton, Button, Am. Ceram. Soc.,**68[2]**,376(1989).

[10] F. Selmi, F. Guerin, X.D. Yu, V.K. Varadan, V.V. Varadan and S. Komarneni, Mater. Lett., **12**, 424(1992).

The Use of an Electric Field as a Processing Parameter in the Synthesis and Densification of Powders

Z. A. Munir

Department of Chemical Engineering and Materials Science
University of California • Davis, CA 95616

Abstract
The imposition of an electric field is shown to activate the synthesis of powders by combustion methods. This makes it possible to synthesize a variety of materials which otherwise cannot be prepared by the self-sustaining combustion method. The use of this method to synthesize ceramics, composites, and solid solutions is described. A modification of this method is used to simultaneously synthesize and densify $MoSi_2$. Products of relative densities up to 99% were synthesized.

1. Introduction

The use of electric fields to influence transport and interfacial phenomena in solids has been well-documented [1]. The novel use of fields in the synthesis of powders by combustion methods is, however, recent [2] and is the focus of this paper. The goal in this case is to demonstrate the use of an electric field as a *processing parameter* in the synthesis of particulates (or dense bodies) of ceramics, intermetallics, and composites. The application of a field *activates* the combustion process and when used with pressure, can result in the simultaneous synthesis and densification of powders.

2. Powder Preparation through Field-Activation
2.1 Synthesis of Ceramics

The use of self-propagating combustion for the synthesis of powders is widely employed [3,4]. This method, however, is limited to those materials whose formation enthalpies are large. This limitation excludes the possibility of synthesis of a large number of important materials such as, SiC, B_4C, most intermetallics, and many composite systems. The limitation has been empirically related to the adiabatic combustion temperature, T_a, such that systems with $T_a \leq 2000K$ are not capable of sustaining a combustion wave and therefore cannot be synthesized by this method. Such systems can be thermally activated by preheating the reactants to raise T_a. However, such an approach is normally unattractive because of the formation of pre-combustion phases due to interdiffusion [5].

In the new approach, ignition of the reactants is effected while a field is present across the sample compact. At fields greater than a threshold value, a self-propagating combustion wave is initiated. The threshold field value is

447

materials-dependent, being influenced by the nature of the reactants and by their particulate characteristics (e.g. relative density, particle size, etc).

Field-activated combustion synthesis (FACS) has been utilized in the preparation of ceramics such as silicon carbide [6], tungsten silicide [7], tantalum carbide [8], and others [2]. In the case of silicon carbide, the product is β-SiC. It should be emphasized that in the absence of an electric field, silicon carbide cannot be synthesized by the self-propagating combustion method. Likewise, the formation of silicides of tungsten is not possible without field-activation. For this system, investigations were made for powder mixtures of W and Si ranging in composition from 6 to 30 wt % Si and encompassing the two silicides of tungsten, W_5Si_3 and WSi_2. A threshold phenomenon was also observed in this case, with the threshold field value being dependent on the sample resistivity [7]. In turn, the resistivity depended on composition (%Si) and relative density. The latter changed roughly linearly from 56 to 65% as the silicon content increased from 6 to 30 wt %. The explicit dependence of the threshold field on wt % Si is shown in Figure 1 (curve a). Curve "b" in Figure 1 is the boundary above which the product is molten. Thus below curve "a" no reaction takes place, between curves "a" and "b" the product is an agglomerate solid, and above curve b the product is a solidified melt. The nature of the product, and thus the mechanism of silicide formation was found to depend on the magnitude of the applied field, i.e. whether the field is below curve b (low field) or above curve b (high field), as shown in Figures 2(a) and (b). When low fields are applied (13-18 V/cm), the phase W_5Si_3 is only present in trace amounts in the product at (or near) the composition corresponding to its stoichiometry, 8.4 wt % Si. In contrast, when high fields are applied ($\sim$ 18-40 V/cm), the product is 100% W_5Si_3 at the corresponding initial composition. The importance of this observation is that the field influences the mechanism of silicide formation and that the product (for the case of W_5Si_3) is consistent with the equilibrium diagram *only* when the synthesis is carried out in the presence of relatively high fields.

2.2 Synthesis of Composites

The benefit of field-activation in the synthesis of composite powders has been demonstrated [9-11]. The advantage of combustion synthesis is clearly demonstrated by the feasibility of preparing composites directly from the elements, or simple reactants. However, for many systems, the addition of reactants to form the second phase, limits the use of combustion synthesis. An example of this is the synthesis of composites containing $MoSi_2$ as the primary phase. The addition of a secondary phase, e.g. SiC is made to enhance its mechanical properties [12]. Although $MoSi_2$ can be combustion-synthesized from the elements, composites of $MoSi_2$-SiC cannot be prepared by this method without field activation. For composites $(1-x)MoSi_2$-xSiC with $0 < x \leq 0.18$, a self-propagating wave can be established without the application of a field but the product is composed of $MoSi_2$ and unreacted Si and C. However, under the influence of a field, the product is $MoSi_2$ and SiC for this x value and for all other compositions up to x = 1.0 (i.e. pure SiC). As in the synthesis of single-phase materials, the synthesis of composites by field-activation requires threshold values to sustain a combustion wave. The threshold field value depended on composition, i.e. x [9].

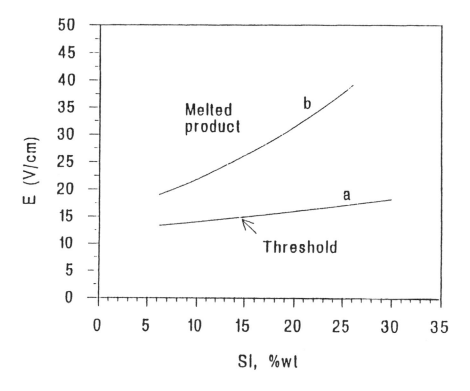

Figure 1. Dependence of the threshold field (curve "a") and the state of the product (curve "b") on silicon content.

Another example is the B_4C - TiB_2 composite. The synthesis of xB_4C - TiB_2 composites directly from the elements can be made without a field if x ≤ 0.5 [13]. However, composites with higher values of x, desired for armor application, can only be directly synthesized with field-activation [11]. It was shown that composites with x values up to 8 can be synthesized directly from the

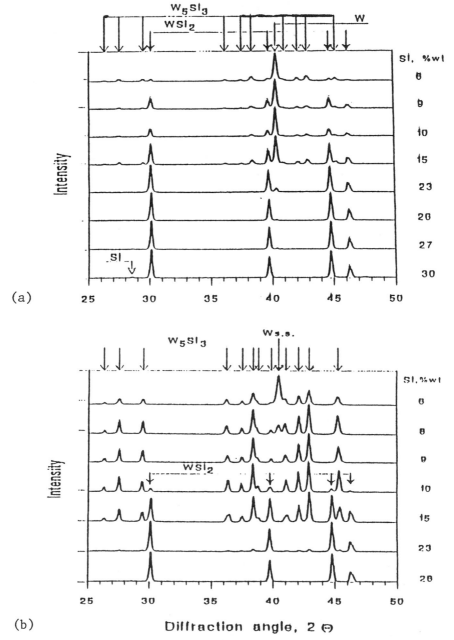

Figure 2.   X-ray diffraction patterns of products in the W-Si system: (a) low field, (b) high field.

elements. It should be noted that in the absence of a field the synthesis of composites requires pre-heating to high temperatures if compositions with x > 0.5 are desired. For example, for x = 1.0 the reactants must be heated to 800K and for x = 2.0 they must be heated to 1200K. The advantage of the field is seen by the energy savings of avoiding pre-heating, and by eliminating undesirable pre-combustion phase formation [5].

2.3 Synthesis of Solid Solutions

An example of the use of field-activation in the combustion synthesis of solid-solution is the system AlN-SiC. Because of attractive high temperature and mechanical properties this solid solution has been the focus of several recent investigations [14-17]. The general approach to the synthesis of this solid solution is to heat mixtures of AlN and β-SiC at relatively high temperatures (~2000°C) for extended periods (~15 hrs) [16]. Through the use of field-activation, a more energy-efficient method to synthesize this material has been recently demonstrated [18]. Synthesis is based on the following powder reaction

$$Si_3N_4 + 4Al + 3C \rightarrow 3SiC + 4AlN \qquad (1)$$

In the absence of a field, no self-sustaining combustion reaction among the reactants powders can be initiated with ignition despite the fact that the adiabatic combustion temperature for Eq (1) is high, 2504K. However, when a field of 8.3 V/cm is applied simultaneously, ignition resulted in a self-propagating wave. X-ray analysis of the product, Figure 3(a), indicated the presence of AlN and $\propto$-SiC (2H structures), Si, and unreacted carbon. Because of the close proximity of the peaks for AlN and α-SiC, only careful measurements can provide the exact identity of the 2H peaks in this figure. It was that these phases were present as solid solutions, each containing substitutional carbon or nitrogen, respectively [18].

Increasing the applied field to 16.7 V/cm resulted in a complete conversion, i.e. the product containing AlN and α-SiC, Figure 3(b). Again, it was shown that these phases are limited solid solutions. However, when the field is increased to 25.0 V/cm, the product becomes a single-phase solid solution as indicated by the x-ray results in Figure 3(c). These observations demonstrate the following: (a) that the formation of a composite of AlN and SiC (see Eq (1)) can be achieved through field-activation. (The entire process now taking minutes (or seconds) as compared to hours in the conventional approach), and (b) that the field influences the nature of the product, a two-phase product at low E values and a solid solution at higher E values.

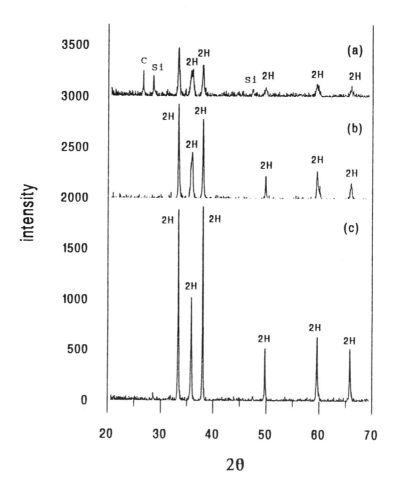

Figure 3.   X-ray diffraction patterns of products of Eq(1):
            (a) E = 9.3 V/cm;   (b) E = 16.7 V/cm;   (c) E = 25.0 V/cm.

3.   Simultaneous Synthesis and Densification

The materials produced by the method described above are typically in the form of a loose or highly agglomerated powder. In the latter case, porosities of as high as 50% are typical and represent the most notable disadvantage of the self-propagating combustion method. Because of the complex nature of the source of porosity in this method of synthesis [19], fully dense materials cannot be synthesized without the application of pressure in one of several forms [20].

It has been recently demonstrated that the simultaneous synthesis and densification can be effected using a modification of the field-activated method [21]. Using a plasma-assisted sintering (PAS) apparatus, high currents are passed through reactant compacts which are simultaneously subjected to a uniaxial pressure. The process involves five stages. In the first stage the system is evacuated and this followed by a pressure application (stage 2). Stage 3 constitutes the application of a pulsed power (30V, 750A, on-off durations of 30 ms). This is followed by resistance heating (stage 4) and subsequent cooling (stage 5). Heating is effected by the application of a 75V potential across the die and the passage of high current (2000-3000A), primarily through the graphite die containing the sample. When powder mixtures of Mo + 2Si were used in this method, the Joule heating resulting from the passage of the current through the reactants led to the formation of $MoSi_2$. With the existence of an applied pressure of 60 MPa, the products were nearly fully-dense silicides. Their relative density ranged from 94.5 to 99.2%, depending on the specific aspects of the process parameters [21]. Analysis of the shrinkage with regards to the various stages of the process (Figure 4) produced the results shown in Table 1. It is seen that nearly half of the total shrinkage (60.5%) occurs before the initiation of the reaction.

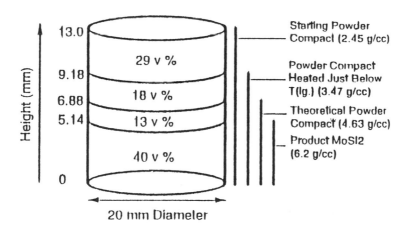

Figure 4. Schematic representation of volume changes during synthesis and densification of $MoSi_2$

Table 1 -- Porosity Changes During Field-Activated and Pressure-Assisted (FAPA) Synthesis of MoSi$_2$

	Initial Sample	Before Ignition	Reactants (theo)	Product (expt)	Product (theo)
Density (g/cm^3)	2.45	3.47	4.64	6.20	6.25
Sample Volume (cm^3)	4.08	2.88	2.16	1.61	1.60
Pore Volume (cm^3)	1.92	0.72	0.0	0.006	0.0
Volume Change (%)	0.0	19.4	47.1	60.5	60.8
Incremental Volume Change (%)	0.0	29.4	17.7	13.4	0.3

Acknowledgments:

The support by the Army Research Office for this work is gratefully acknowledged.

References:
1. Z. A. Munir and H. Schmalzried, *J. Mater. Synth. Process.*, **1**, 3-16, (1993).
2. Z. A. Munir, W. Lai, and K. Ewald, U. S. Patent No. 5,380,409, Jan. 10, 1995.
3. Z. A. Munir, *Amer. Ceram. Soc. Bull.*, **67**, 342-349 (1988).
4. Z. A. Munir and U. Anselmi-Tamburini, *Mater. Sci. Repts*, **3**, 277-365, (1989)
5. D. C. Halverson, B. Y. Lum and Z. A. Munir, Proceeding of the Symposium on High Temperature Materials Chemistry - IV, Z. A. Munir, D. Cubicciotti, and H. Tagawa, Editors, The Electrochemical Society, Pennington, N. J., 1987.
6. A. Feng and Z. A. Munir, *Metall. Mater. Trans.*, **26B**, 587-593 (1995).
7. S. Gedevanishvili and Z. A. Munir, *J. Mater. Res.*, in press, 1995.
8. H. Xue and Z. A. Munir, unpublished results.
9. S. Gedevanishvili and Z. A. Munir, *Scripta Metall. Mater.*, **31**, 741-743 (1994).
10. I. J. Shon and Z. A. Munir, *Mater Sci. Eng.*, in press, 1995.
11. H. Xue and Z. A. Munir, *Metall. Mater. Trans.*, in press, 1995.
12. R. K. Wade and J. J. Petrovic, *J. Amer. Ceram. Soc.*, **75**, 1682-1684 (1992).
13. D. C. Halverson, Z. A. Munir, and B. Y. Lum, in Combustion and Plasma Synthesis of High Temperature Materials, Z. A. Munir and J. B. Holt, Editors, VCH Publishers, N.Y., 1990, pp. 262-272.
14. R. Watanabe, A. Kawasaki, M. Tanaka, and J. F. Li, *Int. J. Refractory Metals and Hard Materials*, **12**, 187-193 (1993-1994).
15. S. Y. Kuo and A. V. Vikar, *J. Amer. Ceram. Soc.*, **73**, 2650-2646 (1990).

16. R. Ruh and A. Zangvil, *J. Amer. Ceram. Soc.*, **65**, 260-265 (1982).
17. M. Landon and F. Thevenot, *Ceram. Int.*, **17**, 97-110 (1991).
18. H. Xue and Z. A. Munir, unpublished results.
19. Z. A. Munir, *J. Mater. Synth. Process.*, **1**, 387-394 (1993).
20. Z. A. Munir, *Solid State Phenomena*, **25 & 26**, 197-208 (1992).
21. I. J. Shon and Z. A. Munir, *J. Amer. Ceram. Soc.*, submitted.

Reaction Sintering: The Role of Microstructure

L. C. De Jonghe
Lawrence Berkeley Laboratory, University of California
Berkeley, CA 94720

M. N. Rahaman
Department of Ceramic Engineering, University of Missouri
Rolla, MO 65401

Abstract
The reaction sintering of powder compacts, in which densification and chemical reaction of the starting powders occur in the same heating cycle, is considered. Results for the $ZnO-Fe_2O_3$, $ZnO-Al_2O_3$ and $CaO-Al_2O_3$ systems indicate that the reaction and any associated volume change are, by themselves, not the dominant factors in the subsequent densification. Instead, the microstructural change produced by the reaction is a major factor in the subsequent densification. The change in the microstructure is itself affected by the mechanism of the chemical reaction. An important advance of the present work is the identification of the relation between the reaction mechanism and the associated volume change and their influence on the subsequent densification. Processes that can be used to modify the microstructural change produced by the reaction and, hence, the subsequent densification, are discussed.

I. Introduction
Reaction sintering, sometimes referred to as reactive sintering, is the term used to describe a particular firing process in which the chemical reaction of the starting materials (the reactants) and the densification of the powder compact are both achieved in a single heat treatment step (Kolar, 1981; Yangyun and Brook, 1985). Depending on the processing conditions such as particle size, temperature, applied pressure and heating rate, reaction and densification can occur in sequence, or concurrently, or in some combination of the two (Yangyun and Brook, 1985). Provided that the reaction is not accompanied by a significant volume change, products with high density and controlled microstructure can best be achieved if densification of the reacting powders is completed prior to the reaction itself. To achieve this, processing should be carried out with fine powders, in a temperature

457

regime chosen to maximize the ratio of the densification rate to the reaction rate. The use of applied pressure (e.g., hot pressing) or high heating rate is also advantageous. A well recognized example of such process control is the preparation of ZrO_2-toughened mullite composites from mixtures of $ZrSiO_4$ and Al_2O_3 (di Rupo et al, 1979; Claussen and Jahn, 1980).

In many systems, however, the reaction is relatively rapid and favorable conditions for the completion of densification prior to the occurrence of the reaction are difficult to achieve. For such systems, it is important to determine whether the reaction sintering route can still be exploited for the production of controlled microstructures. The occurrence of a reaction is expected to produce microstructural changes in the powder compact which will influence the subsequent densification. The factors that control the extent of the microstructural changes during the reaction need to be clarified. A further consideration is whether the microstructural changes produced by the reaction can be modified to achieve adequate control of the subsequent densification. These aspects of the reaction sintering process are considered in the present paper for the $ZnO-Fe_2O_3$, $ZnO-Al_2O_3$ and $CaO-Al_2O_3$ systems in which the reaction occurs prior to densification.

II. Reaction Sintering Systems

The $ZnO-Fe_2O_3$, $ZnO-Al_2O_3$ and $CaO-Al_2O_3$ systems have important similarities and differences which make them useful model systems for the present investigation. In the $ZnO-Fe_2O_3$ and $ZnO-Al_2O_3$, two simple oxides react to form an alloy oxide such that the molar volume of the two reactants and of the product are almost the same. Both systems have seen considerable use as model materials for investigations into the kinetics and mechanism of solid state reactions. The data indicate that the reaction mechanisms in the two systems are different. At temperatures normally used in their sintering (less than ≈ 1300 °C) the reaction between ZnO and Fe_2O_3 occurs by a Wagner counter-diffusion mechanism in which cations migrate in opposite directions and the oxygen ions remain essentially stationary (Linder, 1955; Kuczynski, 1971). In comparison, the reaction mechanism in the $ZnO-Al_2O_3$ system is not as clear. The work of Branson (1965) indicates that the reaction occurs by a solid state mechanism in which the diffusion of Zn ions through the $ZnAl_2O_4$ spinel layer is rate controlling. However, Schmalzried (1974) reported that the rate constant for the gas-solid reaction between ZnO vapor and Al_2O_3 is in excellent agreement with the measured reaction rate. In the $CaO-Al_2O_3$ system, high melting point intermediate phases are formed prior to significant densification (Wu et al, 1995).

III. Experimental Procedure

Powder compacts for the reaction sintering experiments were prepared by conventional ceramic powder processing techniques. The procedure has been described in detail elsewhere and only a brief outline will be given here. For the

ZnO-Fe$_2$O$_3$ system (Rahaman and De Jonghe, 1993) and for the ZnO-Al$_2$O$_3$ system (Hong et al, 1995), equimolar mixtures of the two simple oxide powders were prepared by ball milling. For the ZnO-Fe$_2$O$_3$ system, the average particle sizes of the ZnO and Fe$_2$O$_3$ powders were ≈0.3 μm and ≈1 μm, respectively, while for the ZnO-Al$_2$O$_3$ system, the average particle size of both powders was ≈0.3 μm. After drying, the ball-milled mixture was ground in an agate mortar and pestle and compacted, by die pressing or a combination of die pressing and cold isostatic pressing, to produce pellets (≈6 mm in diameter by ≈5 mm). Sintering was performed at a constant heating rate (1-10 °C/min) in air, in a dilatometer that allowed continuous monitoring of the axial shrinkage. The sinterability of the reaction sintered samples was compared with that for a pre-reacted, single phase powder of the complex oxide (ZnFe$_2$O$_4$ or Zn Al$_2$O$_4$). For the CaO-Al$_2$O$_3$ system (Wu et al, 1985), mixtures containing up to 5 wt% CaO were prepared by dispersing the Al$_2$O$_3$ powder (particle size ≈0.5 μm) in a solution of calcium nitrate. After drying, the mixture was calcined to decompose the nitrate, ground lightly and compacted to form pellets. Sintering was performed in air isothermally at 1330 °C or at a constant heating rate of 5 °C/min.

IV. Sintering and Microstructural Evolution

For the ZnO-Fe$_2$O$_3$ system, Fig. 1 shows a comparison of the densification data for the reaction sintered powder mixture and the single phase ZnFe$_2$O$_4$ powder during sintering at 4 °C/min to 1350 °C. For this system, the densification of the reaction sintered powder mixture is fairly close to that for the single phase powder and final densities in excess of 95% of the theoretical are achieved. Under the conditions used for sintering, the reaction between ZnO and Fe$_2$O$_3$ is completed prior to any significant densification. X-ray analysis indicates that the reaction commences at ≈600 °C and is completed at ≈800 °C. Figure 1 shows that the reaction is accompanied by little change in the volume (< 3 vol%) of the powder compact. Scanning electron micrographs of the sintered compacts (Fig. 2) indicate that the microstructures of the reaction sintered sample and the single phase sample evolve in roughly the same way.

Figure 3 shows the densification data for the reaction sintered mixture and the single phase ZnAl$_2$O$_4$ powder for the ZnO-Al$_2$O$_3$ system during sintering at 4 °C/min to 1400 °C. A distinct feature is the expansion of the reaction sintered sample (starting at ≈800 °C) and, compared to the single phase powder, the inhibition of the subsequent densification. X-ray diffraction indicates that the reaction between ZnO and Al$_2$O$_3$ starts at ≈800 °C and is essentially completed by 1000 °C. The expansion of the reaction sintered sample can therefore be correlated with the reaction. The magnitude of the expansion is ≈25 vol% and this value has been shown by Hong et al (1995) to be fairly insensitive to the green density (in the range of 0.56 to 0.69 of the theoretical) or to the average particle size of the Al$_2$O$_3$ powder (0.3 to 1.5 μm). The microstructures of a reaction sintered sample and a

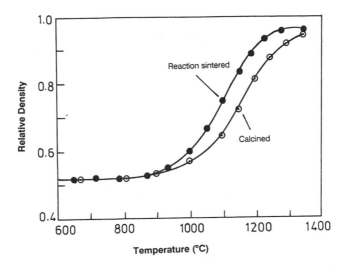

Figure 1 Relative density versus temperature for the reaction sintered ZnO-Fe$_2$O$_3$ powder mixture and for the single-phase ZnFe$_2$O$_4$ powder (referred to as *calcined*).

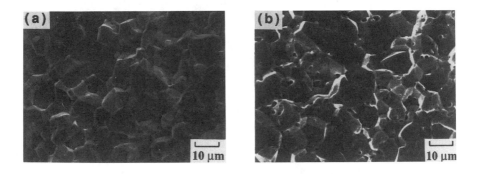

Figure 2 SEM of the fractured surfaces of the sintered compacts formed from (a) the reaction sintered ZnO-Fe$_2$O$_3$ mixture and (b) the single-phase ZnFe$_2$O$_4$ powder.

single phase sample sintered under almost identical conditions (4 °C/min to 1450 °C) are shown in Fig. 4. The reaction sintered sample (relative density ≈0.60) consists of a highly porous network of particles. Furthermore, as shown by Hong et al (1995), the morphology of the particles appears to be a slightly coarsened version of the starting Al$_2$O$_3$ particles used in the powder mixture.

For the CaO-Al$_2$O$_3$ system sintered below the eutectic temperature, Wu et al (1985) have shown that the densification is inhibited by the presence of the CaO.

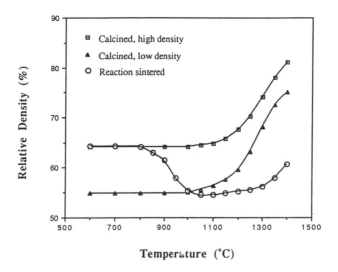

Figure 3 Relative density versus temperature for the reaction sintered ZnO-Al$_2$O$_3$ powder mixture and for the single phase ZnAl$_2$O$_4$ powder (referred to as *calcined*).

Figure 4 SEM of the fractured surfaces of the sintered compacts formed from (a) the reaction sintered ZnO-Al$_2$O$_3$ mixture and (b) the single-phase ZnAl$_2$O$_4$ powder.

Furthermore, inhibition of sintering can be correlated with the formation of high melting point intermediate phases such as CaO.Al$_2$O$_3$ and CaO.2Al$_2$O$_3$.

V. Discussion

In the ZnO-Fe$_2$O$_3$ and ZnO-Al$_2$O$_3$ systems, where the reaction occurs prior to densification and is accompanied by almost no change in molar volume, the reaction sintering behavior of the two systems shows significant differences. In the

ZnO-Fe_2O_3 system, the subsequent densification and microstructural evolution of the reaction sintered sample are fairly close to those for the single phase powder compact. However, in the ZnO-Al_2O_3 system, the subsequent densification is inhibited and the microstructure evolves differently from the single phase $ZnAl_2O_4$ sample. The reaction and microstructural changes in the ZnO-Al_2O_3 system are also accompanied by a fairly large expansion. The expansion is, by itself, not the dominant factor in the inhibition of the sintering. As shown in Fig. 3, a single phase $ZnAl_2O_4$ powder compact with a green density close to that of the minimum value reached by the reaction sintered sample has a significantly higher sintering rate.

In the powder compacts formed from the ball-milled, equimolar mixture of the simple oxides, the volume fraction of each oxide is well above the percolation threshold. It can be assumed that the volume change in the dominant percolative network will be reflected macroscopically by an equal volume change of the compact. For the ZnO-Fe_2O_3 system, in which the reaction occurs by a Wagner counter-diffusion mechanism, the balance of fluxes required to maintain electroneutrality dictates that the ratio of $ZnFe_2O_4$ formed on the ZnO and on the Fe_2O_3 is 1:3. Taking the Fe_2O_3 particulate network, and assuming that 1/4 of its volume leaves the network, the Fe_2O_3 (molar volume = 30.5 cm^3) will react to form 33.7 cm^3 of $ZnFe_2O_4$, thereby increasing its volume by $\approx 10\%$. This expansion is close to values of 2-5 vol% found in experiments by Toolenaar and Verhees (1988) and by Rahaman and De Jonghe (1993). Essentially, in the $ZnFe_2O_4$ system, a dominant percolating network is not formed in the reaction because sufficient material leaves each network to preserve the initial powder packing of the mixture. The subsequent densification and microstructural evolution are not significantly altered by the reaction, as observed in Figs. 1 and 2.

Assuming a reaction mechanism of one-way transport of Zn ions in the ZnO-Al_2O_3 system, the Al_2O_3 will react with ZnO to form $ZnAl_2O_4$ spinel, increasing its molar volume from 25.7 cm^3 to 40.03 cm^3. Based on this increase in the molar volume, the ratio of the volume of the compact after reaction to the initial volume of the compact (equal to $V_{spinel}/V_{alumina}$) is expected to be 1.54. The experimental value, however, is only ≈ 1.2 to 1.3. The lower expansion of the percolative Al_2O_3 network can be accounted for if it is assumed that during the reaction, ≈ 15-20 vol% of the Al_2O_3 leaves the network. This will be the case if the reaction mechanism involves a mutual counter-diffusion of Zn and Al ions. The proposed mechanism for the expansion of the powder compact during the reaction does not involve details of the particle distribution, as long as the dominant percolative network is established. It is consistent with the data of Hong et al (1995) which showed an insignificant influence of the green density and the ratio of the particle sizes on the magnitude of the expansion. The proposed model for the expansion is different from that put forward by Leblud et al (1981), who assumed uniform packing of the particles and the formation of the product on the Al_2O_3 particles at each contact point with the ZnO.

As outlined earlier, the expansion produced by the ZnO-Al$_2$O$_3$ reaction cannot, by itself, account for the depression of the sintering rate. The origins of the reduced sintering must lie in microstructural changes produced by the reaction. The most significant changes are the formation of a fairly rigid network of particles based on the original Al$_2$O$_3$ particle structure and the development of fewer pores which are larger than those in the unreacted green compact. This overall microstructure is well known for its ability to provide serious obstacles to sintering.

The results for the ZnO-Al$_2$O$_3$ system indicate that if the microstructural disruption caused by the reaction is alleviated, then the subsequent densification can be improved. One way of alleviating the disruption is to prevent the formation of the dominant network based on the original Al$_2$O$_3$ particle structure. To test this, Al$_2$O$_3$-ZnO powder mixtures were prepared by coating the Al$_2$O$_3$ particles with ZnO by chemical precipitation (Hong et al, 1995). The coated powders were compacted and sintered under the same conditions as a ball milled mixture of the same composition. Figure 5 shows a comparison of the densification data for the coated powder and the ball-milled mixture in the Al$_2$O$_3$-ZnO system. Following the reaction, the densification of the compact formed from the coated powder is significantly higher than that of the compact formed form the ball-milled mixture.

VI. Conclusions

In the reaction sintering of ceramic powder mixtures, the reaction and any associated volume change are, by themselves, not the dominant factors in the

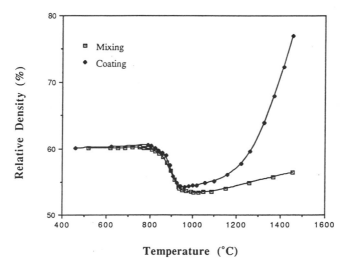

Figure 5 Relative density versus temperature for the reaction sintering of ZnO-Al$_2$O$_3$ compacts formed from the mixed powder and from the coated powder.

subsequent densification. Instead, the microstructural change produced by the reaction is the major factor. The change in the microstructure is, in turn, affected by the reaction mechanism. In the ZnO-Fe_2O_3 system, the reaction mechanism of Wagner counter-diffusion leads to matter transport in which the original microstructure is not severely disrupted. Reaction sintering is a viable fabrication route for this system. Unequal diffusional fluxes in the ZnO-Al_2O_3 system leads to the formation of a dominant percolative network based on the original Al_2O_3 particle structure. This disruption of the microstructure inhibits subsequent sintering but the inhibition can be relieved by modification of the initial powder packing as achieved by the use of coated powders. The formation of high melting point intermediate phases in the CaO-Al_2O_3 system inhibits subsequent sintering.

References

D. L. Branson, "Kinetics and Mechanism of the Reaction Between Zinc Oxide and Aluminum Oxide", J. Am. Ceram. Soc., **48** [11] 591-95 (1965).

N. Claussen and J. Jahn, "Mechanical Properties of Sintered, In Situ-Reacted Mullite-Zirconia Composites", J. Am. Ceram. Soc., **63** [3-4] 228-29 (1980).

E. di Rupo, E. Gilbart, T. G. Carruthers and R. J. Brook, "Reaction Hot-Pressing of Zircon-Alumina Mixtures", J. Mater. Sci., **14** 705-11 (1979).

W.-S. Hong, L.C. De Jonghe, X. Yang and M.N. Rahaman, "Reaction Sintering of ZnO-Al_2O_3", J. Am. Ceram. Soc., (1995). In press.

D. Kolar, "Microstructural Development During Sintering in Multicomponent Systems", Sci. Ceram., **11** 199-211 (1981).

G. C. Kuczynski, "Formation of Ferrites by Sintering of Component Oxides", pp. 87-95 in Ferrites: Proceedings of the International Conference. Edited by Y. Hoshino, S. Iida and M. Sugimoto. University Park Press, Baltimore, MD, (1971).

C. Leblud, M. R. Anseau, E. di Rupo, F. Cambier and P. Fierens, "Reaction Sintering of ZnO-Al_2O_3 Mixtures", J. Mater. Sci. Lett., **16** 539-44 (1981).

J. Linder, "Studies of Solid State Reactions with Radiotracers", J. Chem. Phys., **23** [2] 410-11 (1955).

M. N. Rahaman and L.C. De Jonghe, "Reaction Sintering of Zinc Ferrite During Constant Rates of Heating", J. Am. Ceram. Soc., **76** [7] 1739-44 (1993).

H. Schmalzried, Solid State Reactions. Academic Press, New York, 1974, pp. 101-102.

F.J.C.M. Toolenaar and M.T.J. Verhees, "Reactive Sintering of Zinc Ferrite", J. Mater. Sci., **23** 856-61 (1988).

S. Wu, L.C. De Jonghe and M. N. Rahaman, "Subeutectic Densification and Second-Phase Formation in Al_2O_3-CaO", J. Am. Ceram. Soc., **68** [7] 385-88 (1985).

S. Yangyun and R.J. Brook, "Preparation of Zirconia-Toughened Ceramics by Reaction Sintering", Sci. Sintering, **17** [1] 35-47 (1985).

Processing, Sintering Behavior and Mechanical Properties of Reaction-Bonded Al_2O_3/ZrO_2 Ceramics

Suxing Wu, Hugo S. Caram, Helen M. Chan and Martin P. Harmer
Materials Research Center, Lehigh University
Bethlehem, PA 18015-3194

Abstract

Reaction-bonding of Al_2O_3 (RBAO) is a novel ceramic processing method. It will be shown in this paper that the reaction behavior of RBAO precursor powders can vary depending on the heat-treatment cycle. Since cracking of compacts occurs under some conditions, successful application of the RBAO process requires careful control of the heating cycle. For the RBAO precursor powder produced in this work, a fast heating rate ($5°C/min$) across the melting point of Al ($660°C$) prevented crack formation, whereas a slow heating rate ($\leq 1°C/min$) did not. Reaction-bonded Al_2O_3/ZrO_2 processed using a rapid heating rate exhibited superior microstructures and fracture strengths compared to conventionally processed materials: $<1\mu m$ grain size and >800 MPa have been achieved.

1. Introduction

A novel technique pioneered by Claussen et al., reaction bonding of aluminum oxide (RBAO), has recently been developed for fabricating ceramics to near net-shape (Cluassen et al. 1990; Wu et al. 1991; Wu and Claussen, 1991; Wu et al. 1993; Claussen et al. 1993; Wu and Claussen, 1994; Holz et al. 1994) and has been attracting increased interest (Luyten et al. 1992; Park and Ahn, 1994; Li et al. 1995; Derby, 1995). The RBAO process utilizes the oxidation of powder mixtures containing a substantial amount of metallic Al (30-60vol%). During heat-treatment in an oxidizing atmosphere (usually air), the Al oxidizes to $\gamma-Al_2O_3$ at temperatures below $900°C$, and then undergoes a phase transformation to $\alpha-Al_2O_3$ up until $1100°C$. A net volume increase of 28%, with respect to the final product of $\alpha-Al_2O_3$, partially compensates for the sintering shrinkage, and consequently the reaction-bonded bodies exhibit lower shrinkage than conventionally fired Al_2O_3 ceramics.

RBAO precursor powders prepared from different raw materials and under different milling conditions, can respond differently to the same heat-treatment. For example, we have encountered cracking of compacts of RBAO precursor powders prepared using the same heating cycle that worked successfully for previous work (Wu et al. 1993). The objective of the present work was to investigate the effect of the heating cycle on the microstructure and properties of RBAO ceramics.

2. Experimental Procedure

A powder mixture consisting of 45 vol% Al (~20 μm, >99%, Aldrich Chemical Comp., Inc. USA), 30 vol% Al_2O_3 (AKP-30, ~0.4 μm, Sumitomo Chemical Co., Ltd. Japan) and 20 vol% ZrO_2 (Tosoh-Zirconia, TZ-3Y, Tosoh Corp., Japan) was used in the present work. This was the composition used previously at TUHH (Wu et al. 1993). The mixture was attrition milled in acetone with 2-3 mm TZP-balls for 5, 6 and 7 hrs. After drying, the powder mixture was uniaxially pressed in a graphite die at 44 MPa to form disks of 25.4 mm in diameter and 3 mm in thickness. The disks were then isopressed at 100 - 400 MPa, and heat-treated in a box furnace in air. Two distinct heating cycles (schematically indicated in Fig. 1(A) and 1(B)) were applied to the samples. Both cycles include a heating rate of 1°C/min from room temperature to 450°C. This slow heating step was designed to remove adsorbed moisture from the powder surface. Cycle A consists of only one heating rate of 5°C/min up to the

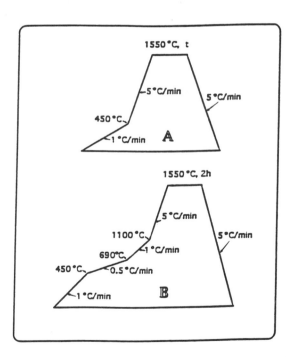

Fig. 1 Schematic representations of heating cycles (A) and (B) used in this study.

sintering temperature (1550°C), whereas cycle B consists of the following three successive stages: 0.5°C/min from 450°C to 690°C, 1°C/min from 690°C to 1100°C and, and 5°C/min from 1100°C to 1550°C. In previous studies cycle B was found to produce homogeneous, crack-free RBAO bodies (Wu et al. 1993).

Weight and dimensional changes of samples milled for 6 hrs and compacted at 300 MPa were characterized by a thermogravimeter (TG, Model 409, Netzsch, Selb, FRG) and dilatometer (Model 502, Bähr Gerätebau, Hülshorst, FRG). Final sintered densities were measured by the Archimedes method. Microstructures of the reaction-bonded Al_2O_3/ZrO_2 ceramics were characterized by scanning electron microscopy (SEM, Jeol 6300F).

Strengths were measured by fracturing sintered disks in biaxial flexure. The disks were loaded on a three-ball support of 9.37 mm in radius using a 2.47 mm radius flat. Prior to testing, the prospective tensile faces were ground and polished to 1 μm finish. Hardness measurements were obtained using a Vickers diamond pyramid by applying a 10 kg load on a polished surface for 15 seconds.

3. Results and Discussion

3.1 Reaction Bonding Behavior

Figure 2 shows the effect of heating cycle on the appearance of as-fired RBAO bodies. Cycle A produced successful reaction bonding, whereas cycle B resulted in cracking. This result was obtained regardless of the milling times (5, 6 and 7 hours) and compaction pressures (100-400 MPa) used in the study.

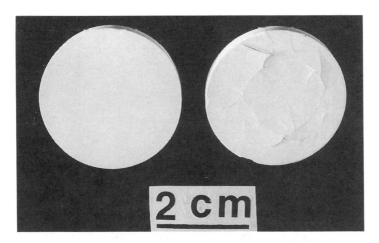

Fig. 2 Typical RBAO bodies identical in green state but subjected to heat-treatment cycles A (left) and B (right) in Fig. 1, respectively.

Weight and dimensional changes of samples milled for 6 hrs, compacted at 300 MPa, and processed using cycle A are shown in Fig. 3. DTA data are also plotted in the uppermost diagram in Fig. 3. As usually observed in the RBAO process, between room temperature and ~380°C the evaporation of fugitive species results in a weight loss. The weight gain from ~380°C in the TG curve is the response of the oxidation reaction of Al. At ~1250°C, the TG curve becomes constant indicating that the Al phase has been fully converted to

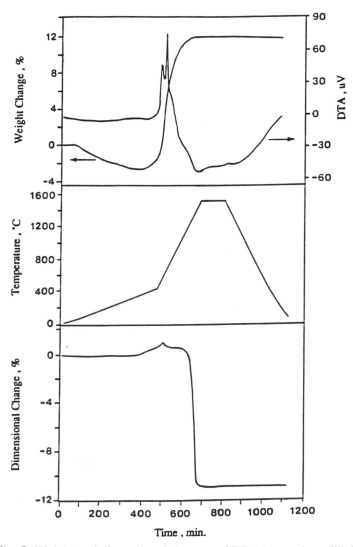

Fig. 3 Weight and dimensional changes of RBAO samples milled for 6 hrs and isopressed at 300 MPa as a function of heating cycle.

Al2O3. Although these observations are analogous to those reported in previous studies (Wu et al. 1993; Holz et al 1994), there are some noteworthy differences. Firstly, in the present study, for heating cycle A, two exothermic peaks were observed in the DTA curve, one at ~520°C and another at ~640°C. This contrasts with previous studies
employing the slower heating rate cycle, where only one such peak was observed (at ~520°C) (Wu et al. 1993; Holz et al 1994). The afore-mentioned exothermic DTA peaks are significant in that their positions indicate periods in the heat-treatment cycle where the reaction rate of the Al —> γ–Al2O3 reaction is at a maximum. Whereas previous studies[4] reported that between 520°C and the melting temperature of Al (T_m=660°C) the oxidation reaction is sluggish, and that the TG curve exhibits a shallow sloped plateau during slow heating, the data in Fig. 3 suggests that a maximum in the reaction rate can occur in the temperature range of 520 - 660°C, with no plateau in the TG curve. Furthermore, there is a sharp expansion peak in the dilatometer curve corresponding to the temperature of the second DTA peak.

Using similar powder compacts, further experiments have shown that a single DTA peak appears at 498°C for cycle B, while double DTA peaks occur at 542°C and 623°C for cycle A (cf. Fig. 1). These results, together with the difference in cracking behavior show clearly that the RBAO process is sensitive to the heating cycle. Further, as evidenced by the differences between the present results and the previous studies[4], RBAO precusor powders prepared under different conditions may respond differently to the same thermal history. Given the complexity of the RBAO reaction, and the large number of processing variables, a definitive explanation for the above observations requires further systematic study. Nonetheless, we can tentatively interpret our observations as follows.

Preliminary measurements show that up to the temperature of the first DTA peak, the fraction of Al that has undergone oxidation is 0.73 for cycle B; whereas it is only 0.26 for cycle A. Up to the temperature of the second DTA peak, this fraction is 0.55 and 0.78 for cycle A and B, respectively. This suggests that the cracking behavior obtained with the slower heating rate is linked to the volume expansion associated with the higher degree of solid/gas oxidation. During the RBAO process, both oxidation of the Al metal, and sintering of the nanocrystalline secondary alumina particles (100-200 nm) are taking place simultaneously. We can postulate that for the higher heating rate, because oxidation is deferred to higher temperatures, a greater degree of particle rearrangement/neck formation can occur. Consequently, the structure is more able to accomodate the high local stresses induced by the oxidation of the Al particles. Further, one can envisage a situation where sintering induced healing of the microcracks takes place concurrently with the rupture process.

The origin of the peaks in the DTA trace is not completely understood. Claussen et al. [12] attributed the single DTA peak which they observed to the rapid oxidation of a large number of Al particles with a radius below that of a certain critical size (r_c). The value of r_c was estimated as ~0.5 μm (Claussen et al. 1995). In the proposed model of the RBAO mechanism (Wu et al. 1993), stresses arising due to oxidation cause rupturing of the oxide skin, thus exposing fresh metal. Note that the volume expansion associated with the Al—>γ-Al$_2$O$_3$ reaction is quite considerable (~ 39 vol%). These workers argued that the resultant stresses increase with decreasing size of Al particle, thus particles below the critical radius can oxidise rapidly and completely at temperatures below Tm.

The fact that it is possible to obtain two separate peaks by varying the heating rate suggests that there are two distinct processes occuring (with different activation energies), which overlap for cycle A, but which are differentiated when cycle B is used. Work is ongoing to determine these mechanisms, but we may speculate that one of the processes could correpond to the rupture model proposed by Claussen et al., whereas the other process could be conventional oxidation limited by solid state diffusion through the oxide skin.

3.2 Microstructure and Mechanical Properties of ZrO$_2$-Containing RBAO Ceramics

Data showing the final densities of samples heat-treated following cycle A (Fig. 1(A)) as a function of sintering time at 1550°C are given in Fig. 4. After only 30 min sintering, final densities of 98.6% TD were achieved. Maximum density (99.7% TD) was obtained after only 2 hrs sintering. Prolonged sintering times (>2 hrs) resulted in a decrease in the final density.

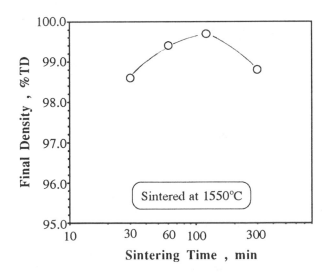

Fig. 4   Final density data of RBAO samples (processed using heat-treatment A) as a function of sintering time at 1550°C.

The microstructure of a sample sintered for 120 min is shown in Fig. 5. It can be seen that the RBAO process results in a very fine-grained microstructure (d < 1 μm), with minimal residual porosity (> 99 % TD). The ZrO_2 particles (~0.2 μm) are uniformly distributed at triple points and grain boundaries, and are clearly effective in suppressing grain growth in the sintering range. The measured fracture strength of the samples were in the range of 800 - 820 MPa. These relatively high intrinsic strength values are attributed to the fine grain size and high density of the RBAO ceramics. Previous work has not detected significant effect of ZrO_2 phase transformation toughening in these materials[5]. The Vicker's hardness of these samples was determined to be ~18 GPa, which is comparable to values commonly cited for Al_2O_3.

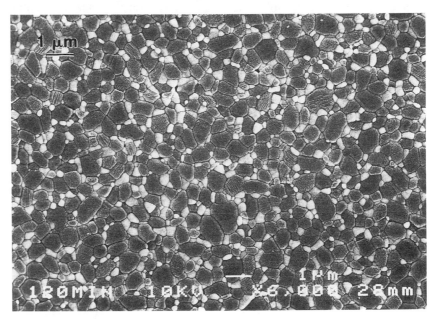

Fig. 5   SEM micrographs showing the microstructure of the reaction-bonded
RBAO samples sintered at 1550°C for 120 min.

4. Conclusions

1. The reaction behavior of RBAO precursor powders is sensitive to the heating cycle. Slow (≤1°C/min) and fast (≥5°C/min) heating rates result in single and double DTA peaks, respectively.

2. For the precursor RBAO powders used in this study, a slow heating rate (≤1°C/min) produced sample cracking, whereas this was avoided using the fast heating rate cycle (~5°C/min).

3. RBAO precursor powders prepared from different raw materials and milling conditions may respond differently to the same thermal history.

Reproducibility of the RBAO process can be achieved by adjusting processing variables, especially the heating cycle.

4. ZrO_2-containing almina ceramics of high density (>99%TD) and fine microstructure (<1μm) are readily produced by the RBAO process.

5. ZrO_2-containing RBAO aluminas exhibit superior fracture strengths cf. conventionally processed materials: >800 MPa was achieved by pressureless sintering at 1550°C for 30 - 120 min.

Acknowledgments

The authors are grateful to the U. S. Office of Naval Research and the Department of Energy for financial support. Special thanks are due to N. Claussen, D. Holz, K. Barmak and S. Gaus for valuable discussions, and B. Badelt and S. Pagel for TG and dilatometer measurements.

References

N. Claussen, N. A. Travitzky, and S. Wu, "Tailoring of Reaction-Bonded Al2O3 (RBAO) Ceramics", *Ceram. Eng. Proc.*, **11**, 806-20 (1990).

S. Wu, A. Gesing, N. A. Travitzky, and N. Claussen, "Fabrication and Properties of Al-Infiltrated RBAO-Based Composites", *J. Euro. Ceram. Soc.*, 7, 277-281 (1991).

S. Wu and N. Claussen, "Fabrication and Properties of Low-Shrinkage Reaction-Bonded Mullite", *J. Am. Ceram. Soc.*, **74** [10] 2460-2463 (1991).

S. Wu, D. Holz, and N. Claussen, "Mechanisms and Kinetics of Reaction-Bonded Aluminum Oxide Ceramics", *J. Am. Ceram. Soc.*, **76** [4] 970-980 (1993).

N. Claussen, S. Wu, and D. Holz, "Reaction Bonding of Aluminum Oxide (RBAO) Composites: Processing, Reaction Mechanisms, and Properties", *J. Euro. Ceram. Soc.* **14** 97-109 (1993).

S. Wu and N. Claussen, "Reaction Bonding and Mechanical Properties of Mullite/SiC Composites", *J. Am. Ceram. Soc.*, 1994. **77** [11] 2898-904 (1994).

D. Holz, S. Wu, S. Scheppokat, and N. Claussen, "Effect of Processing Parameters on Phase and Microstructure Evolution in RBAO Ceramics," *J. Am. Ceram. Soc.*, **77** [10] 2509-17 (1994).

J. Luyten, J. Cooymans, P. Diels and J. Sleurs, "The Effect of Powder Conditions on the Synthesis of RBAO", *Silicates Industriels*, **1992** [7-8] 91-94 (1992).

T. Li, R. J. Brook, and B. Derby, "Reaction Bonding of Cr2O3 Ceramics", pp. 231-235, Ceramic Transactions, Vol. 51, *Ceramic Processing Science and Technology*, edited by Hans L. Meesing nd Shin-ichi Hirano, The American Ceramic Society, Westerville, (1995).

B. Derby, "Reaction Bonding of Ceramics by Gas-Metal Reactions", pp. 217-223 Ceramic Transactions, Vol. 51, *Ceramic Processing Science and Technology*, Edited by Hans L. Meesing and Shin-ichi Hirano, The American Ceramic Society, Westerville, (1995).

N. Claussen, R. Janssen, and D. Holz, "The Reaction Bonding of Aluminum Oxide (RBAO)", *J. Ceram. Soc. of Japan.*, **103** (1995) in press.

THE EFFECT OF TWO STAGE SINTERING ON THE MICROSTRUCTURAL, MECHANICAL AND AGING PROPERTIES OF Y-TZP

Simon Lawson[1], K. Sing Tan[2], J. Malcolm Smith[2], H. Collin Gill[2] and Graham P. Dransfield[3]

[1] British Nuclear Fuels plc, Sellafield, Seascale, Cumbria. CA20 1PG. UK.
[2] University of Sunderland, School of Engineering & Advanced Technology, Chester Road, Sunderland. SR1 3SD. UK.
[3] Tioxide Specialties plc, West Site, Haverton Hill Road, Billingham, Cleveland. TS23 1PS. UK.

Abstract

Yttria-tetragonal zirconia polycrystalline (Y-TZP) ceramics exhibit exceptional mechanical, chemical and wear properties. Y-TZPs manufactured from stabilizer-coated zirconia have an inhomogeneous distribution of the yttria leading to enhanced fracture toughness and superior hydrothermal degradation resistance. This paper discusses work which has been done to optimise the sintering performance of these compositionally zoned ceramics and in particular to assess the effects of the firing cycle on the resultant physical, mechanical and hydrothermal properties.

1 Introduction

Yttria-tetragonal zirconia polycrystalline ceramics (Y-TZP) have attracted intense research efforts internationally in order to exploit their unique mechanical, chemical, thermal and wear properties. It is known that zirconia ceramics can suffer from a relatively low temperature (65-500°C) degradation phenomenon in water or humid environments (Lawson, 1995a) whereby the metastable tetragonal phase can transform martensitically to the monoclinic polymorph resulting in a loss of mechanical, wear and electrical properties. This hydrothermal degradation can clearly limit the potential applications of these materials. The addition of stabilizing oxides (Sato et al,1988 and Boutz, 1993) or the use of sintering aids which promote densification but prevent grain growth (Kimura et al, 1989, Tsukuma & Shimada, 1985 and Lawson et al, 1995b) are established techniques to prevent or retard the transformation.

It has been shown by Ruhle et al (1984) that the purity and powder fabrication route can dramatically affect the phase transformation kinetics. It has been shown recently (Lawson et al, 1994) that powders manufactured with the stabilizer coated on to the monoclinic zirconia particle and subsequently sintered have superior mechanical and ageing properties compared to powders manufactured using other processing routes.   These enhanced mechanical and degradation properties are attributed to compositional zoning of the stabilizer within the sintered Y-TZP.   Fig.1 shows a transmission electron micrograph of a 2.5Y-TZP sintered from these coated powders, revealing a tetragonal shell and monoclinic core.

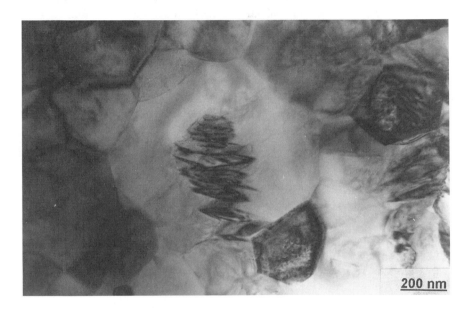

Fig. 1   Transmission electron micrograph showing the distinctive cored nature of coated-Y-TZP grains (courtesy of W.M. Rainforth, University of Sheffield, UK).

This coring effect is the extreme case, as normally the centre of the grain is tetragonal phase but with little or no stabilizer present, a situation confirmed by microanalysis (Lawson et al, 1994).   This is clearly beneficial as the majority of the grain with low yttria content is highly unstable leading to enhanced fracture toughness through transformation toughening, whereas the grain boundary area of the material is yttria-rich resulting in enhanced ageing resistance.

The aim of this work was to investigate the sintering characteristics of the coated material and to optimise the sintering cycle for maximum densification and tetragonal phase retention, without sacrificing the toughness and ageing performance of the Y-TZP.   This work forms part of a larger study investigating the sintering behaviour of large Y-TZP components.

2 Experimental

Zirconia powders supplied by Tioxide Specialties plc are produced using plasma technology and coated with the stabilizer or any other additive (Dransfield, 1993). The die-pressed 2.5Y-TZP powders were isostatically pressed at 200MPa. The firing experiments were divided into three distinct routes: pre-sintering at 1200°C for upto 12 hours and then cooling, pre-sintering at 1250°C followed by cooling, and pre-sintering at 1200 or 1250°C for upto 12 hours and then firing at 1450°C for 2 hours. The aim of the holding experiments was to encourage diffusion of the yttria, at a temperature higher than the monoclinic to tetragonal phase transformation (~1170°C). The sintered discs were characterized by measuring the bulk density, Vickers hardness, indentation fracture toughness (Niihara et al, 1982) and by x-ray diffraction (XRD). The average grain sizes were determined using a linear intercept method (Mendelson, 1969) on micrographs obtained from the scanning electron microscope. The hydrothermal ageing was performed in a PTFE lined autoclave containing superheated water at 180°C (1MPa) for upto 200 hours (Nakajima et al, 1984). The degradation was assessed by monitoring the development of aged surface monoclinic phase. The thickness of which was measured by sectioning the sample and optical microscopy (Lawson & Smith, 1993).

3 Results and Discussion

3.1 Sintering studies

The effect of pre-sintering the samples at 1200 and 1250°C for upto 12 hours is shown in Fig.2.

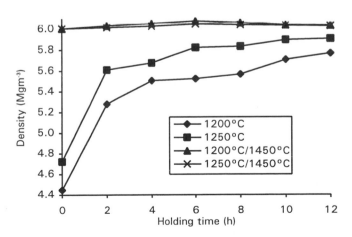

Fig. 2 Density variations as a function of intermediate hold time and temperature for 2.5Y-TZP.

The figure shows that the density increases with time for both temperatures with the 1250°C hold having the greatest effect. Fig. 2 also shows the effect of holding at the lower temperatures and then sintering at 1450°C for 2 hours. For both sintering regimes there is optimum densification after a 6 hour hold. The increase in density with time is due to a reduction in porosity and the fall in density after the peak at 6 hours is attributed to cubic phase formation in larger grains. It is noticeable that the 1200°C hold followed by sintering at 1450°C results in higher densities than the 1250°C pre-sinter.

The effect of the pre-sinter temperature and time on the retained tetragonal phase after sintering are shown in Fig.3.

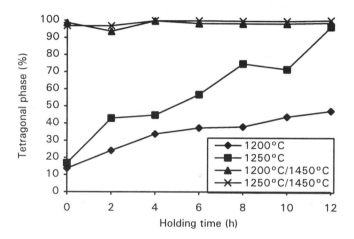

Fig. 3   The tetragonal phase retention of 2.5Y-TZP as a function of intermediate hold time and temperature.

An increase in tetragonal phase with holding time and temperature is observed for the discs sintered at 1200 and 1250°C and then cooled. It is observed that at the lower temperature a gradual increase in tetragonal phase is found. When this holding temperature is increased by 50°C, then over the same time scale >90% tetragonal phase is retained to room temperature. The samples held at the lower temperatures and then sintered at 1450°C all exhibit very high tetragonal phase retention. It is noticeable that the samples sintered at 1450°C with no intermediate hold also exhibited very high tetragonal phase content after firing. The effect of the holding time and temperature of the sintered samples on the grain growth was found to be negligible with the average grain size ranging from 0.25-0.28µm. Therefore, the most significant effect of an intermediate hold during the sintering cycle of coated Y-TZP is to promote densification and encourage the diffusion of yttria with no increase in the grain growth kinetics.

The effect of the pre-sintering stage on the Vickers hardness is shown in Fig. 4. The increase in hardness for the samples sintered at 1200 and 1250°C was rapid and reached some limiting value. Comparison with Fig.2 indicates that this behaviour was mainly due to the porosity of the ceramics. However, when the samples were sintered at 1450°C a relatively constant hardness was achieved for all holding times. This supports the density, tetragonal phase and grain size results which predict that little difference in hardness would occur.

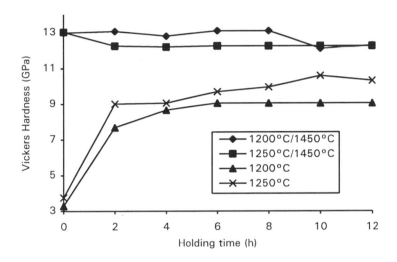

Fig.4   The influence of intermediate holding time and temperature on the Vickers hardness of 2.5Y-TZP sintered at 1450°C for 2 hours.

The enhancement in fracture toughness resulting from an inhomogeneous yttria distribution was discussed earlier. The effect of the pre-sinter on the yttria distribution, and ultimately the fracture toughness, was investigated and the results are shown in Fig. 5. The 1200°C pre-sinter showed a slight increase in fracture toughness with holding time due to the correspondingly slow tetragonal phase formation. The 1250°C pre-sinter resulted in an initial slight increase in fracture toughness, but holding for greater than 6 hours results in a rapid increase in toughness, which slowed again after 8 hours holding time. The effect of the holding time on the sintered samples was not as clearly defined. There was a peak in fracture toughness after a 6 hour pre-sinter hold, and then the fracture toughness fell slightly. The maximum toughness after 6 hours corresponded to the peak in density observed for the samples. The slight fall in toughness with holding time indicates that diffusion of the yttria was promoted, resulting in less transformability of the tetragonal phase.   Additionally the development of cubic phase in the microstructure would have resulted in a lowering of the fracture toughness.

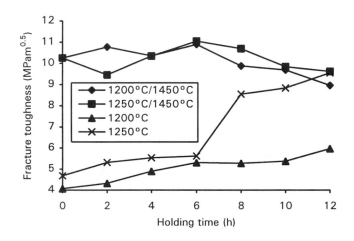

Fig.5   Indentation fracture toughness as a function of intermediate holding time and temperature of 2.5Y-TZP sintered at 1450°C for 2 hours.

3.2   Hydrothermal degradation

It has been shown in section 3.1 that two stage sintering promotes yttria mobility and leads to high densities, phase retention and enhanced mechanical properties.   It is also important to understand how this homogenizing of the stabilizer in the zirconia affects the hydrothermal degradation properties of the ceramic.

The results in Fig. 6 show the effect of pre-sintering at 1250°C for upto 12 hours followed by firing at 1450°C for 2 hours on the surface monoclinic phase formation (the results for the 1200°C pre-sinter were similar).   The resultant phase development was very typical of coated Y-TZPs, in that a rapid increase in monoclinic phase was observed until a plateau region (monoclinic saturation) was reached at 24 hours.   The autoclave environment used was a highly aggressive one and it has been shown (Lawson et al, 1994) that comparable co-precipitated materials are destroyed within 24 hours.   These results do not show any significant changes in the ageing behaviour of the pre-sintered Y-TZPs.   Using XRD is not conclusive however, as the penetration depth of the x-rays is exceeded by the aged layer thickness.   Measurement of the ageing-induced monoclinic layer showed that the holding time during the pre-sinter stage had very little effect on the layer thickness, which was ~173µm. However, the sample held for 6 hours at 1200°C had a thinner aged layer than the other samples, but this is attributed to its higher sintered density.

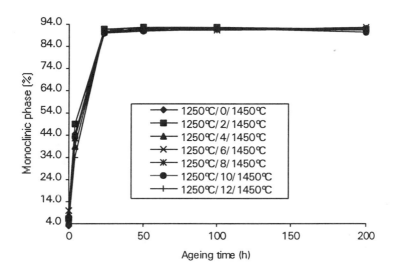

Fig.6 The development of surface monoclinic phase as a function of ageing time during hydrothermal degradation in an autoclave containing superheated water at 180°C (1MPa) for 2.5Y-TZP sintered at 1450°C and subjected to an intermediate sintering stage.

These results show that pre-sintering of coated Y-TZP did not adversely effect the ageing performance of these ceramics and indicates that the yttria enrichment of the grain boundaries was still high enough to retard hydrothermal degradation.

4 Conclusions

1. Two stage sintering of coated Y-TZP ceramics for upto 6 hours resulted in enhanced densification, high tetragonal phase retention and good mechanical properties. Holding for longer times could be detrimental to properties.
2. The inhomogeneous yttria distribution had been maintained in these coated Y-TZPs as the fracture toughness remained high and the hydrothermal degradation resistance was unchanged.
3. The potential for low temperature sintering for these ceramics exists as high fracture toughness and tetragonal phase could be achieved at 1250°C.

Acknowledgements

The authors would like to acknowledge the support of the school of Engineering & Advanced Technology at the University of Sunderland for their support, and like to thank Tioxide Specialties plc for the materials and technical discussions.

References

S. Lawson, "Environmental degradation of zirconia ceramics, part I: review", *J.Eur.Ceram.Soc.*, **15** [6] 485-502 (1995)a.

T.Sato, S.Ohtaki, T.Endo & M. Shimada, "Improvement to the thermal stability of yttria-doped tetragonal zirconia polycrystals by alloying with various oxides", pp.28-38 in Advances in Ceramics, Vol. 24, *Science and Technology of Zirconia III*, edited by S.Somiya, N.Yamamoto & H.Yanagida, The American Ceramic Society, Inc., Columbus, Ohio, (1988).

M.M.R. Boutz, "Nanostructured, tetragonal zirconia ceramics: microstructure, sinter forging and superplasticity", Ph.D. Thesis, University of Twente, Enschede, The Netherlands, (1993).

N.Kimura, S.Abe, Y.Hayashi, J.Morishita & H.Okamura, "Sintering behaviour and anti-degradation property of MO_x-doped Y-TZP (M:Cu,Mn,Co,Ni,Zn)", *Sprechsaal*, **122** [4] 341-3 (1989).

K.Tsukuma & M.Shimada, "Thermal stability of Y_2O_3-partially stabilized zirconia (Y-PSZ) and $Y-PSZ/Al_2O_3$ composites", *J.Mat.Sci.Lett.*, **4** [7] 857-61 (1985).

S. Lawson, H.C. Gill & G.P. Dransfield, "The effects of copper and iron oxide additions on the sintering and properties of Y-TZP", *J.Mat.Sci.*, **30** [12] 3057-60 (1995)b.

M.Ruhle, N.Claussen & A.H.Heuer, "Microstructural studies of Y_2O_3-containing tetragonal ZrO_2 polycrystals (Y-TZP)", pp.352-70 in Advances in Ceramics, Vol. 12, *Science and Technology of Zirconia II*, edited by N.Claussen, M.Ruhle & A.H.Heuer, The American Ceramic Society, Inc., Columbus, Ohio, (1984).

S.Lawson, G.P.Dransfield, A.G.Jones, P.McColgan & W.M.Rainforth, "Enhanced performance of Y-TZP materials through compositional zoning on a nano-scale", presented at 8th CIMTEC, *World Ceramic Congress*, Florence, Italy, (1994).

G.P.Dransfield, "Plasma synthesis and processing of ultra-fine ceramic powders", pp.1-8 in Br.Cer.Proc., Vol. 50, *Engineering ceramics: Fabrication Science and Technology*, edited by D.P.Thompson, Institute of Materials, UK., (1993).

K.Niihara, R.Morena & D.P.H.Hasselman, "Evaluation of K_{1c} of brittle solids by the indentation method with low crack-to-indent ratios", *J.Mat.Sci.Lett.*, **1** [1] 13-16 (1982).

M.I. Mendelson, "Average grain size in polycrystalline ceramics", *J.Am.Ceram.Soc.*, **52** [8] 433-36 (1969).

K. Nakajima, K. Kobayashi & Y. Murata, "Phase stability of Y-PSZ in aqueous solutions", pp.399-407 in Advances in Ceramics, Vol. 12, *Science and Technology of Zirconia*, edited by N.Claussen, M.Ruhle & A.H.Heuer, The American Ceramic Society, Inc., Columbus, (1984).

S. Lawson & P.A. Smith, "A new technique for monitoring ageing in yttria-tetragonal zirconia polycrystals", *J.Am.Ceram.Soc.*, **76** [12] 3170-72 (1993).

Reactive Liquid Phase Sintering in Zinc Oxide Varistor Materials

E. R. Leite, M. A. L. Nobre, and E. Longo
DQ-UFS Car, P.O. Box 676, São Carlos, SP, 13565-905, Brazil

J. A. Varela
IQ-UNESP, P.O. Box 355, Araraquara, SP, 14800-900, Brazil

Abstract

Densification as well as microstructural evolution of the systems $ZnO.Bi_2O_3$, $ZnO.Bi_2O_3.Sb_2O_3$ and transition metal doped $ZnO.Bi_2O_3.Sb_2O_3$ were studied using constant heating rate, high temperature X-ray diffraction and SEM. The presence of Sb_2O_3 retards the densification process of the $ZnO.Bi_2O_3$ system, that otherwise starts at 750°C, due to formation of pyrochlore phase (PY phase). For the $ZnO.Bi_2O_3.Sb_2O_3$ system the decomposition of pyrochlore phase occurs at 900°C. The addition of transition metals (ZBSCCM system) alters the formations and decomposition reaction temperatures of the PY phase and the morphology of $Zn_7Sb_2O_{12}$ spinel phase. After the decomposition of pyrochlore phase into $Bi_2O_3(l)$ and $Zn_7Sb_2O_{12}(s)$ the densification and grain growth process start. Thus sintering occurs by liquid phase with solution-precipitation as dominant mechanism.

I. Introduction

ZnO based semiconducting ceramics devices show highly non linear behavior between voltage and electric current. This non ohmic characteristic is related to the microstructure. Thus a knowledge of microstructural evolution is fundamental to control of the electric properties of ZnO based varistors.

Densification and grain growth in the systems $ZnO.Bi_2O_3$ (ZB) and $ZnO.Bi_2O_3.Sb_2O_3$ (ZBS) have been studied by several authors [1-5]. The formation of a bismuth rich eutectic liquid controls the densification and grain growth in the ZB system [1,2] with phase boundary reaction being the controlling mechanism for Bi_2O_3 concentration up to 0.54 mol%[2,5]. Kim et al [4] showed that the concentration ratio Sb_2O_3 / Bi_2O_3 controls the temperature for liquid phase formation in the ZBS system. For $[Sb_2O_3]/[Bi_2O_3]$ > 1 a Bi_2O_3 rich liquid is formed near to 1000°C due to the reaction between $Zn_2Bi_3Sb_3O_{14}$ and ZnO, as described by the equation:

$$2\,Zn_2Bi_3Sb_3O_{14} + 17\,ZnO \xrightarrow{\;950 - 1050^\circ C\;} 3\,Zn_7Sb_2O_{12} + 3\,Bi_2O_3\,(Liq.)\;(1)$$

For $[Sb_2O_3]/[Bi_2O_3] < 1$ the Bi_2O_3 rich liquid is formed at $750^\circ C$. This corresponds to the eutectic of the ZB system.

There are few works on densification and grain growth kinetics for the transition metal doped ZBS system. Asokan et al [6] reported that the addition of CoO and MnO_2 decreases the Bi_2O_3 evaporation at high sintering temperature and that the addition of Cr_2O_3 and NiO control ZnO grain growth in systems containing Bi_2O_3, Nb_2O_5, Sb_2O_3, CoO and MnO_2. Chen et al [7] reported the influence of CoO on grain growth kinetics of CoO doped ZBS system. Chen et al [7] considered that the Co^{+3} in the ZnO matrix inhibits the effect of bismuth rich liquid during sintering. Both Chen et al [7] and Asokan et al [6] considered a direct effect of additives on ZnO grains.

The objective of this work is to relate densification and grain growth with phase evolution for the ZB, ZBS and CoO, MnO_2, and Cr_2O_3 doped ZBS (ZBSCCM) systems during sintering.

II. Experimental Procedure

ZnO and the additives oxides (Bi_2O_3, Sb_2O_3, Cr_2O_3, MnO_2 and CoO) were mixed in a polypropylene ball mill jar with stabilized zirconia balls using isopropylic alcohol as milling medium. Compositions are listed in Table I. The powder was isostatic pressed in pellet form (210 MPa) reaching a green density of 3.37 ± 0.03 g.cm^3 for all studied compositions.

Pallets were sintered in a Netzcsh dilatometer (Model 402E) using constant heating rates of 2.5 and $10^\circ C/min.$ for temperatures up to $1200^\circ C$. After reaching this temperature the samples were cooled down with no soaking time.

Microstructural evolution was carried out using a Scanning Electron Microscope with X-ray microanalysis (SEM-EDS). For this analysis the polished sintered samples were chemically attacked with NaOH (1M) during several minutes. Phase evolution was characterized by High Temperature X-ray Diffraction (HT-XRD) using a high temperature camera (Model HDK S1) attached to the X-ray diffractometer (Siemens Model D-5000) with CuK_a radiation and Ni filter. Grain size was determined by the intercept method [8] according to the following equation:

$$G = 1.56L \qquad\qquad (2)$$

where G is average grain size and L is the average intercept. Grain sizes were measured in samples sintered up to $1200^\circ C$ using several heating rates.

Table I - Chemical compositions used in this study.

System	ZnO (mol%)	Bi_2O_3 (mol%)	Sb_2O_3 (mol%)	CoO (mol%)	Cr_2O_3 (mol%)	MnO_2 (mol%)
ZB	99.5	0.5	-	-	-	-
ZBS	98.5	0.5	1.0	-	-	-
ZBSCCM	97.4	0.5	1.0	0.5	0.1	0.5

III. Results

Figure 1 shows linear shrinkage rate, $d(\Delta l/lo)/dT$, as a function of temperature for the ZB, ZBS and ZBSCCM systems. The maximum shrinkage rates for theses systems are at $771^{o}C$, $1048^{o}C$ and $1109^{o}C$ respectively. The shrinkage results shows that the addition of Sb_2O_3 retards the densification process of the ZB system. Figure 2 shows linear shrinkage rate as a function of linear shrinkage ($\Delta l/l_o$). This figure shows that the maximum for shrinkage rate depends on the composition of the system. Considering that the densifying mechanisms are dominant up to the maximum shrinkage rate [9], non densifying mechanisms are dominant at lower shrinkage for the ZB system ($\Delta l/l_o$>5%) while in the ZBS system the non densifying mechanisms are dominant for $\Delta l/l_o > 14\%$. The addition of transition metals to the ZBS system decreases the shrinkage for the maximum shrinkage rate ($\Delta l/l_o = 10\%$).

Micrographs of Figure 3 show the microstructure for the three systems sintered to $1200^{o}C$. Two phases, ZnO and a Bi_2O_3 rich phase located at triple and quadruple grain boundary are observed in the ZB system (Figure 3a and 3b). An additional spinel phase with elongated grains surrounded by a rich Bi_2O_3 phase is observed in the ZBS system (Figure 3c). The ZBSCCM system also presents the same three phases observed in ZBS system; however, the $Zn_7Sb_2O_{12}$ (SP) grains are equiaxial and are homogeneously distributed through the microstructure. The EDS analysis of the SP phase shows the presence of Sb, Zn, Mn, Cr and Co. Moreover the EDS analysis of several regions of the microstructure indicated that the transition metal locates preferentially in the SP phase as reported by Olson et al [10].

Table II shows average grain sizes for the systems ZBS and ZBSCCM for different heating rates. Grain sizes are smaller is the ZBSCCM system and should be related with the spinel morphology. The small equiaxial spinel grains act as pinning for grain boundary motion and inhibit grain growth [11]. These results suggest that the transition metal oxide has an indirect effect on ZnO grain growth contrary to the direct effect proposed by Chen et al [7] and Asokan et al [6].

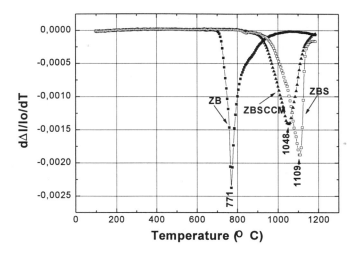

Figure 1. Linear shrinkage rate as function of temperature the ZB, ZBS and ZBSCCM system.

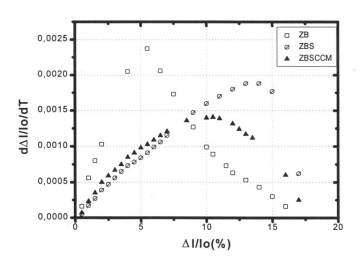

Figure 2. Linear shrinkage rate as function of linear shrinkage for the ZB, ZBS and ZBSCCM systems (heating rate = 10°C/min.).

Table II. Average grain size for the ZBS and ZBSCCM systems after
sintering in a dilatometer up to 1200°C.

System	Average grain size (μm)	
	Heating rate 2.5°C.min⁻¹	Heating rate 10°C.min⁻¹
ZBS	9.2 ± 0.9	6.5 ± 0.7
ZBSCCM	4.3 ± 0.9	2.6 ± 0.1

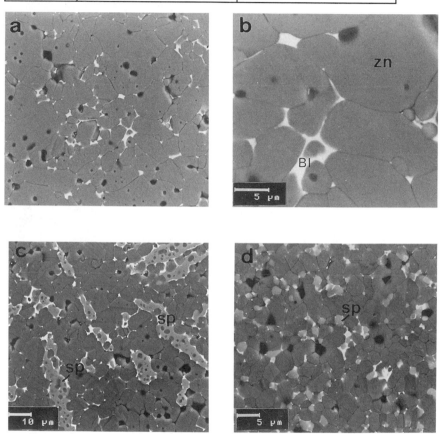

Figure 3. Scanning electron micrograph of sintered samples at constant heating rate
up to 1200°C.a) and b) system ZB (heating rate = 2.5°C/min.); c) system ZBS (heating
rate = 2.5°C/min.); d) system ZBSCCM (heating rate = 5°C/min.).

IV. Discussion

As observed in Figure 1, densification starts at 750°C for the ZB
system which is approximately the eutectic temperature for this system (740°C).

For the nominal composition of 0.5 mol% of Bi_2O_3 the amount of liquid phase formed up to 1200°C is not large as shown in Figure 4. This small amount of liquid cannot explain the fast densification of the ZB system at temperatures between 750°C and 1050°C. Considering that the liquid formed has up to 33 mol% of ZnO in solution [12], the variation of the amount of liquid with temperature must follow a curve with composition of 67 mol% of Bi_2O_3, reaching 100% of liquid at 1050°C as shown in Figure 4. Above this temperature the linear shrinkage is approximately constant for this system as observed in Figure 1. Then the local composition controls the sintering for ZB system, that is, the amount of Bi_2O_3 in contact with ZnO grains. This should promote a differentiated densification and grain growth, leading to abnormal grain growth and the presence of trapped pores (Figures 3a and 3b). These factors   should be   determinant   for densifying mechanism be limited   by low $\Delta l / lo$ values (Figure 2).

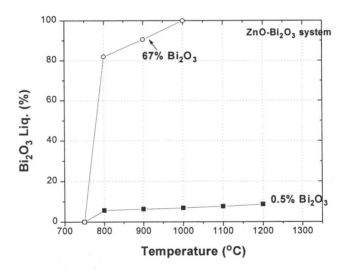

Figure 4. Amount of rich Bi_2O_3 liquid as function of temperature. This plot was made based on the   phase equilibrium diagram of the $ZnO.Bi_2O_3$ system [13].

Regarding the ZBS system, the HT-XRD results described in Figure 5 show that the $Zn_2Bi_3Sb_3O_{14}$ (pyroclore) is formed in the 700-900°C range and the reaction of this phase with ZnO (equation 1), forming SP phase and Bi_2O_3 liquid, occurs in the range of 900 to 1100°C. Thus the formation of liquid Bi_2O_3 at 900°C promotes the densification of this system via liquid phase. The shrinkage for ZBS system starts at 900°C as indicated in Figure 1. These results are in agreement with Kim et al [4] for $[Sb_2O_3]/[Bi_2O_3]>1$. The formation of spinel phase for this system begins at 800°C but this phase starts to decompose at 1050°C.

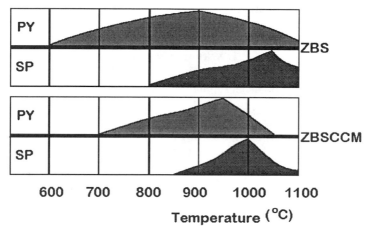

Figure 5. Phase evolution determined by HT-XRD of the ZBS and ZBSCCM systems.

As observed in Figure 5, for the ZBSCCM system the phase evolution is similar to that of the ZBS system, however the formation of PY phase initiates at temperatures above 700°C and the reaction between this phase and ZnO occurs in the 950-1100°C temperature range. Thus the formation of Bi_2O_3 liquid occurs in a narrower temperature range compared with the ZBS system. As a consequence more liquid phase is formed in a narrower temperature range to promote the densification. This result may explain the shrinkage of ZBSCCM system at lower temperature compared with the ZBS system. The spinel phase formation for this system start at 850°C and the decomposition is observed above 1000°C.

The simultaneous process of liquid phase formation and the SP phase formation in the ZBS and ZBSCCM system must inhibit the non densifying mechanism, thus is observed high value of shrinkage rate at high linear shrinkage (Figure 2).

V. Conclusions

The results of this work lead to the following conclusions:

a) The addition of Sb_2O_3 to the ZB system for ratio $[Sb_2O_3]/[Bi_2O_3] > 1$ leads to the formation of pyrochlore phase which retards densification of the system to temperatures above the decomposition of this phase.

b) The addition of transition metal oxides to the ZBS systems (named ZBSCCM) promotes the decomposition of pyrochlore phase in a narrow temperature range (950>T>1100°C) leading to densification at lower temperature as compared with ZBS system.

c) The addition of transition metal oxides to the ZBS system leads to spinel phase formation of equiaxial morphology that should act as pinning and inhibit grain growth. Hence the ZBSCCM system has a finer microstructure compared with the ZBS system.

VI. Acknowledgment

The authors acknowledge CNPq, FINEP/PADCT and FAPESP, all Brazilian agencies, for financial support of this work.

VII. References

1. J. Wong, "Sintering and Varistor Characteristic of ZnO-Bi$_2$O$_3$ Ceramics",*J.Appl. Phys*, $\underline{51}$[8], 4453-59 (1980).

2. T. Senda and R.C. Bradt, "Grain Growth in Sintered ZnO and ZnO-Bi$_2$O$_3$ Ceramics", *J. Am. Ceram. Soc.* ,$\underline{73}$[1], 106-14 (1990).

3. J. Kim, T. Kimura and T. Yamaguchi, "Effect of Bismuth Oxide Content on the Sintering of Zinc Oxide", *J. Am. Ceram. Soc.*, $\underline{72}$[8], 1541-44 (1989).

4. J. Kim, T. Kimura and T. Yamaguchi, "Sintering of Zinc Doped with Antimony Oxide and Bismuth Oxide",*J.Am.Ceram.Soc.*, $\underline{72}$[8], 1390-95 (1989).

5. D. Dey and R.C. Bradt, "Grain Growth of ZnO During Bi$_2$O Liquid Phase Sintering", *J. Am. Ceram. Soc.*, $\underline{75}$[9], 2529-34 (1992).

6. T. Asokan, G.N.K. Iyengar and G.R. Nagabhushana, "Studies on Microstructure and Density of Sintered ZnO Doped non-linear Resistor", *J. Mat. Science*, $\underline{22}$, 2229-36 (1987).

7. Y.C. Chen and C.Y. Shen, "Grain Growth Process in ZnO Varistor with Various Valence States of Manganese and Cobalt", *J. Appl. Phys.*, $\underline{69}$[12], 8363-67 (1991).

8. M.I. Mendelson, "Average Grain Size in Polycrystalline Ceramics", *J. Am. Ceram. Soc.*, $\underline{52}$[8], 443-46 (1969).

9. F.F. Lange, "Approach to Reliable Powder Processing", *Ceramic Transaction* Vol. 1, ed. by G.L. Messing, C.R. Fulles and H. Hausner, Am. Ceram. Soc., Westerville, Ohio, pg. 1069 (1988).

10.G.Olson,G.Dunlop and R. Osterlund, "Development of Functional Microstructure During Sintering of a ZnO Varistor Material", *J. Am.Ceram. Soc.*, $\underline{76}$(1), 65-71 (1991).

11. E.R. Leite, M.A.L. Nobre, E. Longo and J.A. Varela, "Microstructural Development of ZnO Varistor During Reactive Liquid Phase Sintering", Sub. to *J. Mater. Science*.

12. G.W. Morris, "Electrical Properties of ZnO-Bi$_2$O$_3$ Ceramics", *J. Am. Ceram. Soc.*,$\underline{56}$[3], 360-64 (1973).

13. G.M. Safronov, V.N. Batog, T.V. Stephanyuk and P.M. Fedorov, "Equilibrium Diagram of the Bismuth Oxide-Zinc Oxide System", *Russ. J. Inorg. Chem.*, $\underline{16}$[3] 460-61 (1971).

The Influence of Primary Particles and Agglomerates on Pressure Assisted Sintering

I. Bennett, G. de With and P.G.Th. van der Varst

Centre for Technical Ceramics
Eindhoven University of Technology, Netherlands

Abstract

Pressure assisted sintering has been used in the densification of dielectric materials. In this paper, a model is presented showing the effect of agglomerates on the densification rate during pressure assisted sintering. The influence of the agglomerate size, the primary particle size and the processing parameters are discussed.

Introduction

Pressure assisted sintering involves densification of material by the application of a pseudo uniaxial pressure at an elevated temperature in the absence of lateral constraints. This allows control over the shear forces within the material which can break up large, non-sinterable pores into small, sinterable pores (Venkatachari and Raj, 1987), leading to a high final density achieved in a short time. The application of pressure introduces an extra driving force for densification permitting a lower temperature to be used than for conventional sintering, so reducing the amount of grain growth during densification.

The material used in this work consists of calcined stacks of high purity barium titanate films. This material is used for dielectric applications where the density and grain size are important for the performance of the final product (Arlt, Hennings and de With, 1985), making pressure assisted sintering a suitable processing method (Derksen and de Wild, 1991; Keizer and de Wild, 1994). Experimental work is described in a previous publication (Bennett and de With, 1994).

To describe densification, the equations compiled by Wilkinson and Ashby (Wilkinson, 1977; Wilkinson and Ashby, 1978) were used. These authors estimate the contribution of several densification mechanisms to the overall densification rate at various stages during the sintering process. Data for the equations were obtained from literature: diffusion coefficients and

activation energies were taken from work by Wernicke (1976) and
Mostaghaci (1982). Creep constants were found in work by Xue et al. (1990).
Further data were obtained from experimental densification curves and
microstructural analysis. Microstructural analysis also revealed a degree of
agglomeration of the green material, with primary particles under 0.8 μm and
an agglomerate size of several micrometers. In this work, primary particles
are taken to have a size of between 0.1 and 1 μm, with agglomerates between
1 and 10 μm. A limited number of models describing the densification of bimodal powders, with grains of two sizes sintering simultaneously can be
found in the literature. Both Coble (1973) and Chappell and Ring (1986)
emphasize the significant influence of the particle size distribution in the
modelling of the densification rate. Lin et al.(1987) considered densification
of a heterogeneous structure with an approximately bi-modal distribution of
pores. They found an increased densification and creep rate for relative
densities in the range 0.6 to 0.75. A small number of other papers consider
the role of agglomerates in ceramic processing (Reed et al., 1978; Halloran,
1983). These authors primarily address the importance of agglomerates
experimentally. In this work, densification by pressure assisted sintering of
calcined stacks of BaTiO$_3$ films, containing agglomerates made up of primary
particles, is studied.

Theory

A number of approaches can be considered for modelling the
pressure assisted sintering of agglomerates. In the Wilkinson-Ashby equations
the coordination number, increasing with increasing density, plays an
important role. In the first approach all the primary particles within the
agglomerates and within the necks between the agglomerates are assigned a
coordination number according to the Wilkinson-Ashby equations, while the
primary particles at the surface of the agglomerates and at the surface of the
necks can be given, say, half that value. This fraction is chosen as the surface
particles are surrounded by approximately half the number of particles as the
bulk particles. The ratio of bulk to surface particles can be calculated from
the pore size, included in the Wilkinson-Ashby equations, and primary
particle size. Using this ratio, an average coordination number can be
determined.

A second approach is to determine two distinct sintering volumes: a
bulk volume and a shell volume of one primary particle thickness. The
densification rate for the bulk volume would be calculated using the primary
particle size, whilst the shell densification rate would be calculated using the
agglomerate size. The overall densification rate would be obtained from a
combination of these two rates. Due to the smaller size of the primary
particles, the radius of the sphere containing the bulk material would
decrease at a faster rate than the radius of the shell. To maintain the

integrity of the agglomerate, a coupling stress between the two volumes must be calculated. This stress can then be included in the applied stress used to determine the densification rate between the primary particles in the bulk volume.

A third approach would be to calculate a properly averaged radius of curvature from the radius of the agglomerates and the radius of the primary particles, leading to an improved description of the neck geometry between the agglomerates.

This paper reports on a fourth approach where the densification was divided between the primary particles and the agglomerates. Due to their smaller size, it was assumed that the primary particles densify more rapidly than the material as a whole. The primary particles were densified by a predetermined density step, thereby reducing the radius of the agglomerates. The agglomerates were then densified for the same period of time using the new radius. Subsequently, the primary particles were densified by another density step, resulting in a new agglomerate radius, followed by the same densification time for the agglomerates using the new radius. This alternating process is continued until the agglomerates reach full density. Further densification between the agglomerates is continued as for a single grain sized material, with the grain size equal to the current agglomerate size, until full density is achieved for the complete material.

The total volumetric density, ρ_T, of the material can be calculated from:

$$\rho_T = \frac{4}{3}\pi r^3 NM$$

where r is the primary grain radius, N the number of primary grains per agglomerate and M the number of agglomerates per unit volume.

The volumetric density of the agglomerates, ρ_a, can be calculated from:

$$\rho_a = (\frac{r}{R})^3 N$$

where R is the agglomerate radius. Using these equations and knowing the initial values of R, r, ρ_T and ρ_a from experimental measurements, the change in R due to sintering of the primary particles can be calculated and the influence of this change on total densification evaluated.

The Wilkinson-Ashby equations are for isostatic conditions whereas here, the pressing is performed pseudo uniaxially. Although shear deformation is known to contribute to densification, the contribution is assumed to be small and has been neglected.

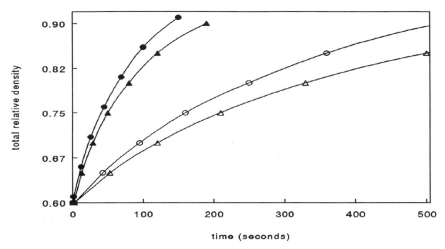

Figure 1. Densification for a primary particle radius of 0.1 μm at 50 MPa, 1460 K, agglomerate size 1 μm (●), 10 μm (▲) and 1410 K, agglomerate size 1 μm (○), 10 μm (▵).

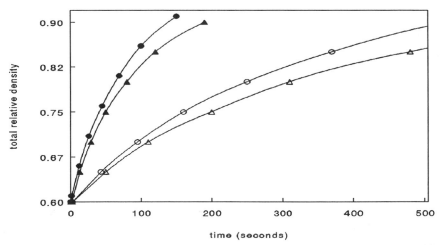

Figure 2. Densification for a primary particle radius of 0.1 μm at 1460 K, 50 MPa, agglomerate size 1 μm (●), 10 μm (▲) and 25 MPa, agglomerate size 1 μm (○), 10 μm (▵).

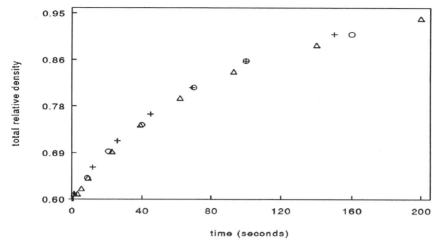

Figure 3 Densification at 1460 K and 50 MPa for an agglomerate radius of 1 μm at various particle radii: 0.1 μm (+), 0.2 μm (○) and 0.4 μm (▵).

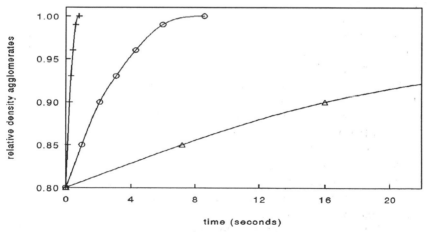

Figure 4. Initial densification of primary particles showing that full density of agglomerates is reached rapidly, primary particle size 0.1 μm (+), 0.2 μm (○), 0.4 μm (▵), 1460 K, 50 MPa.

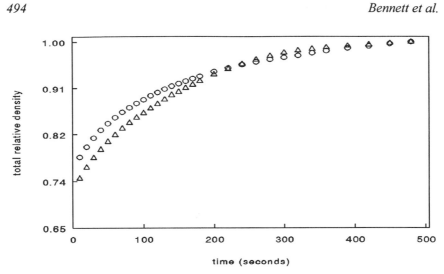

Figure 5. Experimental densification curve at 1460 K and 50 MPa for a primary particle diameter of 0.6 μm (○) and 0.8 μm (▵).

Results and Discussion

For primary particles between 0.1 and 1 μm and agglomerates between 1 and 10 μm, the model predicts that the agglomerate size is a determining factor on the overall densification rate during pressure assisted sintering (figures 1 and 2). The rate of densification between the primary particles is rapid relative to the rate of densification between the agglomerates. For the alternating process used to calculate the overall densification rate in this model, it is the slower densification rate between the agglomerates that dominates. For a specific agglomerate size, the influence of the primary particle size is limited (figure 3). The computational results agree with the assumptions underlying the model.

Densification between the primary particles is rapid due to the small particle size and the low initial porosity (figure 4). The time needed for the agglomerates to reach full density is a small fraction of that needed for densification of the complete material, so the influence on overall densification time is minimal.

The influence of agglomerate size is amplified at lower temperature (figure 1) and lower applied pressure (figure 2). For a given relative density the difference in time between the curves made at the lower temperature and the lower pressure is greater. This effect is caused by the decreased importance of creep mechanisms at low pressures and temperatures. The influence of particle size on these mechanisms is smaller than for the diffusion mechanisms dominating at lower pressures and temperatures.

Figure 5 shows experimental densification curves for barium titanate stacks with two different primary particle diameters (0.8 and 0.6 μm). Both samples were treated in a similar way prior to pressing and can be assumed to have a similar agglomerate size. These plots indicate that the primary particle size has a relatively small influence on the densification rate for materials with a comparable agglomerate size. This agrees, at least partially, with the computative predictions.

Conclusions and Final Remarks

From experimental evidence, it is a well known fact that the agglomeration state of the starting material is a determining factor in the densification rate during pressure assisted sintering. In this work, a simple model is presented which demonstrates this. The degree of influence of the agglomeration state on densification is in turn determined by the processing parameters (pressure and temperature). The results of the model justify the approach chosen to describe the pressure assisted sintering of a material containing agglomerates made up of primary particles.

Further development of the model by incorporation of an average coordination number, densification of two volumes and an average radius of curvature will allow a more complete description of the densification process, permitting more definite conclusions to be drawn. In addition to this, an agglomerate and primary particle size distribution should be considered to model the material as realistically as possible.

References

K.R. Venkatachari and R. Raj, "Enhancement of Strength through Sinter Forging", *J.Am.Ceram.Soc.* **70** [7] 514-520 (1987).

G. Arlt, D. Hennings and G. de With, "Dielectric Properties of Fine-Grained Barium Titanate Ceramics", *J.Appl.Phys.* **58** [4] 1619-1625 (1985).

R.J.A. Derksen and W.R. de Wild, "Quality Control of Thin Dielectric Layers in Multilayer Ceramic Capacitors by means of Pressure Induced Sintering", *5th EuroCARTS*, Munich, Germany, pp.221-228, (1991).

P.H.M. Keizer and W.R. de Wild, "A New Concept for Integration of Passive Components", pp.1045-1053, *Proceedings Electroceramics IV*, edited by R. Waser, S. Hoffmann, D. Bonnenberg and Ch. Hoffmann, (1994).

I. Bennett and G. de With, "Sinter Forging of Barium Titanate", pp.1271-1274, *Proceedings Electroceramics IV*, edited by R. Waser, S. Hoffmann, D. Bonnenberg and Ch. Hoffmann, (1994).

D.S. Wilkinson, "The Mechanisms of Pressure Sintering", Ph.D. Thesis, Cambridge, (1977).

D.S. Wilkinson and M.F. Ashby, "Mechanism Mapping of Sintering under an Applied Pressure", *Science of Sintering* **10** [2] 67-76 (1978).

R. Wernicke, "The Kinetics of Equilibrium Restoration in Barium Titanate

Ceramics", *Philips Res.Rep.* **31** 526-543 (1976).

H. Mostaghaci,"Fast Firing and Hot Pressing of Barium Titanate Compositions", Ph.D. Thesis, University of Leeds, (1982).

L.A. Xue, Y. Chen, E. Gilbart and R.J. Brook,"The Kinetics of Hot-Pressing for Undoped and Donor-doped $BaTiO_3$ Ceramics", *J. Mat. Sci.* **25** 1423-1428 (1990).

R.L. Coble,"Effects of Particle Size Distribution in Initial Stage Sintering", *J.Am.Ceram.Soc.* **56** [9] 461-466 (1973).

J.S. Chappell, T.A. Ring and J.D. Birchall,"Particle Size Distribution Effects on Sintering Rates", *J.Appl.Phys.* **60** [1] 383-391 (1987).

M. Lin, M. N. Rahaman and L. C. de Jonghe,"Creep-Sintering and Microstructure Development of Heterogeneous MgO Compacts", *J.Am.Ceram.Soc.* **70** [5] 360-366 (1987).

J.S. Reed, T. Carbone, C. Scott and S. Lukasiewicz,"Some Effects of Agglomerates in the Fabrication of Fine Grained Ceramics", pp.171-180, *Processing of Crystalline Ceramics*, edited by H. Palmour, R.F. Davis and T.M. Hare, Plenum Press, New York, (1978).

J.W. Halloran,"Role of Powder Agglomerates in Ceramic Processing", pp.67-75, Advances in Ceramics, Vol.9, *Forming of Ceramics*, edited by J.A. Mangels and G.L. Messing, The American Ceramic Society Inc., Columbus, (1983).

Effect of Non-Densifying Inclusions on the Densification of Ceramic Matrix Composites

Samuel M. Salamone and Rajendra K. Bordia

Department of Materials Science and Engineering
University of Washington, Seattle, WA.

Abstract

Since grain growth during sintering is a characteristic feature of polycrystalline ceramics, in this study, we focused on the effect of grain growth on the densification of composites. Rigid dense inclusions that are at least an order of magnitude larger than the average grain size of the matrix material were introduced (5, 10, and 20 vol%) into a commercial $Zr(Y)O_2$ powder with very little grain growth during sintering (ZrO_2-2.5 mol% Y_2O_3). The inclusion particles were made from the same powder as the matrix (ZrO_2-2.5 mol% Y_2O_3) in order to avoid any possible compositional changes. The observed densification behavior for constant heating rate and three different isothermal heat-treatments was examined. The constant heating rate experiments showed a delay in sintering for the composites, however, the final densities were comparable, i.e. > 97 %. The isothermal experiments showed a larger reduction in densification to those of the matrix alone. The microstructure showed no signs of excessive grain growth in both the inclusion free matrix and the composite matrix.

1. Introduction

The role of non-densifying inclusions in a polycrystalline matrix material has been shown to have a detrimental effect on densification during both constant heating rate and isothermal heating schedules. For constant heating rate experiments Fan and Rahaman(1992) added 5-22 vol% of 14 μm ZrO_2 particles in a ZnO matrix and Sudre and Lange(1992) added 9 and 30 vol% of large(38-53 μm) ZrO_2 inclusions in an Al_2O_3 matrix. Both observed a reduction in the densification rate and final density for all composite powders and the effect was more severe as the volume fraction of inclusions increased. Similar results were obtained in isothermal heat treatments by Bordia and Raj(1988) for Al_2O_3 inclusions in TiO_2 matrix and by De Jonghe et. al.(1986) with 10 vol% of 12µm SiC particles in a ZnO matrix.

Recent studies have shown the effect of improved green state processing on the densification of polycrystalline composites. Hu and Rahaman (1992) investigated the sinterability of coating inclusion particles with a porous cladding of the matrix powder. They found that high densities could be achieved and that the problems associated with non-uniform matrix packing and interacting inclusions were alleviated by this technique.

As opposed to the polycrystalline matrix, the effect of non-densifying inclusions on the densification of glass matrix composites has been found to be insignificant unless the inclusion loading exceeds the percolation threshold (Rahaman & DeJonghe, 1987). One significant difference between the glass matrix and the polycrystalline matrix is the concurrent grain growth during densification in polycrystalline matrices. The existing analyses of densification of composites do not account for grain growth explicitly. However, grain growth will increase the viscosity of the matrix influencing visco-elastic analyses (Bordia & Scherer, 1988) and it will also increase the resistance to deformation of the constrained network (Lange, 1987).

The effect of grain growth on densification of composites has not been well studied. Fan and Rahaman (1992) did report on the grain size for both the matrix alone and the composite system. However, in their matrix (ZnO) there is significant grain growth during sintering and a significant detrimental effect of the inclusion was shown. In this study we examine the hypothesis: Will the absence of grain growth in the matrix during sintering alleviate the detrimental effect of non-densifying inclusions. Both constant heating rate and isothermal experiments, for a range of volume fraction of inclusions, have been conducted.

2. Experimental Procedure

As received ZrO_2 with 2.5 mol% Y_2O_3 (Tosoh USA, Inc., Atlanta, GA) were used as the matrix powder. The crystallite size of the powder was .025 µm. The inclusion particles were formed by spray drying (Bowen Engineering) and then heated to $1450°$ C for .5 hours to ensure that they were rigid and non-densifying in the temperature regimes used for sintering experiments. The inclusions range in size from 1-15 µm and are fully dense. The matrix powder was suspended in an aqueous medium at a solids content of 30 volume percent. The pH (7-8) of the suspension was adjusted to cause flocculation in order to prevent the segregation of the inclusions and the matrix particles. The inclusion particles were added as a function of total solid content of 5, 10, and 20 volume percent. The composite suspensions were filter pressed with an applied pressure of approximately 14MPa.

The densification behavior of the matrix and composites was studied during constant heating rate and isothermal experiments. All experiments were performed in air. The linear shrinkage during both sets of experiments was continuously measured using a dilatometer. Final geometric measurements found that the compacts sintered isotropically. The density during sintering was calculated from linear shrinkage data and the final density. Archimedes method in water was used to determine the open and closed porosity. From this the relative density of the compacts was calculated using $\rho_{th} = 6.05$ g/cm^3. The sintered compacts were polished and thermally etched to allow observation of the microstructure/grain size using a scanning electron microscope, SEM (JSM-5200, JOEL). The grain sizes were measured by counting the number of grains intersecting a given diameter overlaid onto a SEM micrograph. At least 50 intercepts were taken for a given sample.

3. Results

3.1 *Densification Characteristics*

Figure 1 shows the densification behavior of pure ZrO$_2$ and ZrO$_2$ matrix composites with a constant heating rate of 5° / min. To account for the small differences in green densities (the same pressure was applied during filter pressing) the densification curves were normalized using the following equation:

$$\Delta\rho/\Delta\rho_{max} = [\rho_m(t) - \rho_m(0)] / [\rho_m(Th) - \rho_m(0)]$$

where $\rho_m(0)$ [t = 0] is the matrix density at the onset of densification, $\rho_m(Th)$ is the theoretical density and $\rho_m(t)$ is the matrix density as a function of time.

In Fig 2, the densification rate $\{1/\rho_m \ (d\rho_m/dt)\}$ has been plotted as a function of temperature to clearly illustrate the effect of inclusions. From both these figures, it is clear that the inclusions delay the densification of the composite to higher temperatures. The peak densification rate is lower and is shifted to higher temperatures as the volume fraction of inclusions increases. Low volume fraction of inclusions (5 vol%) have a negligible effect.

The final densities of all the constant heating rate compacts were > 97 % of theoretical. The inclusions had the effect of delaying densification but they did not hinder the composite from eventually obtaining a high final density. Isothermal densification at much lower temperatures was done to see

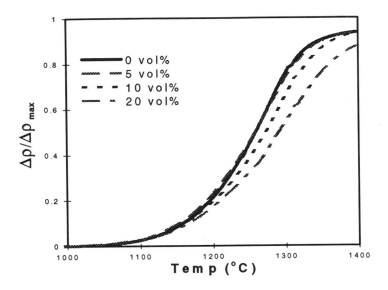

Fig. 1 *Normalized density as a function of time for composites containing 0-20 vol% inclusions at a constant heating rate of 5°C/min.*

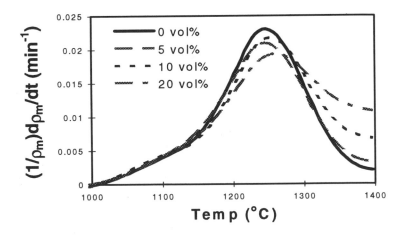

Fig. 2 *Densification rate of the matrix as a function of temperature for data in Fig. 1.*

no sign of grain growth inhibition or rapid growth behavior due to the presence of the inclusions nor does the inclusion free matrix appear to have undergone significant coarsening as has been seen in other systems (Rahaman & De Jonge 1991).

Figure 5 is a micrograph of the fracture surface of the composite. There is no detrimental effect of the inclusion on the matrix microstructure and in particular there are no crack-like defects that have been observed in other systems (Bordia & Raj 1988 and Sudre & Lange 1992).

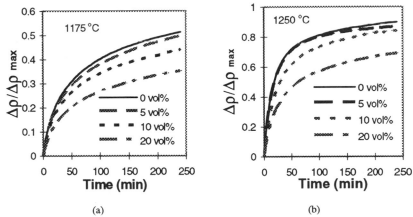

(a)   (b)

Fig 3 (a) & (b) Normalized matrix density as a function of time for composites with different volume fraction of inclusions for isothermally sintered samples.

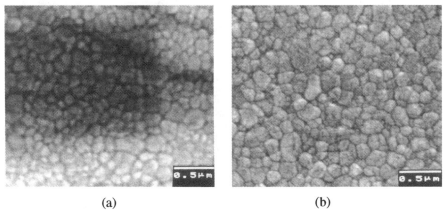

(a)   (b)

Fig. 4 (a) Microstructure of the matrix (inclusion free) (b) Microstructure of 10 vol% composite (Both sintered at 5°C/min to 1400 °C with 1 hour hold)

if the inclusions had a greater effect when the densification kinetics were not as rapid.

Figure 3 shows isothermal densification curves for 1250 °C and 1175 °C. Isothermal densification at 1215 °C was also done, but is excluded due to space constraints. The densification rate of the composite matrix is slowed significantly for the isothermal regimes studied. The densification behavior of the composites are similar to that of the constant heating rate curves where the 5 and 10 vol% curves are offset slightly from each other and the 20 vol% is significantly delayed from the 5 and 10 vol%.

The final density versus volume fraction relationship for the isothermal experiments indicates that as the isothermal temperature is increased, the inclusions have a smaller effect on the final sintered densities of the matrix. This is consistent with the constant heating rate curves that show a larger inclusion effect at low temperatures, from the delayed sintering, but then eventually the densities become comparable to the inclusion free matrix.

3.2 *Microstructural Observations*

The matrix material has a very low grain growth behavior during sintering (the grain size increased from .1μm to .13μm for density changes from 75% to 99%). The grains do not significantly increase in size as the compacts densify. This has been reported elsewhere (Slamovich & Lange 1992) for this system.

The Figure 4 shows that the microstructure of the matrix alone and the matrix in the composite are not significantly different for this system. There is

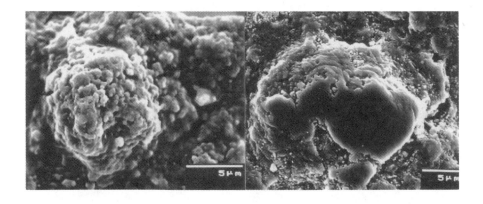

(a)   (b)

Fig 5 *(a) & (b) Fracture surface of 5 vol% composite, showing inclusion/matrix environment.*

4. Discussion & Conclusions

In this study, we report on the densification behavior of a polycrystalline matrix composite during both constant heating rate and isothermal heating schedules. Colloidal processing techniques were used to obtain a homogeneous green body and minimize packing defects around inclusions. It is shown that the effect of rigid inclusions increases as the volume fraction of inclusions increases and as the sintering temperature decreases. In all cases the effect was significantly less severe than what has been reported in the literature for polycrystalline matrix composites. In particular, the densification behavior of a composite with 5 vol% inclusions was almost identical to that of the matrix . Noticeable differences between the pure matrix and the composite, under all heating schedules, were observed only for the composite with 20 vol% inclusions. This is to be expected and has been observed even for glass matrix composites since this loading level is beyond the percolation threshold. For a comparison of the constant heating rate and isothermal heating experiments again only the samples with 20 vol% inclusions showed any deviation. This being due to the more rapid densification kinetics of the CHR experiments which enabled the compact to eventually reach a much higher final density.

The primary difference between this and other systems that have been studied is the grain growth kinetics during densification. For systems that have shown a significant effect of the inclusions, and where grain growth data has been reported, the grain growth during densification is rapid (Bordia, 1986 and Fan & Rahaman, 1992). In this case, the matrix undergoes very little grain growth during sintering. Further, the final grain size of the matrix with or without the inclusions are very similar. Thus a preliminary conclusion can be drawn: The detrimental effect of inclusions on the densification of the matrix will decrease as the grain growth kinetics during densification decrease. This will be the focus of further experimental and modeling studies. In particular the creep and densification viscosities of the matrix and the composites will be determined as a function of microstructural evolution.

5. Acknowledgments

This research was supported by NSF grant number MSS9209775 and DMR9257027.

6. **References**

C.L. Fan and M.N. Rahaman, "Factors Controlling the Sintering of Ceramic Particulate Composites: I, Conventional Processing," J. Am. Ceram. Soc., **75** [8] 2056-65 (1992).

O. Sudre and F.F. Lange, "Effect of inclusions on Densification: I, Microstructural Development in an Al2O3 Matrix Containing a High Volume Fraction of ZrO2 Inclusions," J. Am. Ceram. Soc., **75** [3] 519-24 (1992).

R.K. Bordia and R. Raj, "Sintering of TiO2-Al2O3 Composites: A Model Experimental Investigation," J. Am. Ceram. Soc., **71**[4] 302-10 (1988).

L.C. De Jonghe, M.N. Rahaman, and C.H.Hsueh, "Transient Stresses in Bimodal Compacts During Sintering," Acta Metall., **34** [7] 1467-71 (1986).

C.L. Hu and M.N. Rahaman, "Factors Controlling the Sintering of Ceramic Particulate Composites: II, Coated Inclusion Particles," J. Am. Ceram. Soc., **75** [8] 2066-70 (1992).

M.N.Rahaman and L.C.De Jonghe, "Effect of Rigid Inclusions on the Sintering of Glass Powder Compacts," J. Am. Ceram. Soc., **70** [12] c-348-51 (1987).

R.K.Bordia and G.W.Scherer, "On Constained Sintering-II. Comparison of Constitutive Models," Acta Metall., 36 [9] 2399-2409 (1988).

F.F. Lange, "Constrained Network Model for Predicting Densification Behavior of Composite Powders," J. Mater. Res., **2** [1] 59-65 (1987).

E.B. Slamovich and F.F. Lange, "Densification of Large Pores: I, Experiments," J. Am. Ceram. Soc., **75** [9] 2498-508 (1992).

M.N.Rahaman and L.C.De Jonghe, "Sintering of Ceramic Particulate Composites: Effect of Matrix Density," J. Am. Ceram. Soc., **74** [2] 433-36 (1991).

R.K. Bordia , "Ph.D. Thesis " Cornell University (1986).

Rate Controlled Sintering of Cofired Multilayer Electronic Ceramics

W. S. Hackenberger, T. R. Shrout, and R. F. Speyer[*]

Intercollege Materials Research Laboratory, The Pennsylvania State University •
University Park, PA 16802
[*]School of Materials Science and Engineering, Georgia Institute of Technology •
Atlanta, GA 30332

Abstract

The effect of constant heating and shrinkage rate on flaw generation during cofiring of multilayer electronic substrates was investigated. The substrates studied consisted of a low temperature cofireable ceramic (LTCC) insulator material and one of three silver conductor materials: a fritted silver powder, a fritless silver powder and a fritless silver flake. Rate controlled sintering (RCS) was used to impose constant shrinkage rates on the sample substrates. Delaminations, conductor film discontinuities, and conductor film thickening were all observed for constant heating rate sintering and alleviated to varying degrees by RCS. Multilayer shrinkage anisotropy and conductor film thickness measurements indicated that the RCS profiles tended to increase the overlap in conductor and insulator densification rates during the early stages of cofiring.

1 Introduction

In this work the effect of total shrinkage rate on flaw generation in an LTCC electronic substrate was investigated using rate controlled sintering. Since stresses that result from differential densification depend on strain rate (Scherer and Garino, 1985; Evans, 1982; Bordia and Raj, 1985), it is postulated that controlling the strain rate via RCS may be a way of influencing the transient stress during cofiring. Previous work (Hackenberger, Shrout, and Speyer, 1995) has shown that LTCC substrates densified using RCS have significantly fewer flaws than substrates sintered at constant heating rates. The two types of defects affected were conductor/insulator delamination and conductor film discontinuities. In this paper the effect of RCS on differential densification rates is investigated to determine a mechanism for flaw reduction in silver/LTCC multilayers. The role of silver conductor morphology is also addressed.

505

2 Experimental Procedure

The construction and sintering of commercial multilayer substrates with fritted silver conductor films are described elsewhere (see previous citation). To more completely understand the cofiring behavior of the constituent materials, model multilayers were constructed using a tape-cast glass-alumina composite insulator (68 wt% Pb-Ca-aluminoborosilicate glass and 32 wt% alumina) and either fritless silver powder or flake thick film inks. The use of fritless conductors prevented conductor-substrate adhesion during film free strain rate measurements and did not significantly affect the relationship between heating rate and differential densification.

Multilayers were constructed by first screen printing silver onto 2.54 cm square pieces of tape so that one side of the tape was completely covered by silver. Twenty printed layers of tape were stacked with 10 layers facing up and 10 facing down so that the multilayer had mirror symmetry across the center plane parallel to the printed surfaces. This was done to prevent warping so that shrinkage anisotropy could be used as an indication of the transient stress level. The samples were laminated at 55 MPa and 75 °C for 15 min. The multilayers were then diced into 4 mm x 4 mm squares followed by binder burnout at 400 °C for 1 hour in a box furnace with flowing air.

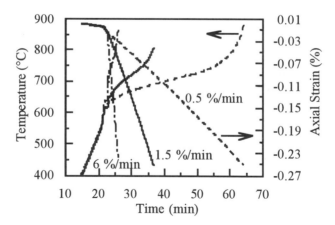

Fig. 1. Rate controlled sintering temperature and shrinkage profiles for model multilayers with silver powder films.

The sintering experiments were performed with the previously described RCS dilatometer and infrared imaging furnace (Hackenberger and Speyer, 1994) which provides a broad range of heating and shrinkage rates (≤ 1 - 500 °C/min. and ≤ 0.1 to 6 %/min., respectively). Multilayers were sintered at constant heating rates of 10 and 50 °C/min. and constant shrinkage rates of 0.5, 1.5, and 6 %/min. For the RCS trials, the samples were heated to the sintering onset temperature (~ 550 °C) at 25 °C/min. (see fig. 1). Shrinkage was monitored in the axial direction, and some samples were partially sintered so that intermediate radial shrinkages and microstructures could be observed.

The free strain rate behavior of the multilayer constituents was evaluated at constant heating rates of 10 and 50 °C/min. For the insulator material this was done by fabricating monolithic laminates and sintering them in the dilatometer. Free radial strain rates were determined for fritless silver films that had been presintered to 600 °C on a rigid substrate. This was done to simulate the constraint on multilayer films before insulator densification. After presintering, the films were peeled off the substrate and reheated to 900 °C at either 10 or 50 °C/min. in the hot stage of an optical microscope. The images were recorded on video tape and later analyzed for radial shrinkage with image analysis software.

3 Results and Discussion
Three main types of sintering defects were observed in the LTCC multilayers: 1) conductor/insulator delaminations, 2) conductor film discontinuities, and 3) conductor film thickening and cavitation. The causes of each of these defects and the portions of the sintering schedule critical to their formation are discussed in the following sections.

3.1 Contributions to Transient Stress
The resultant multilayer radial strain rate includes both the driving force for stress development and stress relaxation. The difference in the resultant and free strain rates for the constituent materials governs the net stress level in each material. This is shown graphically in fig. 2 for multilayers with silver powder films sintered at a CHR of 10 °C/min. The vertical line is at the cross-over point in the film and monolith free strain rates (point of zero stress). The film and substrate are each in radial tension to the left and right of this line, respectively.

Before the onset of sintering in the glass-alumina composite, the resultant multilayer strain rate was zero; therefore, the silver film was rigidly constrained. Above the cross-over point, the film strain rate rapidly decreased as the silver approached its theoretical density. However, even after the silver had densified, there was still a large resultant strain rate indicating that the dense silver films can relax the compressive stress imposed on them by the sintering insulator. Because the films densified before the insulator, the radial stress in the multilayers decreased the total radial sintering strain compared to the free sintering of insulator monoliths as shown in fig. 3. Besides affecting the shrinkage anisotropy, the transient stress can also lead to a number of defects which are discussed below.

3.2 Delamination

In the previous work on a commercial LTCC system, delaminations were found to occur at heating rates greater than or equal to 25 °C/min. (fig. 4). Microstructure observations of partially sintered multilayers revealed that

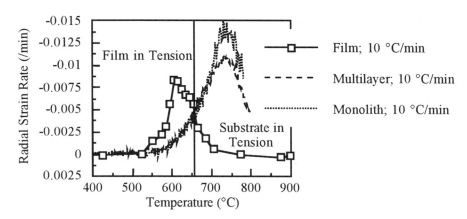

Fig. 2. Film and insulator monolith free strain rates and resultant multilayer strain rate for model multilayer with silver powder films sintered at a CHR of 10 °C/min.

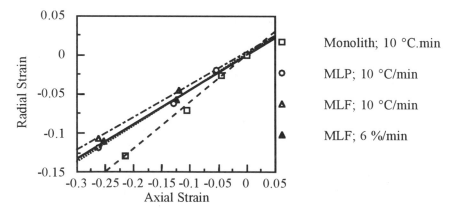

Fig. 3. Sintering strain anisotropy for model multilayers and insulator monoliths. MLP = multilayer with silver powder and MLF = multilayer with silver flake. Note the reduction in anisotropy ($\varepsilon_z/\varepsilon_r$) for rapid RCS of MLF.

delaminations initiated at about 800 °C. By this temperature the films were usually dense and the insulator was at or near its peak densification rate. Furthermore, delaminations were only observed at triaxially constrained electrode edges (see fig. 4).

Since the films were nearly dense by the time delamination occurred, the defect probably resulted from the compressive stress applied to buried film tips by the sintering glass-alumina composite. This compression could have caused the film and adjacent insulator layers to buckle. Because the silver was dense, the sintering schedule would not be expected to have much influence on delamination. However, previous work revealed that an RCS shrinkage rate of 1.5 %/min. dramatically reduced the incidence of delamination over CHR sintering at 25 °C/min. The implication is that in the early stages of cofiring RCS enhances the interface strength so that it can withstand buckling stresses that occur at the end of cofiring. The mechanism by which this strength enhancement occurs is not yet known.

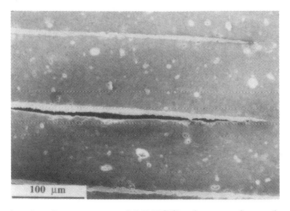

Fig. 4. Delamination in a commercial LTCC substrate sintered at 25 °C/min.

3.3 Film Discontinuities

Glass filled breaks in the conductor film cross sections were observed at low heating and shrinkage rates for both fritted and unfritted silver powder films (fig. 5) and to a lesser extent in films comprised of silver flake. The origin of these defects was probably green state inhomogeneities such as large pores. However, the lower incidence of these defects with increased RCS shrinkage rate suggested that the transient stress on the film played a role in healing the discontinuity. This could be done by matching the initial film and insulator free strain rates so that the film was less constrained or by increasing the compressive load on the film after its strain rate decreased below the resultant multilayer strain rate.

Fig. 6 is a comparison of the silver powder and glass-alumina free strain rates at 10 and 50 °C/min. and of the resultant multilayer strain rate for an RCS shrinkage rate of 1.5 %/min. As the heating rate increased, there was more overlap in the film and insulator free strain rates. At 50 °C/min. the cross-over point was approximately 25 °C lower than for 10 °C/min., and after the cross-over, the silver free strain rate remained at a finite value rather than going to zero. The RCS resultant strain rate followed a path between these two extremes, potentially matching the silver and insulator free strain rates in the early stages of cofiring. This would have reduced the transient tensile stress on the silver films leading to enhanced densification and rapid closure of large pores. Matched free strain rates may have also have strengthened the interface by subjecting it to less shear stress. This in turn could account for the success of the 1.5 %/min. rate in avoiding delaminations, as discussed previously.

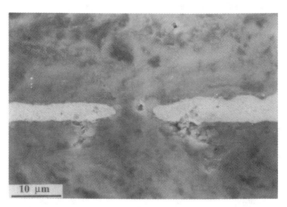

Fig. 5. Film discontinuity in a commercial LTCC substrate.

3.4 Film Thickening

Average conductor film thicknesses, measured from SEM micrographs, are shown in fig. 7 for multilayers with silver powder and flake films. The points refer to the film thicknesses after burnout, at the midpoint of multilayer densification, and at the end of cofiring. The standard deviation on these measurements is $\pm$ 1 μm and is due mainly to roughness and non-uniform screen printing. Since the films were not constrained in the axial direction, they responded to the radial compressive stress in the later stages of cofiring by thickening. The silver powder material thickened very little, but the final flake thickness was greater than it value after burnout.

SEM observations of exposed surfaces of silver flake films revealed that the films reached nearly full density by the midpoint of cofiring then cavitated under further transient compressive stress. The sintering schedule did not effect

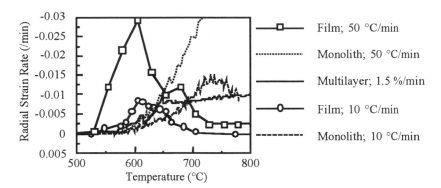

Fig. 6. Comparison of film and insulator free strain rates at 10 and 50 °C/min. and multilayer strain rate at 1.5 %/min.

the final film thickness very much, but the microstructural evolution path the films took on the way to their final thicknesses was changed. In fig. 7, multilayers sintered at RCS constant shrinkage rates of 1.5 and 6 %/min. had higher intermediate film thicknesses than the slower rates. This would occur if the higher rates pushed the film densification to higher temperatures so that the film did not completely densify before the compressive stress from the sintering insulator forced it to thicken. Furthermore, the drag on the insulator layers would decrease because they would be working against a porous film rather than a dense one. This, in fact, showed up as reduced sintering strain anisotropy for the whole multilayer (see fig. 3).

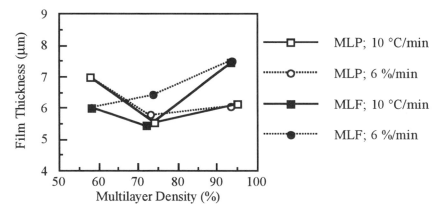

Fig. 7. Average silver film thickness for model multilayers during cofiring.

RCS had less effect on the evolution of powder film thickness. The density of the powder films after burnout was lower than that of the flake (65 and 75 %, respectively). As a result porosity probably persisted longer in the powder than in the flake, and cavitation was not needed to achieve the same amount of stress relaxation. However, the densification of the films was still affected by sintering schedule. The intermediate film thicknesses were again larger for RCS schedules of 1.5 and 6 %/min. compared to the slower rates. Both the powder and flake data verify that RCS creates more overlap between the film and insulator densification rates as suggested in section 3.

4 Conclusions

The effect of rate controlled sintering at constant shrinkage rates was investigated for a commercial LTCC electronic substrate and using model multilayers with glass-alumina composite insulator layers and fritless silver powder or flake conductor layers. For the commercial system, an RCS schedule was found that reduced the incidence of conductor/insulator delamination and conductor film discontinuities over that found in samples conventionally sintered at constant heating rates. Densification strain anisotropy and conductor film thickness measurements revealed that multilayers densified according to RCS profiles had more overlap between the insulator and conductor densification rates than existed during CHR sintering. This would have reduced the tensile stress in the conductor and shear stress on the conductor/insulator interface leading to a reduced incidence of defects.

References

G.W. Scherer and T. Garino, "Viscous Sintering on a Rigid Substrate", *J. Am. Ceram. Soc.*, **68** [4] 216-220 (1986).

A.G. Evans, "Consideration of Inhomogeneity Effects in Sintering", *J. Am. Ceram. Soc.*, **65** [10] 497-501 (1982).

R.K. Bordia and R. Raj, "Sintering Behavior of Ceramic Films Constrained by a Rigid Substrate", *J. Am. Ceram. Soc.*, **68** [6] 287-292 (1985).

W.S. Hackenberger, T.R. Shrout, and R.F. Speyer, "The Effect of Differential Shrinkage Rate on the Microstructure and Electrical Properties of a Cofired Electronic Substrate", under review for publication in *J. Am. Ceram. Soc.*

W.S. Hackenberger and R.F. Speyer, "A Fast-Firing Shrinkage Rate Controlled Dilatometer Using an Infrared Image Furnace", *Rev. Sci. Instrum.*, **65** [3] 701-706 (1994).

Modelling Macroscopic Behaviour and Residual Stresses during Pressure Sintering of Barium Titanate Sheets

P.G.Th. van der Varst, G. de With* and I. Bennett

Eindhoven University of Technology, Centre of Technical Ceramics,
5600 MD, Eindhoven, The Netherlands.
* Also affiliated with Philips Research Laboratories,
Eindhoven, The Netherlands.

Abstract

The densification process of pressure densified barium titanate sheets and the residual stress therein were calculated using a finite element analysis based on the expansion of the Helmholtz energy in the elastic strains. A simple description of the tractions as well as of the kinetic equations was used. In the analysis the volume elements were considered to be able to shrink freely. After removing the applied loading, the residual stress was calculated by restoring the structural integrity of the material. The model yields a significantly less densified rim at the edge of the sheets, as was observed experimentally. The inclusion of shear tractions modifies the residual stress situation qualitatively. Therefore, further attention should be given to the experimental description of the tractions applied to the sintering material.

1. Introduction

For dielectric materials a strong control of grain size and density (Arlt et al., 1985) and size of the sintered compacts (Bennett and de With, 1994) is frequently required . Pressure sintering can be a solution. Indeed pressure sintering appears useful for producing complex components (Keizer and de Wild, 1994) at a relatively low temperature ($\approx$ 1450 K). The sheet-like components are densified between two dies without side constraint. Although the stress state is globally uniform and uniaxial, locally large deviations from this state may be present. It is to be expected that, due to the non-uniformity of the stress state and the resulting non-uniformity of the density, residual stresses are generated. From experimental work done on this system so far (Bennett and de With, 1994) on thin sheets ($\approx$ 1.5 mm thick and 25 mm wide), the following conclusions can be drawn.

- A height reduction predominantly occurs although the lateral dimensions also increase.

- A small ($\approx$2 mm), less densified rim at the edge of the sheet is present.

- Although the lateral sides bulge somewhat there is no visible trace of stick or drag phenomena at the interface. From these three observations we conclude that the normal stress is dominating.

- Above 1485 K increased densification occurs for any applied pressure, indicating limited influence of plasticity. The free sintering rate is negligible at this temperature.

- Using the strain rate equation $\dot{\epsilon} = k.\sigma^n. \exp(-Q/RT)$, where σ denotes the applied stress, n the stress exponent, Q the activation energy, k a temperature and microstructure dependendent proportionality factor, R and T the gas constant and temperature respectively, results in a range of n-values from 0.84 to 1.6 and Q-values from 300 to 600 kJ/mol, indicating the simultaneous presence of more than one mechanism.

The aim of the study was to model the macroscopic behaviour of the densifying sheets, including the build-up of residual stresses. For this purpose a description of the material behaviour is not sufficient. Also the stress transfer at the interface between dies and sheets needs to be considered.

2. Modelling

As is well known, sintering deals with a wide variety of length scales. They range from microscopic (atomic) to macroscopic. In fact, the microstructural changes result in a macroscopically detectable shape and density change. The basic idea is to introduce an intermediate length scale, the meso-scale. Averaging the microscopic equations over the representative volume element (the meso-cell), a process known as homogenisation, results in constitutive equations which can be used in a macroscopic model. All geometrical microstructural changes lead to shape and density changes of this meso-cell and these changes are described by a field variable: the permanent strain tensor α. The microstructural parameters, e.g. grain size distribution, neck size distribution, pore size distribution and total porosity, are typical examples of internal variables in the sense of Bridgman: variables that one can measure but not control directly. Some microstructural changes result in a macroscopically detectable shape and density change. To describe the macroscopic shape change and the densification the simplest choice for the internal variables will be the permanent strains of the meso-cell because functions of microstructural parameters can also be used as internal variables, This idea is applied here by expanding the Helmholtz energy density of a sintering compact as a function of the elastic strains.

In figure 1, stages a, b and c represent the initial, the stress-free reference and the actual state. The transformations I, II and III represent the permanent, elastic and total deformation tensors $\mathbf{F}_p$, $\mathbf{F}_e$ and $\mathbf{F}_t$ respectively, and $\mathbf{F}_t = \mathbf{F}_e.\mathbf{F}_p$. The total (compatible) strain is $\mathbf{E} = (\mathbf{F}_t^c.\mathbf{F}_t - \mathbf{I})/2$ (superscript c denotes transposition). The elastic strains ϵ and the permanent strains

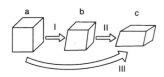

Figure 1: *Schematic representation of the deformation.*

α are defined accordingly. In the small strain approximation $\epsilon = \mathbf{E}-\alpha$. Using this relation after expanding the Helmholtz energy F as a quadratic function of the elastic strains, yields a function of the temperature, the total strains and the permanent strains. The permanent strains act as internal variables dependent on the microstructural features. Specifically:

$$F = F_o(\alpha, T) + \frac{1}{2}(\mathbf{E} - \alpha) : L(\alpha, T) : (\mathbf{E} - \alpha) + ..,$$

where $L(\alpha, T)$ denotes the density and temperature dependent fourth order tensor of elastic constants and the symbol ":" is the double product of tensors (summing over two indices).

The forces corresponding to the variables $\mathbf{E}$ and α (Maugin, 1992) are the mechanical stress tensor σ and the tensor of sintering forces $\mathbf{A}$:

$$\sigma = \partial F/\partial \mathbf{E} = L : (\mathbf{E} - \alpha),$$
$$\mathbf{A} = -\partial F/\partial \alpha = D\,\partial F_o/\partial D\,\mathbf{I} + \sigma + (D/2)\sigma : \partial L^{-1}/\partial D : \sigma\,\mathbf{I},$$

with $D(\alpha) = D_o/\det(\mathbf{F}_p) = D_o/[[\det(\mathbf{I} + 2\,\alpha)]^{1/2}]$ because $\det(\mathbf{F}_p) = [\det(\mathbf{F}_p^c.\mathbf{F}_p)]^{1/2}$ and $\alpha = (\mathbf{F}_p^c.\mathbf{F}_p) - \mathbf{I})/2$ (D_o: initial density). Taking the logarithm of D, differentiating and expanding the results in a power series in α yields $\partial \log D/\partial \alpha = D^{-1}\partial D/\partial \alpha = -\mathbf{I} + \cdots$. From this the approximate relation $\partial D/\partial \alpha \approx -D\,\mathbf{I}$ follows. It was also used when deriving $\mathbf{A}$. The first term in the expression for $\mathbf{A}$ is the driving force for free sintering and can be neglected here since free sintering was observed not to occur at the present temperature. The third term is also negligible under the present conditions.

To calculate the density distribution in a sintering compact the following set of equations has to be solved:

$$\text{div}\,\sigma = 0 \quad \text{(mechanical equilibrium)},$$
$$\sigma.\mathbf{n} = 0 \text{ at the free surface and } \sigma.\mathbf{n} = \mathbf{t} \text{ at the contact surface},$$
$$\dot{\alpha} = \mathbf{G}(\sigma, \alpha, T) \quad \text{(kinetic equation)},$$
$$\sigma = L : (\mathbf{E} - \alpha) \text{ (Hooke's law) and } D = D_o/[[\det(\mathbf{I} + 2\,\alpha)]^{1/2}],$$

with **n** the outward normal and **t** the applied traction at the contact surface. The constraint $0 < D \leq 1$ should be met also.

For the kinetic behaviour, in principle a meso-model, averaging the kinetics over a representative volume should be used. Here, the kinetic equations as described by Besson's model, were used (Abouaf et al., 1988; Besson et al., 1991) because this model could be fitted to our data quite well. The model contains parallel coupled diffusion controlled and interface reaction controlled transport mechanisms and uses a density dependent equivalent stress. The model contains two density dependent functions $c(D)$ and $f(D)$ and also two additional parameters depending on grain size, temperature and activation energy. The functions c and f were taken the same as in the Besson model but the two parameters were determined by fitting the model to the scaled experimental rate defined by $\dot{D}/(3\,D^2)$ and the scaled experimental variable $\sigma\,(c(D)+f(D))^{1/2}$ (figure 2). The latter variable depends on the density and the applied stress but is denoted as scaled density.

Since only the data gathered at 1485 K at 7, 15, 25 and 50 MPa were used, no activation energies are required. Note that this description of the densification in the mesocell does contain neither the influence of agglomerates (Bennett et al., 1995), nor the influence of grain size distribution.

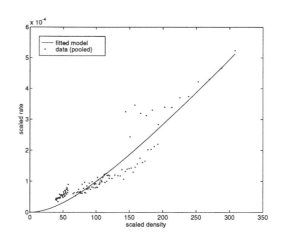

For the tractions in principle true contact elements should be used. To simplify the modelling we used here three simple loading situations. For reference, uniform normal loading was used. This type is denoted as the ideal

Figure 2: *Comparison Besson's model with data for 7, 15, 25 and 50 MPa. Temperature: 1485 K.*

loading. To simulate the contact situation more realistically, normal and shear loading, as detailed by Unksov (1961) was used. It is dependent on a friction coefficient μ and is reasonably well experimentally verified for large plastic deformations of metals. For values of μ above about 0.3, the exact value of μ is unimportant. It yields a high normal pressure in the centre gradually decreasing towards to the edge. The shear traction is zero in the centre, increases up to a certain radius and is afterwards approximately constant. The direction of the shear traction was taken inward. This type of loading is referred to as the Unksov loading. Finally, we considered also the case that only the normal component of the Unksov loading is present (modified Unksov loading).

Neither the choice for the kinetic equations nor that of the tractions is essential for the thermodynamic part of the model. The present choice for the traction was mainly dictated by reasons of simplicity. However it has a large effect on the calculated results.

For the finite element analysis, the densifying sheets were modelled as an axisymmetric system (initial diameter and thickness of 25 mm and 1.6 mm respectively) using 8-noded isoparametric elements with 9 integration points per element.

The interpolation functions were chosen as recommended by Abouaf et al. (1988). A fully implicit numerical integration scheme appeared to be essential. Further details on the procedure will be given later (van der Varst and de With, 1995). The computational procedure was as follows: first the stress state is calculated. This is at t=0+.

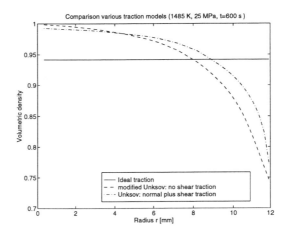

Figure 3: *Density as function of the radius.*

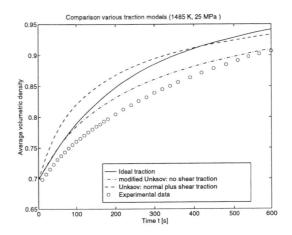

Figure 4: *Density as function of time.*

Next the sintering process was switched on thereby allowing each point to shrink as if it were isolated from the rest of the material and at the end of the process (t=600 s) the tractions were removed and the connections restored. This is a first order, decoupled approach.

3. Results

In figure 3 the density, averaged over the height, is plotted as function of the radius. The figure applies to the case that the average applied stress is 25 MPa. The time is 600 s, i.e. just after the load was removed. Non-uniform loading results in a less densified region at the edges. This is in agreement with experiment. As required, uniform loading results in a uniform density. We will not discuss the uniform loading further; it fulfills all expectations (uniform densification and height reduction, no residual stresses) yielding trust in the numerics used.

The volumetric density (averaged over the sheet) as function of pressing time resembles qualitatively the experimental data (figure 4).

Both types of non-uniform loading result in non-uniform axial displacement as a function of radius (figure 5).

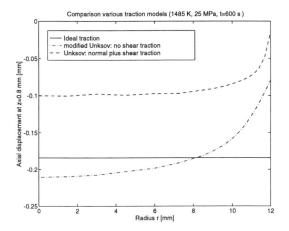

Figure 5: *Axial displacement*

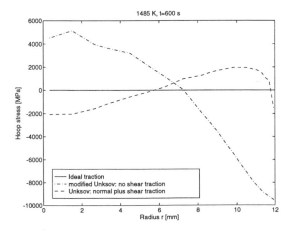

Figure 6: *Hoop stress (25 MPa)*

The displacement shown in the figure is actually half the estimated height reduction. Note the strong influence of the shear traction. The absence of a shear loading leads to a radial expansion whereas the presence of shear loading may lead to radial shrinkage (plots not shown).

For the hoop stress (figure 6), the presence of normal plus shear loading (both non-uniformly distributed) results in tensile stresses in the outer part of the sheet while only normal loading, but also non-uniformly distributed, results in tensile stresses in the centre.

For the radial stress the non-uniform normal plus shear loading results in compressive stress throughout the sheet while only a non-uniform normal loading results in tensile stresses (figure 7). Obviously, the radial stress should always be zero at the outer edge because of the boundary condition.

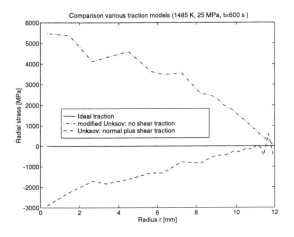

Figure 7: *Radial stress.*

4. Conclusions and final remarks

Presently only microstructural changes leading to densification and shape change were considered. To include other, microstructural changes e.g. grain growth internal variable theory theory can also be used. Obviously, the list of internal variables should be increased in that case.

The actual loading situation has a strong influence on the densification. Spatial nonuniformity is primarily determined by the normal traction. The presence of shear traction generally leads to higher rates of densification. The non-uniformity in normal traction leads to non-uniformity in the final height. The presence of the shear stress mitigates this effect but leads simultaneously to a decrease of the height reduction that can be achieved. Dependent on the value of the shear traction, tensile stresses arise in the outer part or inner part of the sheet, possibly affecting the structural integrity of the sheets.

An improvement upon the present method is as follows. After calculation of the stress at t=0+, the interaction between the volume elements is always taken into account, i.e the connectivity constraint is always present. At the end of the process the tractions are removed and the residual stresses calculated (coupled approach). The implementation of the coupled approach is presently pursued.

In both the decoupled and coupled approach, stress relaxation in the cooling stage can occur. This was not taken into account so far.

Based on the absence of cracking in the experimental situation, it is expected that the coupled approach will result in significantly lower stress values, but not in a qualitatively different stress situation. Although a coupled analysis is in many aspects similar to a creep analysis, the influence of the density upon the various parameters, particularly the modulus of elasticity, makes the solution much more difficult.

Finally, we should remark that, since the stress distribution in the sheet is critically dependent on the interface stress, a better understanding of the stress situation at the interface is required for reliable, quantitative modelling. Experimental determination of the residual stresses probably will supply additional information about magnitude and nature of the actual interface traction.

References

M. Abouaf, J. L. Chenot, G. Raisson, P. Baudin, "Finite element simulation of hot isostatic pressing of metal powders", *Int. J. for Numerical Methods in Engineering,* **25**, pp 191-212, (1988).

G. Arlt, D. Hennings, G. de With, "Dielectric properties of fine-grained barium titanate ceramics", *J. Applied Physics,* **58**, pp 1619-1625, (1985).

I. Bennett, G. de With, "Sinter forging of barium titanate", *Electroceramics IV,* (eds.) R. Waser, S. Hoffmann, D. Bonnenberg, Ch. Hoffmann, Augustinus Buchhandlung, Aachen, pp 1271-1274, (1994).

I. Bennett, G. de With, P. G. Th. van der Varst, "The influence of agglomeration on pressure assisted sintering", Poster presented at the conference *Sintering 1995,* Pennsylvania State University, September 1995.

J. Besson, M. Abouaf, F. Mazerolle, P. Suquet, "Compressive creep tests on porous ceramic notched specimens", *Creep in structures, IUTAM symposium Cracow Poland 1990,*(ed.) M. Zyczkowski, Springer Verlag, Berlin, pp 45-53, (1991).

P. H. M. Keizer, W. R. de Wild, "A new concept for integration of passive components", *Electroceramics IV,* (eds.) R. Waser, S. Hoffmann, D. Bonnenberg, Ch. Hoffmann, Augustinus Buchhandlung, Aachen, pp 1045-1053, (1994).

G. A. Maugin, "The Thermomechanics of Plasticity and Fracture", Cambridge University Press, Cambridge, pp 276-282, (1992).

E. P. Unksov, "An Engineering Theory of Plasticity", Butterworth Publishers Ltd, London, pp 208-213, (1961).

P. G. Th. van der Varst, G. de With, "Thermomechanics of pressure assisted sintering", in preparation, 1995.

Index